CIS Energy and Minerals Development

The GeoJournal Library

Volume 25

The titles published in this series are listed at the end of this volume.

CIS Energy and Minerals Development

Prospects, Problems and Opportunities for International Cooperation

edited by

JAMES P. DORIAN
East-West Center and University of Hawaii, Honolulu, U.S.A.

PAVEL A. MINAKIR
Institute of Economics Research, Khabarovsk, Russia

and

VITALY T. BORISOVICH
Moscow Geological and Prospecting Institute, Moscow, Russia

SPRINGER SCIENCE+BUSINESS MEDIA, B.V.

Library of Congress Cataloging-in-Publication Data

CIS energy and minerals development : prospects, problems, and opportunities for international cooperation / edited by James P. Dorian and Pavel A. Minakir and Vitaly T. Borisovich.
p. cm. -- (GeoJournal library ; v. 25)
Includes index.
ISBN 978-94-010-4889-7 ISBN 978-94-011-2012-8 (eBook)
DOI 10.1007/978-94-011-2012-8
1. Mines and mineral resources--Former Soviet republics. 2. Mineral industries--Former Soviet republics. 3. Energy industries--Former Soviet republics. I. Dorian, James P. II. Minakir, P. A. III. Borisovich, V. T. IV. Commonwealth of Independent States. V. Series.
TN85.C52 1993
338.2'0947--dc20 93-19530

ISBN 978-94-010-4889-7

Printed on acid-free paper

CONTENTS

PART TWO: REGIONAL ISSUES AND AFFAIRS

PART THREE: INTERNATIONAL TRADE AND RELATIONS

LIST OF TABLES

LIST OF FIGURES

PREFACE AND ACKNOWLEDGMENTS

Given the dramatic changes which have taken place in recent years within the former Soviet Union, the editors felt it critical to compile a book documenting the impacts of those changes on the world's largest mining industry. The chapters in this book are based on papers presented at the Conference on Commonwealth Mining Development: Prospects, Problems, and Opportunities for Cooperation with the Pacific Basin held at the Institute of Economics Research in Khabarovsk, Russia, June 1992. The meeting was sponsored jointly by the East-West Center and the University of Hawaii, both in Honolulu, the Moscow Geological and Prospecting Institute, and the Institute of Economics Research of the Russian Far East. It brought together Commonwealth planning and industry officials in the energy and mineral fields with their counterparts in academia and industry from Australia, Canada, China, Japan, the United States and other major producing and consuming countries. The Commonwealth participants in the conference came from all relevant sectors of the energy and minerals industries and included central government planners and policymakers.

Putting together a manuscript containing papers authored by scholars from ten countries would not have been possible without the assistance of numerous people. Indeed, the editors wish to express their appreciation for the generous support from the East-West Center in Honolulu, Hawaii, notably, Julia Culver-Hopper, Technical Editor, Dee Baroman and Lillian Shimoda, Program Secretaries, Lois Bender, Publications Assistant, and Russell Fujita, Graphic Artist. As a Soviet scholar, Ms. Culver-Hopper was instrumental in readying the manuscript for publication. D. Lyn Henegan, also of the East-West Center, provided substantial assistance in the preparation of tables and graphics used in the book. None of the above should be held responsible in any way for any errors or mistakes evident in what follows.

James P. Dorian
East-West Center
Honolulu, Hawaii

Pavel A. Minakir
Khabarovsk Territory
Administration and Institute
of Economics Research
Khabarovsk, Russia

Vitaly T. Borisovich
Geological and
Prospecting Institute
Moscow, Russia

LIST OF CONTRIBUTORS

Yury A. AGABALYAN is Director of the Department of Mining at the Yerevan Polytechnic Institute, Yerevan, Armenia.

Yuri M. ARSKY is First Deputy Minister of Ecology and Environmental Control of the Russian Federation and a Professor of Economics, Moscow Geological and Prospecting Institute, Moscow, Russia.

Madenyat A. ASANOV is Head of the Kazakhstan Geological Fund, Ministry of Geology and Preservation of Kazakh Minerals Wealth, Alma-Ata, Kazakhstan.

L. M. BAGDASARYAN is a Research Analyst at the Institute of Economics, Yerevan, Armenia.

Yuri I. BAKULIN is Director of the Far East Institute of Mineral Resources and DAL Pacific Minerals, Khabarovsk, Russia.

John H. BEAGLEHOLE is an Associate Professor, Department of Politics, the University of Waikato, Hamilton, New Zealand.

Valery G. BELITZKY is Laboratory Head of the All-Russian Scientific Research Institute of Fuel and Energy Problems and Professor of Economics, Moscow Geological and Prospecting Institute, Moscow, Russia.

Vitaly T. BORISOVICH is Head of the Economics Department, Moscow Geological and Prospecting Institute, Moscow, Russia.

Youhua CHEN is an Engineer with the Techno-Economic Research Center of the China National Nonferrous Metals Industry Corporation (CNNC), Beijing, China.

Allen L. CLARK is a Program Director at the East-West Center and an Affiliate of the Graduate Faculty at the University of Hawaii, Honolulu, United States.

James P. DORIAN is a Research Associate at the East-West Center and an Affiliate of the Graduate Faculty at the University of Hawaii, Honolulu, United States.

Alaster EDWARDS is Principal Geologist with BHP-Utah Minerals International, Victoria, Australia.

Christopher FINDLAY is Coordinator of the Minerals and Energy Forum, Pacific Economic Cooperation Conference, and Professor of Economics at the Australia-Japan Research Centre, the Australian National University, Canberra, Australia.

Chennat GOPALAKRISHNAN is a Professor in the Agricultural and Resource Economics Department of the University of Hawaii, Honolulu, United States.

Pavel V. KARAULOV is an Expert in the Research Institute of Geology of Foreign Countries (VZG) and a Representative at Landmark Sovgeo, Ltd., Moscow, Russia.

Boris V. KOCHETKOV is Department Head at Volfram Russian Joint Stock Company and Professor of Economics, Moscow Geological and Prospecting Institute, Moscow, Russia.

Ping-Sun LEUNG is a Professor in the Agricultural and Resource Economics Department of the University of Hawaii, Honolulu, United States.

Stepan K. MEKHAKIAN is an Associate Professor in the Department of Mining at the Yerevan Polytechnic Institute, Yerevan, Armenia.

Pavel A. MINAKIR is Vice Governor of the Khabarovsk Territory Administration and President, Institute of Economics Research, Khabarovsk, Russia.

Stewart MURRAY is Chief Executive of Gold Field Mineral Services, Ltd., London, United Kingdom.

Keun-Wook PAIK is a Ph.D. Candidate in the Departments of Politics and Industrial Relations and Economics at the University of Aberdeen, Aberdeen, Scotland.

Vsevolod I. POTAPOV is a Professor of Economics, Moscow Geological and Prospecting Institute, Moscow, Russia.

Douglas RITCHIE is General Manager, Commercial, CRA Exploration Pty., Ltd., Melbourne, Australia.

Genady V. SEKISOV is Director of the Institute of Mining Affairs, Khabarovsk, Russia.

Terry SHEALES is Manager of the Mineral Economics Branch, Australian Bureau of Agricultural and Resource Economics, Canberra, Australia.

Vitaly T. SHISHMAKOV is Director General and Professor of Economics of the Far Eastern School of Interbusiness, Khabarovsk, Russia.

Vyrene SMITH is a Research Analyst in the Mineral Economics Branch, Australian Bureau of Agricultural and Resource Economics, Canberra, Australia.

Hiroshi TAKAHASHI is a Research Analyst at the Institute for Russian and East European Economic Studies, Tokyo, Japan.

Valentin N. TARASOV Institute of Mineral Economics (UIEMS), Moscow, Russia.

Nikolai I. TSVETKOV is Director of the Far East Management Cadres Center, Khabarovsk, Russia.

John F. YANAGIDA is Head of the Agricultural and Resource Economics Department of the University of Hawaii, Honolulu, United States.

Shakarim F. ZHANSETOV is Deputy Chairman of the Production Factors Council, Khazakh Academy of Sciences, Alma-Ata, Kazakhstan.

Chapter 1

INTRODUCTION AND BACKGROUND: A NEW BEGINNING

James P. Dorian

CREATION OF A COMMONWEALTH

The former Soviet Union possessed some of the world's largest reserves of hydrocarbons and minerals. With the breakup of the country in 1991, the former Soviet republics are now exercising complete control over their resources and related industries. The new Commonwealth of Independent States (CIS), made up of Armenia, Azerbaijan, Kazakhstan, Kyrgyzstan, Belarus, Russia, Tajikistan, Turkmenistan, the Ukraine and Uzbekistan, consists of many new countries which are well-endowed in mineral and energy resources. Russia, Kazakhstan, and the Ukraine, for example, contain large mining industries with an abundance of products. Other states possess substantial geologic potential, indicating that they may become important mineral producers in the years ahead.

While the specific nature of the new Commonwealth will be debated for some time, the alliance will, of necessity, involve some degree of economic interdependence, as long as the former republics have the economic means to continue relations. The full text of the new "Commonwealth Agreement," released in Moscow on December 9, 1991, defined the nature of the loosely-fitted group of nations and the termination of the Union Treaty of 1922 which had formally established the Soviet Union. The Commonwealth Agreement, executed by the Republic of Belarus, the Russian Federation, and the Ukraine, consisted of 14 brief articles. In them, the authors described the basis for the formation of the Commonwealth as being the historic economic, social, and cultural ties already established among the members. The agreement outlined the rights and freedoms guaranteed for Commonwealth citizens, prospects for economic, political, military, and cultural cooperation among members, methods for dispute resolution, and obligations for the fulfillment of international treaties and agreements.

Today, ten independent states are voluntarily part of the Commonwealth, whose members include all former Soviet republics except the Baltic states,

J.P. Dorian et al. (eds.), CIS Energy and Minerals Development, 1–8.

Moldova, and Georgia.The Commonwealth by itself does not possess the legal status of a state, while at the same time, its members do not display all of the attributes of fully independent nations, such as viable national currencies. Notwithstanding its loosely-fitted nature, the member nations of the Commonwealth have been admitted to the United Nations, International Monetary Fund, and other global organizations. Clearly, many of the Commonwealth states have gained widespread international recognition as autonomous countries with self-ruling governments and legal systems.

According to the original founding accords, the Commonwealth was to have its headquarters in Minsk, the capital of Belarus, and a government structure composed of Council Heads of State and Government and numerous ministerial committees, which would serve to coordinate cooperation in such key areas as defense, economic and foreign policy, and communications. The framework for cooperation was to be modeled after the International Economic Committee (IEC).

In reality, however, the CIS represents a transitional structure with a primary objective of minimizing the disruptions caused by the dissolution of the Soviet Union and promoting a peaceful transformation. Most Commonwealth members, many of which were independent states in the past, are wary of the creation of a strong central body which could possibly act as a legal successor to the Soviet Union. As such, member countries are independently managing their own affairs and building economic, political, and legal regimes to govern domestic activities. Their economies do, however, remain strongly interdependent and it will be years before many of the reliances will be overcome. Though in the past independence-minded republics had often maintained that they could survive and prosper on their own, economic reality suggests otherwise. The Soviet model of economic development and centralized planning facilitated a closely interwoven, highly specialized societal framework. The country possessed countless state monopolies and concentrated the suppliers of essential natural resources and basic goods. The Uralmash Machine Tool Works in Sverdlovsk, Russia, for example, produced nearly all of Soviet Union's strip mining excavators, specialized drilling rigs, hydraulic presses, blast furnaces, rolling mills and slab-casting machines (Claudon, 1991). The individual republics simply cannot develop sufficient specialized production capabilities to satisfy their total domestic requirements in the short term, particularly in light of growing budget deficits and increased production costs. Moreover, the republics will have difficulty moving their products onto world markets as the quality of manufacturing is not, for the most part, internationally competitive.

The interdependent nature of the economies of the former Soviet republics is particularily exemplified in the coal industry. According to recent figures from the State Statistics Committee, Russia imports 13 percent of the amount of coal it consumes and exports seven percent of what it produces; for the Ukraine, the figures are 11 percent and six percent, respectively; Georgia imports 51 percent and exports 33 percent, while Kazakhstan imports 14 percent and exports 40 percent (Romanyuk, September 1991). Azerbaijan, Lithuania, Moldova, Latvia,

Armenia, Estonia and Turkmenistan import 100 percent of their coal and do not export any. The situation with rolled ferrous metal products is very similar. As for resins and plastics, Armenia imports 87 percent; Estonia, 91 percent; Latvia and Moldova, 78 percent; Lithuania, 75 percent; and Turkmenistan, 100 percent.

With such a degree of economic integration, most of the former Soviet republics would face severe economic difficulties and even collapse without importing energy, raw materials, chemical products and machinery. If the intricate trading and economic relations between the republics of the former Soviet Union fell completely apart, desperate economic and social times would likely follow. The economic ties among the republics are considered stronger than those among the countries within the European Community (EC), as interrepublic transactions amounted to 20.5 percent of Soviet gross national product in 1990, compared to 16 percent in the EC (Romanyuk, September 1991). Additional evidence of the high degree of economic integration was the relatively small share of exports abroad in the republics' net material product when valued in current domestic prices. The share of foreign imports in the Soviet national income was similarly low. While the Baltic republics, Russia and the Ukraine could possibly survive economic independence from an alliance, the other former Soviet republics would indeed find it nearly impossible to sustain their economies in the short term without some ties to at least the Russian Federation.

MINERAL INVESTMENT OPPORTUNITIES AND CONCERNS

Much of the new Commonwealth's energy and minerals endowment, both developed and underdeveloped, lies within the borders of the Russian Federation. It is by far the wealthiest new independent state, accounting for 90 percent of the former Soviet Union's oil output, and a majority share of gold, diamonds, platinum and base metals production. After Russia, the Ukraine and Kazakhstan are the two former Soviet republics best endowed in energy and mineral resources.

Notwithstanding the minerals abundance in Russia, large quantities of resources do occur outside the Russian Federation, including, for example, more than 90 percent of the former nation's manganese production, which is derived from mines within the Nikopol and Chiatura basins of the Ukraine and Georgia, and a substantial share of coal from the large sedimentary basins of Donetz and Kuznetz in the Ukraine and Kazakhstan (*Mining Journal,* February 1, 1991). Armenia produced about 40 percent of Soviet molybdenum output in 1990, while Kazakhstan accounted for nearly half of the country's copper production. While Azerbaijan is today a primary service and supply center for the Commonwealth oil industry, in the early part of this century it was an important center for crude production. Uzbekistan's output of gold, until now veiled in secrecy, ranked second behind that of the Russian Federation among the former Soviet republics. The republic produced one-quarter of the former country's output of the precious metal.

Having been effectively closed to foreign investors for decades, the former Soviet republics are now consciously trying to attract such investment. State leaders desire to expand their minerals and energy sectors and improve the efficacy of use of raw minerals in the former Soviet republics by introducing capital, technology, machinery and science from abroad. The potential for further mineral discoveries in the new Commonwealth is immense, with the most attractive areas occurring in the Russian Federation (Far East region and Siberia), Kazakhstan and the Caucasus region.

Opportunities for foreign participation in the mining activities of the former Soviet republics are wide ranging and include enhancement of the production performance in the nonferrous metals industry; use of advanced prospecting techniques for the discovery of blind ore deposits; adoption of improved recovery methods to rework gold tailings and waste rock dumps; and cleaning up of environmental damage (Dorian, January 1992). State mining officials are also seeking overseas guidance to improve their managerial and accounting skills.

While opportunities for investing in the new Commonwealth are becoming plentiful, the legal regimes within the new states have not kept pace with the rapid development of economic and social reform. In most former Soviet republics, rules and regulations that protect investors' interests are largely absent or untested. Clearly, the former Soviet republics will require sufficient time to develop legal foundations capable of guiding international business activities in their regions. Until such environments are firmly established, overseas enterprises will need to comprehend fully the terms under which they are investing and the details of any transactions.

In the minerals-rich Russian Federation, a law on foreign investment was approved in August 1991. It concerns all types of property and investment in the form of both joint ventures and wholly foreign-owned enterprises, including 100 percent foreign-owned affiliates established in the form of joint stock companies. The Russian law broadened the rights of investors in comparison with the previous Soviet legislation and is geared to simplifying relationships between Russian and foreign partners in joint investments and generally making the new state more attractive for foreign investors.

While Russia and the other independent states are creating investment laws and provisions, many are also considering establishing policies to guide mineral exploration and development activities within their boundaries. Russia has issued two laws which govern mineral exploration and development activities—the Law on Subsurface Resources, adopted February 21, 1992 (sometimes referred to as the Law on Mineral Resources), and the Regulations on Procedures for Licensing the Use of Mineral Resources, adopted July 15, 1992. The Law on Mineral Resources outlines three important subjects: (1) the State Mineral Fund; (2) licensing of mining activity; and (3) a system of payments and fees for the use of mineral resources. The State Mineral Fund comprises all the mineral resources of the Russian Federation. Before the dissolution of the Soviet Union, all explored and identified reserves were transferred to the relevant ministries for industrial exploitation. Today all minerals wealth will stay permanently in the State Mineral Fund, while parts of it are handed over to various juridical and physical entities to use. The transfer is to be accomplished through a system of licensing. The government body that supervises the State Mineral Fund is the State Geological Committee, or, more accurately, State Committee for Geology and Rational Use of Resources.

For most states, there is emphasis on devising and implementing mineral strategies that will meet economic development objectives and improve social welfare by using natural resources as an impetus to growth. Mineral policies in these former Soviet republics will most probably resemble those in market economies, since in both the former Soviet republics and market economies, resources are viewed as national goods which should be developed for the well-being of the entire society. The means by which the policies achieve their objectives will not, however, be similar.

With independence, each new Commonwealth state is today attempting to guide all phases of the mining industry, including production, sales, and trade. The former Soviet republics are having to devise strategies independently in accordance with their own economic policies and objectives without receiving direction from a central government. While the states are moving through an unprecedented transitional phase, foreign companies are expressing eagerness to invest in mining projects, particularly those related to oil and natural gas development. However, inadequate infrastructure, inefficient management, outdated technology, immature legal systems, the rise of ethnic nationalism, and political and economic instability are hampering both the Commonwealth of Independent States and its prospective partners. Nonetheless, the Commonwealth remains perhaps the world's last great frontier for the exploration and development of energy and mineral resources.

BOOK OVERVIEW

This book contains a comprehensive overview of the mineral and energy industries of the CIS, whose vast resources include much of the world's remaining oil reserves and large deposits of gold. While the book concentrates on the 10 member-nations of the Commonwealth, the other former Soviet republics are also discussed owing to their close historical ties with the Commonwealth states. The objective of the volume is to introduce the many critical issues surrounding the mining industries of the new Commonwealth, and to examine prospects for international cooperation and trade in metals, nonmetals, and hydrocarbons. Opportunities for joint ventures and technology transfers in mining are identified, along with the numerous obstacles limiting needed foreign investment. The authors have varied backgrounds in the energy and mineral fields, and include individuals in academia, government, and industry from throughout the Commonwealth, Asia, the United States, and Europe. The book is the first of its kind to include detailed information on the individual mining industries of the former Soviet republics and an accurate assessment of the many problems governments must overcome before their nations' energy and mineral sectors can foster economic growth.

CIS Energy and Minerals Development: Prospects, Problems and Opportunities for International Cooperation contains twenty chapters and is conveniently divided into three parts, namely, Commonwealth Mining Industry, Regional Issues and Affairs, and International Trade and Relations. Part I provides an overview of the minerals and energy base of the Commonwealth, a review of its newly developing gold industry, and an assessment of joint venture and technology transfer opportunities in mining. Five chapters are presented in Part I, including one by V. T. Borisovich, G. B. Belitzky and V. I. Potapov of Moscow which examines the energy status of various CIS members with a focus on oil and gas resources. In the chapter, the authors reveal the severity of the problems plaguing the Commonwealth energy industry, including an uneven distribution of oil, gas, and coal resources throughout the region, use of antiquated equipment, and lack of capital funds for renovation and expansion. The authors divide the Commonwealth into three categories of nations depending on their energy status—states with a positive energy balance, including Russia, Kazakhstan, and Turkmenistan; states which can only partially satisfy their demands for fuel by domestic sources, including Azerbaijan, the Ukraine, Kyrgyzstan, Tajikistan, and Uzbekistan; and states possessing little or no domestic energy reserves, including Belarus, Moldova, and Georgia.

Stewart Murray, General Director of Gold Fields Mineral Services, Ltd., in London, authored Chapter 5, which evaluates the dramatically changing gold industry of Russia and other former Soviet republics. Murray's analysis includes a description of the organization of gold producers in Russia, a complicated structure composed of "zolotos," or state-owned enterprises, "goks," or mining operations, and "priisks," or alluvial production units. Adhering even more loosely to the operating companies are entities known as "artels," which supervise their own affairs and pay their own workers, but in actuality depend on the state amalgamations for supplies and equipment and physical infrastructure. The artel movement is acknowledged as being the most productive throughout the Commonwealth, and particularly in Russia.

Part II examines present and proposed mineral development activities in Armenia, Kazakhstan, and the Russian Far East, and describes the tumultuous economic conditions of the minerals-rich Far East before, during, and after the dissolution of the Soviet Union. Like many other former Soviet Republics, Armenia and Kazakhstan are well-endowed in mineral resources, yet their mining industries are struggling to overcome a host of technological, economic, and environmental problems. Both nations are reshaping their mining industries in an attempt to facilitate economic growth. Foreign investment in joint ventures is being actively pursued.

The Russian Far East, considered by many to be a significant frontier for minerals exploration and development, is the focus of Chapters 9, 10, 11, and

12. In their chapter on the geology and minerals potential of the Far East, Clark and Sekisov point out that the region "encompasses approximately seven percent of the Earth's land mass, larger in area than the United States." The Russian Far East possesses more than 120 types of metallic, nonmetallic, and energy resources, and the region is endowed with large reserves of antimony, boron, brown coal, diamonds, fluorite, gold, platinum, silver, tin, and tungsten—minerals which are not as abundant in other areas of the Commonwealth. The authors present a partial compilation of known mineral deposit types which occur in the Russian Far East, and predict the types of geologic deposits which will likely be developed in the latter 1990s and beyond.

Part III addresses international trade and cooperation issues, including the emerging economic ties between the Commonwealth of Independent States and Australia, the People's Republic of China, Japan, and New Zealand. In their chapters, Douglas Ritchie (Chapter 14) and Alaster Edwards and Christopher Findlay (Chapter 17) provide valuable insight into the decision-making criteria industry employs in choosing whether to invest in mineral-related joint ventures in the Commonwealth. In addition to geologic potential, mining companies will assess laws and regulations governing minerals exploration and development, as well as possible constraints on access to international markets. Principal investment criteria for the minerals sector include political and economic stability, law and order, attitude towards foreign investors, bureaucratic efficiency, restrictions on equity participation and employment, and tax rules.

Chapter 16 of the final part provides an evaluation of energy cooperation prospects in Northeast Asia, and suggests a "Northeast Asia Energy Charter" may be possible, similar in structure to the European Energy Charter. A significant improvement of bilateral energy relationships in the region has raised real prospects for multilateral cooperation in oil and gas exploration and development in the Russian Far East, particularly the Sakhalin offshore and Yakutia areas, and China's Tarim basin and East China Sea areas.

In the years ahead, the world's mining industry will be influenced strongly by mineral and energy exploration and development activities in Russia, Kazakhstan, the Ukraine, and other newly-independent Commonwealth states. For decades relatively little was known about the physical, economic, and political characteristics surrounding the Soviet mining industry, despite its number one ranking globally. This book provides a much-needed overview of the critical issues facing today's mining industries in the various Commonwealth states, and highlights developments which are likely to occur in the future. Many questions are raised in the text which will only be answered with the passage of time. The transformation of the former Soviet republics into sovereign states will continue to have adverse effects on mining activities until their political, legal and economic regimes are well established and operating. Such impacts will be monitored closely by the mining companies as well as government and academic officials throughout the world.

REFERENCES

Claudon, Michael P., 1991, "Don't Give Up On Gorbechev Yet: Economic Interdependence Could Lead To Stability," *Geonomics,* pp. 2–3.

Dorian, James P., 1992, "Joint Mineral Ventures in the Former Soviet Republics: A Slow Beginning," *Mining Engineering,* Vol. 44, No. 1, Society for Mining, Metallurgy, and Exploration, Inc., Littleton, Colorado, pp. 54–57.

Mining Journal, February 1, 1991, "Russian Roulette," Vol. 316, No. 8107, Mining Journal Publications, Ltd., London.

Romanyuk, V., September 9,1991, "Republics Declare Independence," *Izvestia,* p. 2, Moscow (in Russian).

PART ONE: COMMONWEALTH MINING INDUSTRY

Chapter 2

ENERGY AND FUEL RESOURCES IN THE COMMONWEALTH OF INDEPENDENT STATES: MAJOR DEVELOPMENTS AND PROSPECTS

Vitaly T. Borisovich, Valery G. Belitzky, and Vsevolod I. Potapov

BACKGROUND

The Commonwealth of Independent States (CIS) is considered to be an energy-rich industrialized country. Covering more than 15 percent of the world's land area and inhabited by approximately 5.5 percent of its population, the CIS possesses 20 percent of proved global natural gas reserves and 15 percent of coal reserves, and it produces 21 percent of the world's fuel resources (Table 2.1). Huge energy reserves provided the basis for substantial development of energy and fuel resources within the Soviet Union, which since the beginning of the 1980s was the world leader in development of these resources. In the 1970s, the Soviet Union was first in world crude oil and condensate production, while from the beginning of the 1980s it led in natural gas production as well.

While the CIS has vast energy and fuel resources, shortages have occurred in various parts of the new Commonwealth because of declining oil production and the use of outdated oil refining technologies, resulting in lower output of light oil products (Table 2.2). The deterioration of the energy industry has led to diminished supplies to industry and consumers, as well as shortfalls in fuel resources in wintertime.

A principal problem plaguing the CIS energy industry is the uneven distribution of reserves throughout various states. The European part of the Commonwealth (including the Urals region) accounts for more than 75 percent of total CIS energy consumption while it produces only 20 percent, requiring supplies from the eastern regions of up to 1.2 billion tonnes annually. The severity of the problem is revealed in the following statistics:

J.P. Dorian et al. (eds.), CIS Energy and Minerals Development, 11–28.

Table 2.1. Commonwealth of Independent States: Energy Production, 1991

	Coal		Oil, Including Condensates		Natural Gas	
	Total production (in millions of tonnes)	Percentage change compared to 1990	Total production (in millions of tonnes)	Percentage change compared to 1990	Total production (in billions of cubic meters)	Percentage change compared to 1990
Azerbaijan	—	—	11.7	-6.0	8.6	-13.0
Kazakhstan	130	-0.7	26.6	3.0	7.9	11.0
Russian Federation	353	-11.0	461	-11.0	643	0.4
Turkmenistan	—	—	5.4	-3.0	84.3	-4.0
Uzbekistan	5.9	-8.0	2.8	0.8	41.9	3.0
Ukraine	136	-18.0	4.9	-6.0	24.4	-13.0

Source: Statistical Committee of the CIS, information compiled in International Monetary Fund, April 1992a.

Table 2.2. Energy Balance for the Former Soviet Union and Russia, 1990–1992

	USSR Territory		Russia		
	1990	1991	1990	1991	1992[a]
OIL		(in millions of tonnes)			
Production	569.8	515.5	516.2	461.1	404.7
Consumption[b]	421.7	422.5	...	...	...
Russia	227.4	241.5	227.4	241.5	218.4
Other	194.3	181.0	...	...	...
Exports	159.0	93.0	155.8	91.1	90.0
Of which:					
Crude oil	109.0	52.0	106.8	51.0	60.0
Oil products		50.0			
		41.0	49.0	40.2	30.0
Deliveries to other republics/states[c]	...	...	163.0	158.2	120.8
Of which:					
Crude oil	...	...	123.0	126.4	96.0
Oil products	...	...	40.0	31.8	24.8
Deliveries from other republics/states	...	...	30.0	29.7	24.5
Of which:					
Crude oil	...	...	19.0	18.7	16.6
Oil products	...	...	11.0	11.0	7.9
Imports	10.9	—	—	—	—
NATURAL GAS		(in billions of cubic meters)			
Production	815.0	810.0	640.6	642.9	653.6
Consumption[b]	708.0	708.0	...	...	...
Russia	547.6	415.9	547.6	415.9	453.8
Other	160.4	292.1	...	...	...
Exports	109.0	105.0	95.0	91.0	103.5
Deliveries to other republics/states[c]	...	...	...	139.0	99.3

Sources: Goskomstat of the USSR; Goskomstat of the Russian Federation; and Ministry of Fuel and Energy. Information compiled in International Monetary Fund, April 1992b.

a. Projection

b. Consumption has been derived as a residual of domestic production and exports, thus implicitly assuming no change in stocks

c. The projections for 1992 are based on the trade agreements between the republics of the former Soviet Union.

- much of the Commonwealth's power supply (60 million kilowatts (kw) or 17.6% of the total) is provided by out-of-date, wasteful, and already exhausted equipment;
- in the oil-refining industry, wear and tear of equipment exceeds 50 percent;
- more than 50 thousand kilometers (km) of natural gas pipelines need to be replaced; and
- in the coal industry, 80 percent of the shafts were put into production more than 30 years ago, and half of these are in need of rebuilding.

Consequently, the production capacity does not meet overall requirements.

NEW ENERGY PROGRAM

To solve existing energy problems within the Commonwealth, a draft of a New Energy Program has been formulated. To avoid imposing strict guidelines, the Program contains forecasts based on the participation of all the sovereign republics. The Program, which contains projections from the present to 2010, takes into account the following paramount guidelines:

- residential dwelling space shall be increased from 15.8 square meters (m^2) per person to 24.4 m^2;
- food products output shall be expanded by 2–2.6 times;
- nonnutrient production shall rise by 4–4.5 times;
- cultural and public health and education services with facilities will be expanded by three times;
- automobile production will increase by 2.3–2.5 times (with the number of private cars rising by three times);
- labor conditions will improve, especially in the rural areas, due to an increase in automatization by 3–3.5 times.

To achieve the economic status of most developed countries, it will be necessary to expand national income in the next 20 years by five percent per year which will occur primarily as the result of development of the light, agricultural and food processing, construction, and chemical industries. However, present conditions indicate that such growth in national income is unlikely to occur, especially during 1991–1995. If national income were to grow at a slower rate of three percent per year, national income would increase by 1.6–1.8 times during the 20-year forecast period.

In light of current economic and environmental problems, it is clear that an adequate increase in fuel and energy production can hardly be achieved. Development of energy and fuel resources will require about 20 percent of current total capital investments, while the costs of fuel production are increasing continuously. Between 1985 and 1990, energy and fuel industries required a 30 percent increase in capital investment.

Under such conditions, a primary strategy of the CIS should be production of environmentally sound energy. About one-third of currently consumed energy is considered to be environmentally sound. As a result of utilization of environmentally sound technology, it is believed that the reduction in environmental damage will prevent losses of tens of billions of rubles to the national economy by 2000.

During the first stage of Commonwealth energy development, particularly during the period 1991-1995, the national economy will not yet undergo structural changes or require considerable capital investment and material resources. Already tested environmentally sound technologies, not requiring hefty expenditures and providing quick returns, will therefore be used. In the second stage, during 1996–2000, structural transformation of the energy industry will occur by further development of equipment and increased utilization of advanced technologies. Altogether in the next 20 years, it is anticipated that 65 percent of fuel demand will be met by environmentally sound and more effective technologies, including 55 percent during the first 10 years and 75 percent in the following 10 years.

Therefore, about one-third of the additional CIS demands for fuel and energy, taking into account environmentally sound technologies, will require development of new energy reserves. During the first 10 years the increase will amount to approximately 300–400 million tonnes, while over the 20-year period 500–600 million tonnes of additional energy resources will need to be produced with the annual increase averaging 1–1.2 percent. As a result, the annual per capita consumption should increase from 6.8 to 8 tonnes as compared to the current 11 tonnes in the United States, 6.1 tonnes in Germany, and 4.1 tonnes in Japan.

In the near term, little additional petroleum output is expected from enhanced oil recovery, while power from nuclear power stations is likely to expand only during the second stage after 2000, though only if measures are taken now. Renewable resources, including nontraditional resources, will be of only small importance during the period under review, and then only at the local level.

Under such conditions, before 2000 the main source of new energy will likely be natural gas. However, due to several problems in the gas industry the production of natural gas may decline slightly, which, in turn, will make nuclear and coal industry development more important.

COMMONWEALTH ENERGY DISTRIBUTION

Unevenness

Different regions within the former Soviet Union are characterized by their own problems. The extent of perestroika in various parts of the Commonwealth will certainly influence production patterns as well as energy resource distribution. Generally speaking, the European regions of Russia will experience shortages of fuel. Consequently, securing adequate supplies of fuel is believed to be the paramount problem to be resolved in European Russia. A partial solution involves the construction of new nuclear power stations and obtaining natural gas, coal and power supplies from the East.

Based on available fuel resources, the sovereign states of the Commonwealth and the Baltic republics can be divided into three categories:

(1) states with a positive balance of fuel reserves—the Russian Federation, Kazakhstan, and Turkmenistan;
(2) states which can partially provide their demand for fuel—Azerbaijan, the Ukraine, Estonia, Kyrgyzstan, Tajikistan, and Uzbekistan; and
(3) states possessing almost no reserves or less than 5-10 percent—Lithuania, Latvia, Belarus, Moldova, and Georgia.

In general, reserves under the Soviet resource classification scheme include quantities of economic mineral material identified, explored, and evaluated in place within a specific area which has been subjected to geologic exploration and drilling as well as mining activities. Reserves are considered economically mineable given existing technology and market conditions.

Fuel resources in the republic

The Russian Federation is by far the major supplier of fuel resources in the Commonwealth. At present, its potential exceeds its own needs by 3,545 percent, especially in oil and gas, which is an encouraging feature for a newly independent economy.

Russia's energy and fuel industry contains several regional fuel bases or complexes. The share of these bases in future republican production will be: natural gas, approximately 90 percent; oil, 75-80 percent; and coal, 60 percent. In developing the Western Siberian oil and gas complex, the Timano-Pechora and Kansko-Achinskiy energy complexes, and the Kuznetskiy and Vorkuta coal basins, it will be necessary to expand capital investments in social and industrial infrastructure. Protection and reclamation of the environment in the abovenamed regions will also require additional funds.

The amount of fuel to be exported to other states within the CIS and abroad is expected to increase from 650 million tonnes in 1990 to 800-830 million tonnes in 2010. At the same time, Russia receives supplies of natural gas as well as oil products from Central Asia and from Ekibazstuz. Total imports are projected to be 145-160 million tonnes in 2010.

Energy reserves are distributed unevenly throughout Russia. The European parts of Russia and the Urals run deficits of fuel. Meeting projected demand will require 230-250 million cubic meters of gas from Western Siberia by 2010. However, taking into consideration recent trends in gas industry development, it will also be necessary to develop nuclear energy.

Eastern Siberia and the Russian Far East are almost self-sufficient in energy, primarily because of significant demonstrated and inferred reserves of coal (Figure 2.1). Demonstrated reserves are proven or probable while inferred reserves are possible based on geologic analogy. Up to 75-80 percent of the energy requirements in the eastern part of the former Soviet Union will be met in 2010.

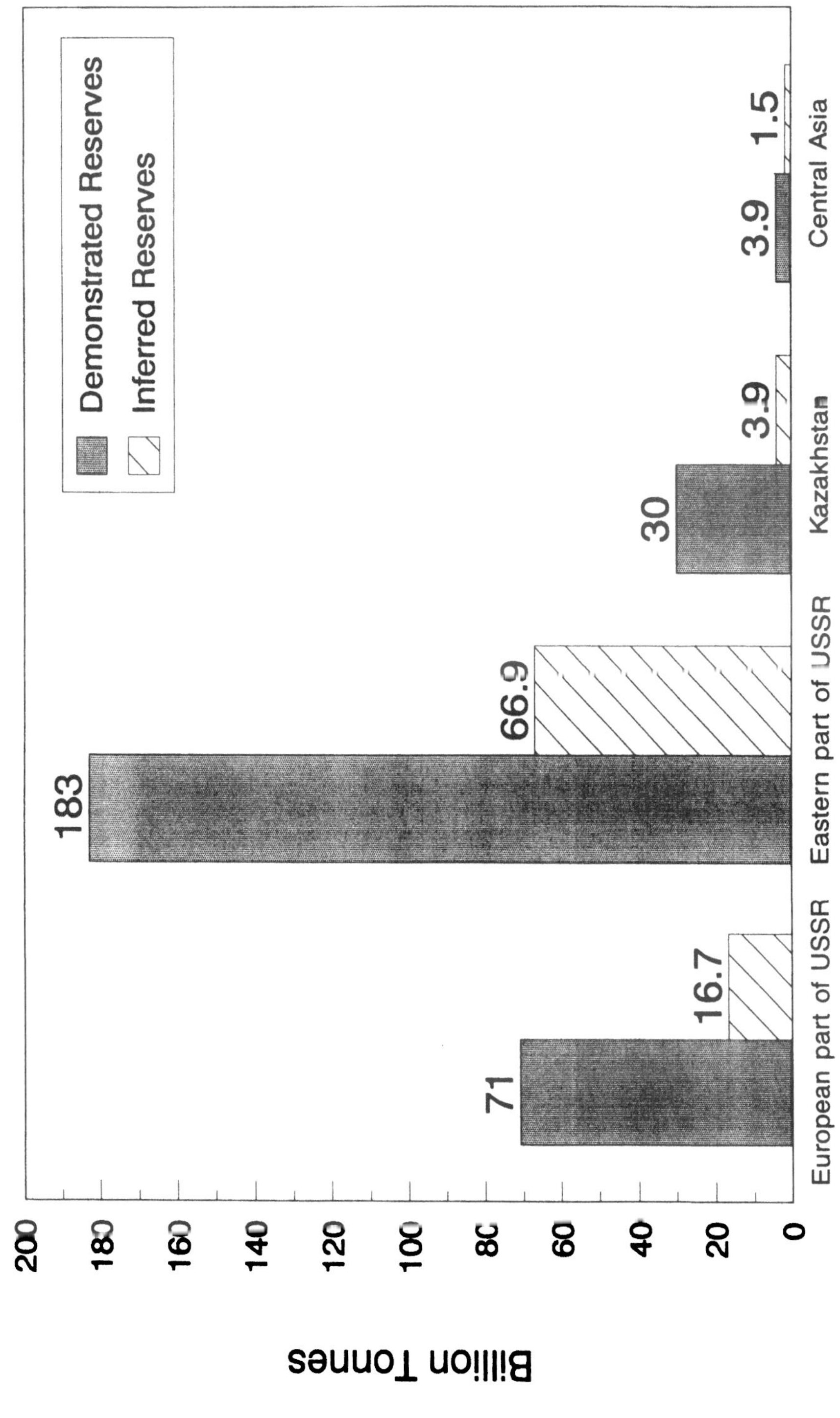

Figure 2.1. Coal Reserves by Region of the Former Soviet Union

Additional supplies of fuel may also be available as a result of the Sakhalin offshore gas field development. Further growth in energy production may also result from increases in labor productivity due to technology upgrading and reduction of material and energy resources consumption.

About 70 percent of the CIS total natural gas reserves are located in the Tyumen region, namely the Nadir-Pur-Tazovskiy, Yamal, and Gyudan deposits. A considerable raw materials base also occurs in the Near Volga and Urals regions, where reserves as of January 1, 1991 amounted to four trillion cubic meters (tcm), or 9.2 percent of CIS total reserves.

The Russian Far East has tremendous natural gas reserves amounting to 1.5 tcm, of which 0.5 billion cubic meters (bcm) are offshore. The natural gas reserves under production are expected to increase by 21.9 tcm during 1991–2010.

The CIS gas industry is expected to undergo massive expansion over the next two decades. Gas output will reach 858 bcm by 1995, 970–980 bcm in 2000, and 990–1000 bcm in 2010. Such production increases will require development of gas fields on the Yamal Peninsula and offshore in the Okhotsk and Barents seas. By 1995, a new pipeline from the northern part of the Tyumen region to central Russia will be constructed which will require imports of three million tonnes of pipes with 1,420 mm diameter and construction and pumping equipment costing a total of about two billion rubles. Under a pessimistic scenario, given problems with Yamal field development, gas production will be 100–150 bcm lower than planned in 1995.

Further development of the CIS oil industry will be constrained by the following conditions:

- currently producing oil fields are mostly exhausted and declining in production;
- extracted oil comprises only 35 percent of proved reserves;
- reserves are distributed over a number of small deposits with low productivity;
- proved reserves in the explored fields are difficult to develop;
- present-day extraction technology requires excessive drilling and special equipment;
- development of oil fields has become more complex because of increasing remoteness of development areas and a deepening of the wells;
- further development of the oil industry will, therefore, concentrate on in-place reserves.

The maximum rates of oil and gas condensate production (435 million tonnes in 1995, 410 million tonnes in 2000, and 300 million tonnes in 2010) can be achieved only if the proved reserves in Western Siberia can be increased by more than three billion tonnes by 1995. The primary region of oil development will continue to be Western Siberia. Its share shall increase to 77.4 percent of the production of the Russian Federation by 2010. However, because of decreasing productivity and declining oil grades, the number of productive wells is expected to rise by only 1.9–2 times in 2010 as compared to 1990 levels, while in Tyumen the number of wells will increase by 3.4 times and in Komi, 1.9 times.

Low-producing fields, with expected yields of 13–15 million tonnes of oil, can be put into operation with additional capital investment. Forty-five million meters of well-drilling is anticipated to put into production up to 14,500 wells. An increase in production will require imports of upgraded equipment and chemical agents, as well as the establishment of higher prices and lower taxes for secondary oil recovery.

Geological work in oil exploration will become more productive and efficient in the near future. Labor utilization in reconnaissance drilling is to be reduced by 1.5–2.5 times as a result of the introduction of third-generation telemetric systems for seismic applications, and broad use of computer-aided processing and interpretation and storage of data.

In the field of oil transportation, power consumption by pumping is expected to be reduced by 1.5–2.0 times, while utilization of labor is to decrease by 2–2.5 times through the widespread use of advanced technology in transporting gas condensates and high viscosity oil and upgraded devices for pipeline control.

Over the long term, the Russian oil-refining industry will be characterized as follows: first, economic requirements for light oil products will increase (especially for motor fuel and petrochemical feedstocks); second, there will be further declines in oil production because of depletion of fields; and finally, there will be considerable increases in production costs. These factors reveal the importance of boosting the share of the more valuable oil products in Russia. This can be achieved if heavy oil residues are upgraded. The proposed development of the oil-refining industry (with throughput of 297.2 million tonnes in 1995, 296.4–298.8 million tonnes in 2000 and 292.4–298.2 in 2010) necessitates heavy upgrading.

According to one scenario, the production of diesel and motor fuel is expected to rise by 13–16 percent owing to increased upgrading of refineries. Production of reactive fuels and benzene as a chemical feedstock will also increase.

The share of heating fuel in product output is to decrease from 32.1 percent in 1990 to 12.9 percent in 2010, while the share of ship bunker fuel is to drop by 18 percent due to lower demand, declining to 2.1 million tonnes in 2010. Domestic furnace fuel production is projected to fall 19.5 percent from 4.1 million tonnes in 1990 to 3.3 million tonnes in 2010, thus making available resources for increased diesel fuel production.

In general, requirements for gasoline shall not be supplied only from oil, but also from alternative fuels. In the gas-processing industry, construction of several small gas-based petrochemical plants is anticipated. Between 1991 and 1995, gas-processing plants with a capacity of 20 bcm per year are expected to be built in the Tyumen region, along with pumping stations with a capacity of 26 bcm of gas per year, gas pipelines and other projects. All capacities will be based on low-temperature complex processes.

Additional plans include reorienting the Otradnensky, Groznensky, and Neftekumsky gas-processing plants to produce gasoline, polyethylene, polypropylene and other endproducts. The Astrakhan gas-processing plant is expected to be rebuilt, while a new gas-based petrochemical complex is to be built in New Urengoy, in addition to a number of such complexes in the northwestern and northeastern parts of the European regions.

Condensate utilization will increase within the industry, led by an increase to 94 percent in Western Siberia by 1995. About 95 percent of extracted condensates will be stabilized in the gas-processing plants. Approximately 20 percent of condensates will be subjected to extensive upgrading in the plants, providing output of a number of oil products. The largest amount of motor fuel is expected to be produced in the Astrakhan gas-processing plant. Increased output of motor fuel has already been achieved in 1986–1990 at the Surgut plant, and a number of minirefineries have begun contributing to production. Expansion of processing capacity will allow for an increase in output of condensed gas by 2.8 times, ethane by 6.2 times and sulfur by three times within the period from 1991 to 2010.

As for the coal industry of Russia, production is expected to reach 375 million tonnes in 1995, 435–450 million tonnes in 2000, and 530–540 million tonnes in 2010. Because of worsening geological conditions, advances in open-pit mining are required, while producing shafts will be overhauled. The most acute need in the industry is an improvement in machinery and technology in order to raise output. Increased output will result from the development of Kuzbass, KATEK and Far East deposits, while production of the European deposits will decline and only provide 91 percent of 1990 output in 2010 (Figure 2.2).

In recent years, almost all technical and economic indices of the coal industry have deteriorated, resulting largely from the disruption of the entire Soviet economy. Requirements for imported advanced technologies and machinery will cost an estimated 2.5 billion rubles (in 1991 prices).

In order to supply power stations in Siberia, the Far East, and the Urals, as well as European Russia with the Kansko-Achinskiy coal, construction of a 400-km railway is needed in addition to railway stations and about 200 km of connecting railways, as well as 300 km of power lines. Total costs of the railway project are estimated at 700–750 million rubles for the railway and 180–200 million rubles for infrastructure.

Kazakhstan possesses tremendous potential in fuel resources. Kazakhstan's coal production is able to currently satisfy 133 percent of the republic's fuel requirements, which will increase to 158 percent by 2010 as a result of development at the Ekibastuz basin and Pre-Caspian deposits. Coal reserves in the republic are 38.9 billion tonnes, with 62 percent of that suitable to strip-mining. At present, the Ekibastuz, Borlinsky, Kuu-Chekinsky, Karaganda, Maykubensky and Shubarkolsky coal basins have been developed and are producing. An open-pit mine, with an annual capacity of 500 thousand tonnes, is being developed at the Preozerskoye deposit within the Turgay basin. Deposits under exploration, development and production in the republic contain 500 million tonnes of resources; thus, production of coal can be increased up to 3.8 times the 1990 levels. In accordance with the CIS New Energy Program, the annual output of

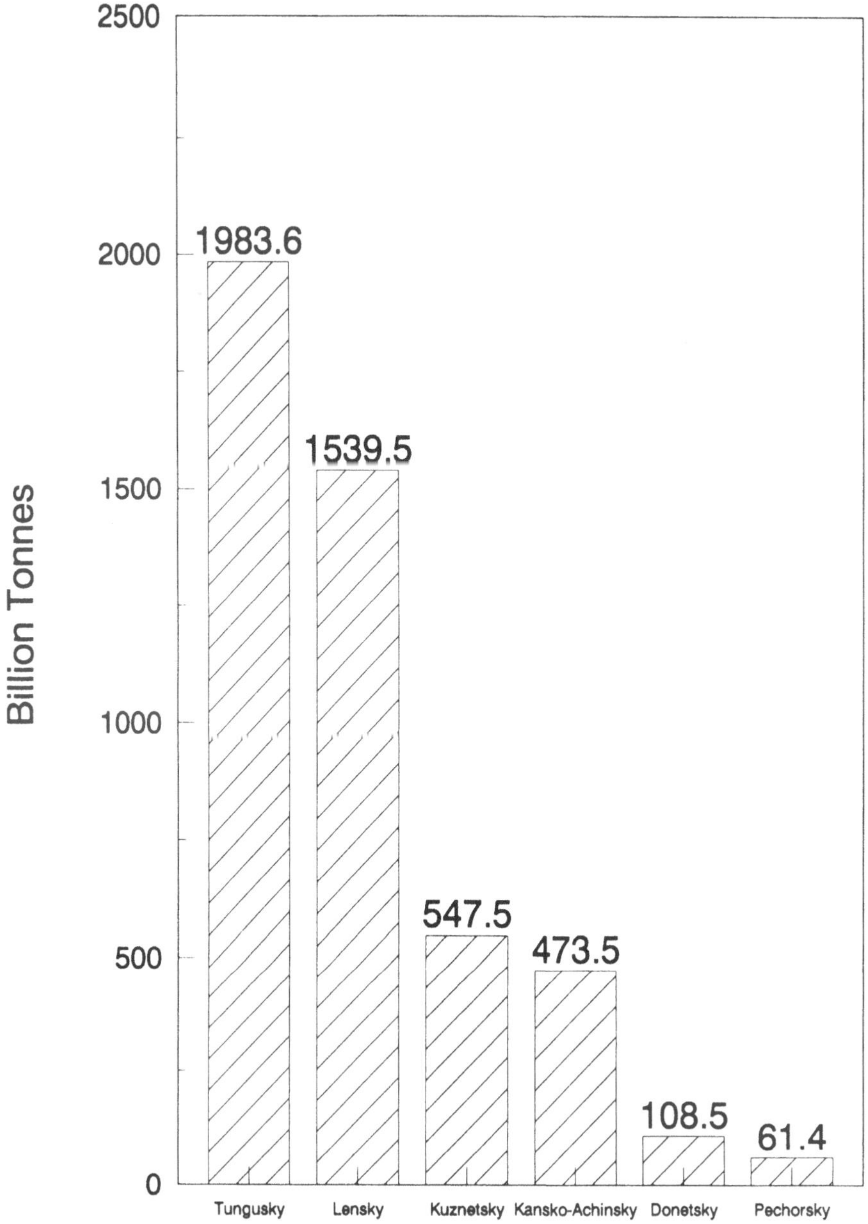

Figure 2.2. Coal Reserves by Field in the Former Soviet Union

coal is projected to be as follows: 1990, 141 million tonnes; 2000, 177 million tonnes; 2005, 182 million tonnes; and 2010, 200 million tonnes. The major increases will be achieved in the open-pit mines of the Maykuben, Ekibastuz and Shubarkol basins.

In terms of oil reserves and production, Kazakhstan is second in the CIS. Currently 115 oil fields are under exploration in the republic. The ten most accessible fields contain about 88 percent of oil reserves and 86 percent of production in the republic. The majority of in-place reserves of the Pre-Caspian fields (90% of crude oil, 98% of gas, and 100% of condensates) are in underground salt formations at depths of five to seven kilometers. Potential hydrocarbon reserves in Kazakhstan are second only to those of Western Siberia, while in terms of sulfur reserves it is the Commonwealth leader.

Exploration in Kazakhstan is actively continuing. Large fields already discovered include the Tenghiz field (Guriev region), which contains 33 percent of the republic's oil reserves, and the Karachaganak oil and gas condensate field of the Urala region which comprises 77 percent of the republic's hydrocarbon resources.

Today, there is considerable interest in the offshore North Caspian areas adjacent to the Tenghiz field. Large oil fields may be discovered there at a depth of around four kilometers. Seismic exploration is to continue in these areas through 1995, followed by more environmentally sound drilling activities from 1996 onward. In addition, a new oil-rich region has been discovered in the South-Turgay Depression that, in turn, may increase the republic's potential, particularly in Kazakhstan's southern regions.

It is forecasted that by 2010 oil production in Kazakhstan will rise to 15 million tonnes and gas output to 29 bcm, or 3.7 and 4.7 times, respectively, the 1990 levels. Increased production will include output from the Karachaganak field (15 million tonnes and 28 bcm, accordingly.)

In 2010, 32 bcm of gas will be processed in Kazakhstan, or 68 percent of the expected gas production. Twenty billion cubic meters of this will be associated gas from the Tenghiz field, and 10 bcm will be Karachaganak natural gas.

Capacities of the Guriev, Pavlodar, and Chimkent oil refineries are expected to be expanded. There are plans to build oil refineries at Mangyschlak (1995, 6 million tonnes), Tenghiz (2000, 12 million tonnes), and in the Urals (2000, 6 million tonnes). Consequently, while now only 4.6 million tonnes (14%) of oil is refined in the republic, in 2010 it will be 32 million tonnes (38%).

Natural gas from the Karachaganak gas field is to be processed and utilized to sustain pressure in the Karachaganak oil and gas-condensate field. Commercial natural and associated gas also can be used as furnace and heating fuel. Annual output for natural gas is estimated to be: 1.7 tcm in 1990; 4.2 tcm in 1995; 8.0 tcm in 2000; 9.7 tcm in 2005; and 11.6 tcm in 2010. Heavy oil output is expected to reach: 5.2 million tonnes in 1990; 5.0 million tonnes in 1995; 8.6 million tonnes in 2000; 9.3 million tonnes in 2005; and 8.0 million tonnes in 2010.

Despite Kazakhstan's vast energy resources, the republic suffers from natural gas and coking coal shortages, as well as shortages in heavy fuel oil for boiler use. Gas requirements will likely expand to 50 million tonnes in the near future, while heavy fuel oil needs will reach two to three million tonnes. Surplus fuels available for export include ordinary coal and light oils.

Turkmenistan is the only state in Central Asia to have a positive energy balance due to significant gas production. Only 25 percent of the state's gas output is used domestically, while the remaining supplies serve customers in Uzbekistan,

Kyrgyzstan, Tajikistan and southern Kazakhstan, as well as Russia by pipeline.

At the beginning of 1990, 14 oil fields were under production in Turkmenistan, with three of these offshore in the Caspian Sea. The largest fields—Leninskoye and Barsa-Gelmes—provide 71.9 percent of the oil output in the republic. Nearly 94 percent of Turkmenistan's recoverable reserves are in the fields already under production, with 60 percent of these deposits nearly completely worked out. This provides strong evidence of near exhaustion of Turkmenistan's existing energy formations.

During the last decade, about 100 wells were put into production annually in Turkmenistan, while average well depths increased by 3.3 percent, causing a decline in economic indices. According to the state balance of reserves of the former Soviet Union in January 1990, Turkmenistan's proved oil reserves (categories $A+B+C_1$) were 2,916.5 bcm and C_2 reserves (inferred) were 879.7 bcm for a total of 91 oil fields. Proved reserves imply the resource has been thoroughly identified and measured using up-to-date geological and geophysical techniques. The reserves of natural gas were 2,849.5 bcm. Eighteen large oil and gas fields account for 89.1 percent of Turkmenistan's oil reserves and 85.5 percent of its recovery of natural gas.

Less than one-half of Turkmenistan's explored fields are operating; these contain 83.3 percent of the republic's gas reserves. Oil reserves in Turkmenistan are categorized as follows: ready for exploitation, 10.9 percent; under exploration, 4.6 percent; and reserved for future exploration, 1.2 percent. Nearly 30.6 percent of proved oil reserves are in-place, while 30.1 percent are exhausted.

Natural gas production in Turkmenistan in 1990 reached 87.8 bcm. During the Twelfth Five-Year Plan (1986–1990), output continued in all fields operated during the preceding Eleventh Five-Year Plan (1981–1985), including the Bagadja and Sovetobad fields. The Bagadja field yielded an output of 230 million cubic meters (mmcm) in 1990.

Large quantities of Turkmenistan's recovered gas were transported to the central regions within the former Soviet Union by the "Central Asia-Center" pipeline. In 1990, 79.9 bcm of gas were allocated outside the republic. To meet its local needs, Turkmenistan consumed only 8.4 bcm, or 9.3 percent of total gas output.

The pace of gas production exceeds the rate of exploration in Turkmenistan, hence the quantity of accessible gas reserves is on the decline. The effectiveness of exploration is low due to intricate geological conditions experienced while drilling resulting from salt and sulfur aggression, high temperatures, anomalous formation pressures, and deep-seated formations.

Most wells in Turkmenistan's highly promising areas such as South Iolotan, Yashlar, Molodeznaya, and Osman are still being drilled, and, therefore, have not yet yielded any considerable results. Fields such as Achak, North Achak, Naip, South Naip, North Naip, Gugurtly, and Mollaker were to stop producing after 1990, while 10 other fields, including Kirpkichli, Beurdeshik, North Balkui, West and East Shatlyk, Tedjen, Uchadjy, Seyrab, Shorkel, and Bayram-Ali, are entering a period of declining output. Gas production is anticipated to decline after 1992. Output from operating fields in 1995 is projected to be 68.2 bcm.

Because of an absence of coal reserves, Turkmenistan receives supplies of around 0.8–1.5 million tonnes of coal from Kyrgyzstan, Tajikistan, and Uzbekistan. Similarly, Turkmenistan's oil industry requirements are satisfied by only 30 percent of its production, thus the shortfall is expected to be made up of seven to nine million tonnes of oil from Western Siberia. Overall, however, Turkmenistan's energy resources exceed required imports due to its large natural gas reserves.

Like Turkmenistan, Azerbaijan is well-endowed in energy resources. It presently satisfies its requirements for energy from domestic production, as well as imports. Heating fuel is provided partially by heavy oil, while natural gas is derived from Azerbaijanian fields, as well as imported sources (approximately 50%). Azerbaijan and the other Caucasian republics receive six to seven million tonnes of crude oil for refining into motor fuel from Russia and Kazakhstan.

Recently, production of oil and gas from both offshore and onshore deposits declined in Azerbaijan. Exploration activities have indicated that there is little possibility of increasing onshore production. Onshore fields are generally depleted and characterized by low daily output and high water content.

In contrast to the onshore resources, offshore areas are highly promising for development. Azerbaijanian deep-sea areas are considered to be the most important. Forty deep-sea platforms are planned for construction by 2005 for oil and gas recovery at depths of around 200 m. Given the planned construction activities, oil and gas output in the next two decades is estimated to be: 12.5 million tonnes in 1990; 13.5 million tonnes in 1995; 15.0 million tonnes in 2000; 16.5 million tonnes in 2005; and 16.5 million tonnes in 2010. For gas, future production levels are projected to approach: 9.9 million tonnes in 1990; 7.5 million tonnes in 1995; 8.0 million tonnes in 2000; 8.0 million tonnes in 2001; and 8.0 million tonnes in 2010.

Oil and gas production in the Ukraine is declining. Self-sufficiency in energy resources will drop from 48 percent in 1990 to 35 percent in 2010. This is primarily the result of declines in oil, gas and coal production, and an uncertain future for nuclear power. A nuclear power moratorium recently adopted in the Ukraine ensures that the state will likely be short of energy in the early part of next century because of deficient supplies in fossil fuels. Power deficiencies are expected to increase and will reach 25–30 billion kwh in 2000 and 45–50 billion kwh in 2010.

The Ukraine is also faced with shortages of funding for coal development and coal-related environmental problems in Donbass. It may be necessary to build several power stations based on coal, including plants in Donetsk, aimed at more efficient use of the fuel.

To alleviate the fuel deficit in the Ukraine, 150–160 bcm of gas from Western Siberia will be supplied. In addition, an estimated 55–56 million tonnes of crude oil from Russia will be provided to the Ukraine's oil refineries. All energy requirements of the Ukraine can not be fulfilled, however, by gas and oil supplies from Russia.

Estonia currently is able to satisfy up to 75 percent of its fuel requirements, primarily with shale oil. Without construction of new shafts, however, shale recovery is likely to decline by one-half in 2005 compared to the 1990 production level; production is to drop to 13–14 million tonnes. Currently, electric power generation exceeds needs by 1.9 times. After 2005, though, Estonia's electric power supply will not be adequate to meet the republic's needs, which are estimated to reach around 13–14 billion kwh.

To meet the republic's requirements in fuel and energy, it will be necessary to support new development of shale mining. Newly constructed shafts will ensure fuel for power supplies at the level of 16–17 billion kwh. The Pre-Baltic and the Estonian power stations are to be restored. Nuclear power stations are also forecasted to be built in the country.

Kyrgyzstan provides 52 percent of its fuel and energy needs, mostly due to hydropower and coal output. The republic is forecasted to be able to meet 64–88 percent of its requirements of coal between 1996 and 2010. Mainly high quality coal is to be supplied from the Kuznetsky and Karaganda basins. Natural gas is routinely supplied from Uzbekistan in the amount of 6.5 bcm per year.

Kyrgyzstan's geography is favorable for hydropower development. Planned construction of a hydroelectric power station on the Naryn River will double the republic's energy output, making it possible to supply power to Central Asia's other republics as well as Kazakhstan.

In Tajikistan, coal resources satisfy 40–62 percent of its energy requirements. An almost two-fold increase in energy consumption during 1996–2010 is forecasted, with supplies of coal expected from Karaganda, Kyrgyzstan, and Uzbekistan, gas from Central Asia, and heavy oil from Uzbekistan and Turkmenistan.

The development of electric power in Tajikistan is to be based on new hydropower stations constructed on the Vakhsh River, as well as on a number of coal-fired power stations to be commissioned. The result will be an expansion of power generation capacity by 2.2 times, while production will increase from 15.5 billion kwh in 1990 to 25.3 billion kwh in 1995 and 39.4 billion kwh in 2010. Such increases in output will ensure not only a 1.5-fold increase in power consumption in the republic, but will necessitate energy imports as well at levels of six to eight billion kwh.

The amount of proved energy reserves in Uzbekistan is estimated at 3,500 million tonnes. Two-thirds of these resources are represented by natural gas. Nearly three-quarters (74%) of Uzbekistan's main oil and gas reserves occur in the Gissar area. Only small portions of these resources are located in the Bukharo-Khiva, Fergana and Surkhandarya areas. A small quantity of gas reserves was explored on the Usturt plateau.

As for coal, more than 97 percent of the republic's reserves are concentrated in the Angrenskoye deposit area. Minor reserves, representing approximately 2.3 percent of total reserves, are in the Shargunskoye and Baysunskoye deposits. Uzbekistan's coal resources include mainly brown coal (97%) and smaller quantities of hard coal (3%). Almost 60 percent of brown coal is available for open-cast mining. The remaining 40 percent of brown coal and all the reserves of hard coal are suited to underground mining.

Only one-half of Uzbekistan's energy needs are provided by coal, due largely to a delay in construction and introduction of new capacity in the Angrenskiy open-pit mine. It is believed that upgrades of the mine will improve Uzbekistan's energy balance, satisfying 64 percent of its domestic needs. The coal deficit shall be alleviated by coal supplied from Kuznetsk, Karaganda, Maykuben and Shubarkol.

Uzbekistan's gas supplies are limited and amount to around 45.5 bcm. With growth in consumption, self-sufficiency will drop from 95 percent in 1995 to 65 percent in 2010. The increasing shortfall in gas supplies is expected to be partially alleviated by additional supplies from Turkmenistan.

Given further evaluation of condensate reserves in the Kokdumalak field, Uzbekistan's oil and condensate production is projected to grow from two to 5.5 million tonnes by 1995, though it is expected to be followed by a decline to 3.3 million tonnes by 2010. Energy self-sufficiency will be somewhat more than 50 percent. Therefore, the republic's oil-refining industry will, as before, be mostly oriented towards crude imports from Western Siberia.

In 1990, Lithuania provided 24 percent of its own energy and fuel consumption. The country has no coal, gas or oil reserves. To satisfy the demand for heating fuel, it is expected to import Silesian coal from Poland and from Kuznetsk, as well as gas supplies from Western Siberia. However, gas consumption is limited because of the relatively small capacities of the industry.

When calculating the supply/demand balance of electric power in Lithuania, it becomes clear that the republic's power supply will have a surplus until 2010, although excess production will decline from 13.2 billion kwh in 1990 to six billion kwh in 2010. The growth in electric power production will be accounted for by the Ignalinskaya nuclear power station, which will have a capacity of 17 billion kwh.

Lithuania's oil refineries will continue to depend on about 9–12 million tonnes of oil from Russia. The republic will continue to have a surplus in the light oil products supply.

In Latvia, local sources of fuel are scarce, except for wood, whose use is limited because of strict requirements in the construction industry. Hydroenergy resources are also not abundant. Hydropower production potential is expected to approach 2.8 billion kwh, taking into account reconstruction.

Latvia's power supply only meets half of its requirements. At the same time, it is expected that power consumption will increase from 10 billion kwh in 1990 to 15–18 billion kwh in 2010. Under such conditions, the new Riga power station to be put into operation will require more fossil fuels. Taking into consideration the fact that gas production is declining, one of the options to be considered is surplus coal supplied from the Moscow basin. There are also updated plans to build a nuclear power station.

Armenia, one of the Caucasian republics, has virtually no discovered energy resources. It is supplied with gas from Central Asia and coal from Donetsk.

After the nuclear power station was shut down in the mid-1980s, Armenia's energy industry suffered several fuel shortages. No hydropower potential exists in the republic. A doubling of fuel and energy supplies is planned based on imported gas, nuclear power development using the safest designs, and the republic's participation in the recovery and transportation of energy from other regions in the Caucasian republics and the CIS.

The Belarus economy depends on energy resources from the Commonwealth's other regions. Oil and gas is supplied from Western Siberia, coal from the Ukraine, Russia, and Poland, and electric power, at least in part, from Lithuania. Local energy production is mainly limited to heavy oil produced from imported crude.

In 2010, the republic's needs will be satisfied by local production by only 8.3 percent, compared to 12 percent in 1990. This will coincide with a comparable decline in the annual rate of increase in energy consumption from 2.5 percent in 1990 to 0.3–0.5 percent in 2010.

Belarus does have a well-established oil-refining industry; however, the amount of crude processed will drop to 3,035 million tonnes in 2010. For heating fuel, gas resources are to be developed, with supplies rising from 14 bcm in 1990 to 30 bcm in 2010.

A considerable shortfall exists in Belarus's power supply. Even in 1990, the electricity shortfall was 12.3 billion kwh. This shortfall will continue to 2010 at a level of 10–12 billion kwh; nevertheless, new generating capacity will be put into operation. To alleviate the power deficit, power supplies are expected from the adjoining power grids of the northwestern and central regions of the CIS.

Like Armenia, Moldova has almost no energy reserves. The republic provides less than 3.2 percent of its own energy requirements, therefore it depends heavily on imported fuel. It needs up to 4.5 million tonnes of coal from Donbass and 1–8 million tonnes of natural gas from Western Siberia.

Moldova's local needs for electric power are basically satisfied by its own generation. After 2000, however, electric power will become deficient, while in 2010 the shortfall will exceed 1.5 billion kwh. Under such circumstances, there is a need to build new power plants based on coal, as well as safe nuclear power stations.

Georgia produces only 15 percent of its required fuel and energy resources. One possible source of improvement in the republic's energy situation appears to be greater hydropower utilization and use of nontraditional sources of energy such as hot springs, solar and wind power. The hydropower stations' output will increase from 7.6 billion kwh to 12.0 billion kwh in 2010, meeting 38 percent of domestic requirements. To fully satisfy its energy requirements the republic will have to build new plants based on fossil fuels, including construction of power stations in Kutaissi and Tbilisi, and reconstruction of one plant in Tbilisi.

In general, the republic's requirements for heating fuel and gas are likely to increase; these supplies will be provided from other former Soviet republics and other countries. In order to increase fuel supplies it will be necessary for the republic to develop an oil-refining capacity. Therefore, Georgia is expected to overhaul the Batumi refinery and construct a new plant.

SUMMARY

The energy industry of the former Soviet Union is large, complex, and unevenly distributed. Major problems exist in the oil and coal industries, while the natural gas industry is also in decline. It will be important in the years ahead to introduce environmentally sound, advanced technologies for coal-based power plants. The process of acquiring and utilizing these technologies will remain slow due to a lack of hard currency and incomplete policies.

Energy demand in the CIS will continue to rise, while the republics' self-sufficiency will likely continue to decline. Under such circumstances it will be necessary to take measures to conserve energy, while energy trade relations between the republics will need to be based on energy needs and deficiencies and the rational use of supplies.

REFERENCES

International Monetary Fund, April 1992a, *Common Issues and Interrepublic Relations in the Former USSR,* Economic Review, Washington, D.C., 55 p.

———, April 1992b, *Russian Federation,* Economic Review, Washington, D.C., 115 p.

Chapter 3

COMMONWEALTH MINERAL RESOURCES AND ECONOMIC DEVELOPMENT

Yuri M. Arsky, Vitaly T. Borisovich, and
Boris V. Kochetkov

OVERVIEW

Mineral resources are of great importance to the development of the new Commonwealth of Independent States (CIS) national economy and a major provider of hard currency earnings. The CIS is the world leader not only in proved reserves of hydrocarbons, coal, iron ore, nickel, cobalt, tin, tungsten, titanium, potassium, apatites, asbestos, and barite, but also in the production of a variety of other minerals. At present, there are about 7,000 mines in operation. Raw materials production accounts for about 20 percent of the total value of minerals development. During the last 30 years, minerals production in the former Soviet Union increased by 12 times.

Domestically produced raw materials currently account for 70 percent of Commonwealth minerals resources consumption. Domestic resources account for 95 percent of the fuel consumed, 90 percent of heavy industry inputs, 17 percent of consumer goods production, as well as 70 percent of exports. In general, proved reserves have been increasing faster than production over the past three years; bauxite, nickel, copper, iron, molybdenum and lead ore reserves increased by 1.7–2.0 times, while tungsten and asbestos reserves rose 1.1–1.2 times.

As one of the foundations of the national economy, the mineral resource base has a significant effect on the country's economic development and the employment of its labor force. From 20 to 40 percent of the country's capital investments are concentrated in the mining sector, which employs up to 20 percent of the labor force. During the past five years, the exploitation of 39 major primary minerals has, on average, accounted for around seven percent of the total national product, with annual output valued at approximately 131 billion rubles. In world prices, the annual value of the commodities produced from minerals is almost US$180 billion (Table 3.1).

J.P. Dorian et al. (eds.), CIS Energy and Minerals Development, 29–48.

Table 3.1. Value of Mineral Commodities Produced in the Former Soviet Union, 1991

Commodity Grouping	In Soviet 1991 prices (billion rubles)	In World 1991 prices (billion US$)
Total	130.9	180.9
Fuel and energy resources	87.4	144.2
Oil and gas	71.7	118.4
Coal	15.7	25.8
Ferrous metals	6.0	16.9
Nonferrous metals and rare metals	8.5	10.2
Precious metals and diamonds	24.6	6.5
Nonmetals	4.2	2.2

Note: Figures are rounded.

From 1976 through 1990, 24 billion tonnes of major types of minerals, as well as 9.3 billion tonnes of crude oil and 7.8 trillion cubic meters of natural gas were produced. Compared with output in 1976, the extraction of most major types of minerals had increased by 1990; for instance, output of metal and nonmetal ores rose by 10–30 percent; chemical ores and agricultural minerals, 45–65 percent; coal, 6.5 percent; and natural gas, 180 percent.

The former Soviet Union earned substantial amounts of hard currency from sales of minerals and their endproducts. In 1989 these earnings were 34.5 billion rubles, including 26.3 billion rubles collected from sales of hydrocarbons.

Geological exploration during the last 20 years was mainly aimed at further development and upgrading of the mineral reserve base. At the same time, the increase in proved reserves surpassed the rate of production. Exploration costs have been increasing by about 25 percent every five years. The country's major mining regions were surface mapped at a scale of 1:5,000. Geological surveys and mapping at a scale of 1:200,000 have been completed for almost all of the territory of the former Soviet Union.

Nearly 14,000 mineral deposits exist in the Commonwealth, containing reserves already evaluated. Close to 5,000 deposits are currently under production, while approximately 9,000 deposits are included in the mining industry balance of resources. Mineral industrial development does not exceed 30 to 50 percent of available resources.

In general, the former Soviet Union is self-sufficient in minerals; however, due to several reasons, it also imports certain types of ores, concentrates and metals. The value of imports amounts to around 7.6 billion rubles. In 1989, imports of fuel and energy sources were valued at about 2.2 billion rubles, while nonmetal ore imports cost 189 million rubles. The deficiencies in raw materials and products were caused by the nonrational and inefficient exploitation of reserves and inadequate capital investments in mining industry development.

While possessing a huge minerals base, the former Soviet Union is not well integrated in the world commodity market, and mineral resource sales to foreign markets do not match its tremendous potential.

Even if economic development is accelerated, the new Commonwealth will still lag behind most developed countries. An improvement of living conditions in the CIS will require an intensification of mineral raw materials development. The mineral resources industry is a promising area for international cooperation with many opportunities for joint ventures.

REGIONAL DISTRIBUTION OF MINERAL DEPOSITS AND DEVELOPMENT PROJECTS

Minerals are distributed unevenly throughout the former Soviet republics. The richest regions include Russia, the Ukraine, Kazakhstan, and Uzbekistan, while the Baltic states, Belarus, and Moldova are extremely poorly endowed in minerals. Only Russia possesses substantial quantities of all types of reserves (Table 3.2). Its share in the former Soviet Union's reserves of coal, crude oil, ferrous and nonferrous metal ores, diamonds, and nonmetal ores ranges anywhere from 50 to 100 percent. It is only in manganese and chromium ores, bismuth, germanium, phosphorite, sulfur, barite and kaolin that the Russian Republic's share is less than 20 percent of the former Soviet Union's total proved reserves. Russia accounts for less than 10 percent of the development of manganese and chromium ores, mercury, titanium, sulfur and barite in the former Soviet Union.

The total annual value of minerals production in Russia is estimated at US$150–180 billion, which represents only 1.5 percent of the value of proved reserves. The total value of proved reserves in the republic exceeds 13 trillion rubles (more than 68% of the value of proved reserves in the former Soviet Union). The bulk of the value of minerals reserves in the CIS is comprised of energy reserves which are valued at 7.8 trillion rubles.

Kazakhstan has the second largest mineral resource base in the former Soviet Union. Its mineral resources have been instrumental in the republic's economic development. Mineral industry activities include the extraction and processing of chromium (99.6% of the former Soviet Union's total proved reserves and almost 100% of those under production), iron ore (9.7% of reserves and 7.5% of production), copper (28.4% and 29.6%), lead (38.5% and 64.3%), zinc (35.3% and 56.1%), bauxite (22.1% and 36.4%), tungsten (53% and 3.8%), and molybdenum (29.3% and 5.4%), as well as nonferrous metals, coal (11.9% and 18.6%), petroleum condensates (27.8% and 17.4%), phosphorite (67.4% and 65.4%), barite (81.7% and 83.6%), chrysotile asbestos (20.1% and 19.5%) and others.

At 1.5 trillion rubles, the value of Kazakhstan's proved reserves is third among the republics of the former Soviet Union, representing eight percent of the value of All-Union reserves. The value of energy resources is estimated at 0.7 trillion rubles.

Table 3.2. Distribution of Inferred Mineral Reserves[a] Among Former Soviet Republics (%)

Commodity	Former Soviet Republic								
	Armenia	Azerbaijan	Georgia	Kazakhstan	Kyrgyzstan	Russia	Tajikistan	Ukraine	Uzbekistan
Aluminum	—	—	—	47.5	—	52.5	—	—	—
Apatite	—	—	—	3.0	—	87.0	—	10.0	—
Antimony	—	—	—	5.0	14.4	23.0	52.3	—	—
Chromium	—	—	—	38.6	—	61.4	—	—	—
Copper	1.6	4.5	1.8	32.2	—	52.5	—	—	7.4
Iron	—	—	—	1.8	—	86.2	—	10.2	1.5
Lead	0.3	1.6	0.4	57.6	—	23.9	8.3	0.6	7.3
Manganese	—	—	1.4	21.1	—	70.8	—	6.7	—
Mercury	—	—	—	—	38.6	50.6	—	4.8	6.0
Molybdenum	1.7	—	—	27.1	—	68.3		2.9	—
Nickel	—	—	—	5.0	—	92.7	1.4	0.9	—
Phosphate	—	—	—	33.2	—	26.4	—	2.0	24.0
Tin	—	—	—	4.6	3.0	84.9	3.2	—	1.4
Tungsten	—	—	—	30.0	3.0	61.0	2.0	—	4.0
Titanium	—	—	—	2.0	—	82.0	—	16.0	—
Zinc	0.4	1.2	0.7	49.0	—	37.2	4.8	0.7	6.0

a. Approximately equivalent to the Soviet categories of probable resources, or P_1, P_2, and P_3. Probable or projected resources of basins, areas, and fields are estimated by analogy with similar and explored resources elsewhere. They are subdivided into three categories: projected resources of explored deposits or those under exploration (P_1); projected resources of undiscovered deposits thought to exist on the basis of evidence from geologic surveys, prospecting, and geophysical and geochemical tests (P_2); and projected resources of potentially promising areas where new deposits may be discovered (P_3).

The Ukraine accounts for almost one-half (45.6%) of the former Soviet Union's iron ore production and nearly one-third (30.6%) of its proved reserves. In general, there is an optimistic outlook for iron ore production over the next 40–50 years, although mining operations in Krivoy Rog are often short of ore. The republic's shares of All-Union manganese ore reserves (73.7%) and extraction (78.6%) are also considerable.

The Ukraine is also the Union leader in titanium ore reserves (42.4%) and production (90.1%). This role will continue in the near future with development of the Stremigorod deposit. However, several placers that supply the Irshin ore-processing plant will stop production, while the Upper-Dnepr plant will limit its production and the general quality of processing sands will decline.

The Ukraine is the former Soviet Union's only region that produces potassium sulfate. All other types of minerals are produced by the other republics. The Ukraine also accounts for 16.3 percent of the former Soviet Union's proved coal reserves and 22.3 percent of the coal production. The value of the Ukraine's proved mineral reserves is 1.7 trillion rubles (9% of that of the former Soviet Union), including 1.1 trillion rubles worth of energy resources.

Uzbekistan's minerals base consists mainly of nonferrous and rare metal ores, gold, iron and chemical ores, and natural gas reserves. Currently, the republic produces raw materials for copper and fluorite-processing plants and will have adequate reserves to continue production until 2005. The value of the republic's proved reserves is 0.6 trillion rubles (or 3% of the value of the former Soviet Union's reserves), which includes fuel and energy reserves valued at 0.1 trillion rubles.

These four republics, Russia, Kazakhstan, the Ukraine and Uzbekistan, account for the bulk of the new Commonwealth's proved mineral raw materials reserves and production. There are only a few minerals in other republics that are of importance to the former Soviet mining industry.

Tajikistan, for example, has sizable deposits of antimony and mercury which represent 34.4 and 11.5 percent of former Soviet reserves and 24.7 and 4.8 percent of production, respectively. The republic also has considerable resources of lead (12.7% of reserves, 7.1% of production), zinc (4.9% and 2.6%), other nonferrous and rare metal ores and nonmetal minerals. The potential value of proved reserves is 0.1 trillion rubles.

Mercury and antimony also prevail in Kyrgyzstan, providing almost 55 and 9.5 percent, respectively, of the production of the two metals in the former Soviet Union. The major problems of the Kyrgyzstan mining industry are associated with the extraction of these minerals. The republic has also made significant progress in gold and tin excavation.

The Turkmenistan minerals base consists mostly of natural gas (5.6% of proved reserves and 11.8% of production of the former Soviet Union), with resources second only to Russia, as well as nonmetal ores. Geological exploration in eastern Turkmenistan resulted in the discovery of two deposits of strontium, namely Ariskoe and Sakyrtminskoe.

A wide spectrum of nonferrous ores, coal, barite, manganese and other deposits comprises the mineral resource base of the Caucasian republics. However, only some of them make up significant shares of the former Soviet Union's reserve base. These are the manganese ore deposits of Georgia (7.6% of former Soviet proved reserves and 19.6% of production) and the molybdenum deposits of Armenia (24.0% and 31.8%, respectively). Some of Armenia's molybdenum deposits have already been depleted, causing the Agarskiy processing plant to periodically fall short of ore supplies. Azerbaijan contains already tapped oil fields (1.9% of reserves and 2.2% of production), and lead and zinc deposits.

In Belarus, potash salt deposits (18.8% of former Soviet reserves) provide half of the Commonwealth's total output. The traditional method of extracting and processing potash salt has had considerable environmental impact on large areas around mining operations.

The minerals potential of the Baltic republics and Moldova is represented only by common minerals. Only Estonia possesses noteworthy reserves of oil shales (59.5%) and phosphorites (13.0%), providing almost 0.5 and 1.5 percent, respectively, of the former Soviet Union's output of the resources.

Until recently, the Soviet Union's mineral industrial base had been developing as an integrated sector. It will be difficult for the republics to meet their mineral needs if the supply links of the pre-Commonwealth system are ruptured. Even Russia, which possesses almost all of the minerals its economy requires, still needs supplies of several types of ore, including manganese, chromium, mercury, and titanium, from other republics.

LEVEL OF INDUSTRIAL DEVELOPMENT OF PROVED RESERVES

The shares of proved mineral deposits that are currently under exploitation are as follows: iron ore, more than 80 percent; manganese, 83 percent; chromium, 80 percent; nonferrous and rare metals, 30–70 percent; potash salt, 76 percent; apatite, 50 percent; phosphorite, 59 percent; barite, more than 30 percent; and asbestos, 80 percent. The percentage of proved reserves not yet under development is still rather high and for some types of minerals reaches 25–40 percent.

There are a total of 4,700 undeveloped deposits with reserves, including iron ore deposits in Kursk and Krivoy Rog; magnetite deposits in Siberia and the Far East; bauxite deposits of the northern Onega and the southern Tyumen regions; lead and zinc deposits in central Kazakhstan; tin ore deposits in Yakutiya, the Magadan region, and Kyrgyzstan; tungsten stockworks in Kazakhstan; and stockwork deposits of molybdenum in Belarus and the Chita area. The undeveloped deposits have characteristics similar to those under production, but exploitation will require considerable capital investment for construction of infrastructure because of their locations in remote areas with severe geographical conditions.

A number of deposits of several types of minerals were identified by the State Commission of Reserves 20–30 years ago, even though they have not yet been claimed by industry. These include large-scale deposits of iron ore, dozens of nonmetal ore deposits, and more than a hundred coal deposits suitable to open-pit mining. Thus, between 1966 and 1990 several hundred million rubles were spent for geological exploration on deposits that are not scheduled for development for another 20 years or so. In addition, a considerable number of mineral deposits accounted for in the state budget are anticipated to be put into production after 2015. The reserves of the abovementioned unexploited deposits calculated as percentages of the all-Union total are as follows: iron ore, 30 percent; copper, 23 percent; lead and zinc, 14 percent; bauxite, 39 percent; tungsten, 24 percent; molybdenum, 25 percent; mercury, 40 percent; titanium, 31 percent; phosphorite, 32 percent; fluorite, 16 percent; and coal, 45 percent. Detailed studies of these deposits are to be completed shortly in order to estimate more accurately the actual mineral supply situation.

Quality of ores and ease of processing

The quality of a number of deposits with proved reserves in the former Soviet Union is inferior to that in other countries, although this may partially be the result of differences in the definition of reserves internationally. The average iron content in Soviet ore is 1.2–1.5 times lower than that in well-known deposits in other countries, while for manganese it is 1.8 times lower; copper, 1.2; nickel, 2.2; lead, 1.8–2.7; zinc, 2.3–2.7; tungsten, 1.4–3.0; molybdenum, 1.5; apatites, 1.3; and phosphorite, 1.6 times lower.

In recent years, newly identified deposits have exhibited increasingly lower grades of ore. For example, copper content in ore has declined by 30 percent over the past few decades. When compared to the situation in 1960, iron content has dropped 23 percent; manganese by 14 percent; tungsten, 24 percent; molybdenum, 30 percent; tin, 25 percent; antimony, 15–20 percent; mercury, 30–40 percent; and phosphorite anhydrides in apatite, 39 percent.

The percentages of easily processed ores in proved reserves were evaluated on January 1, 1989 to be as follows: iron ore, 11 percent; manganese, titanium, and tin, 17–20 percent; tungsten, 14 percent; and molybdenum, 22 percent. In contrast, the shares of ores that are difficult to process were: iron ore, 12 percent; manganese, more than 40 percent; bauxite, almost 50 percent; tin, approximately 40 percent; molybdenum, 17 percent; phosphorite, 25 percent; and barite, more than 50 percent. The shares of ores that are difficult to process are anticipated to increase in the near future by 10–15 percent for ferrous metals and 22–25 percent for nonferrous metals.

Rational and complex development of minerals

The deterioration of ore quality in the former Soviet Union necessitates more sophisticated processing technologies which will increase the costs of the endproducts. The exhaustion of the mineral reserves now under exploitation, the worsening quality of ores, and the use of outdated and inefficient technologies for development and processing have led to considerable losses in the output of main products and by-products, as well as nonrational and incomplete development of the mineral resources. Total losses in production and concentration are estimated to be as high as 26 percent for iron ores; 30 percent for manganese; and 15 percent for chromium. Extremely high losses occur in nonferrous metals production. For example, during extraction and processing, losses for major minerals are as follows: copper, 25 percent; lead, zinc, aluminum, tin, tungsten, antimony, titanium, vanadium, 35–40 percent; zirconium, 45 percent; cobalt, 64 percent; potash salt, 62 percent; barite, 68 percent; apatite, 20 percent; and phosphorites, 31 percent.

At present, only 10 percent of the 15 billion tonnes of rocks mined annually are processed. In general, minerals losses are as high as 50 percent during mining and processing. Every year more than 205 million tonnes of ferrous, nonferrous and rare metals are lost, while chemical ore and coal losses are valued at 12.4 billion rubles. With more integrated development, the efficiency of production could be improved by at least 15 percent and costs could be reduced by 20–30 percent. There are plans, however, to reduce capital investments by 40–50 percent and mining of rock by 15–20 percent.

In evaluating the minerals base of the new Commonwealth, it is necessary to mention that significant amounts of mining wastes containing ferrous, nonferrous and rare metals are dumped or stockpiled. Consequently, these secondary raw materials can and should be considered as an additional source of minerals. The technologies for processing secondary raw materials are much more advanced in the rest of the world than in the former Soviet Union, especially for nonferrous metals production. A preliminary estimate of the share of secondary raw materials in nonferrous metals production is as follows: lead, 30 percent; tantalum, 21 percent; copper, 12 percent; nickel, cobalt, tin and mercury, 6–9 percent; niobium, 28 percent; and zinc, 1.5 percent.

Minerals in the former Soviet Union can be categorized in accordance to their abundance. The first group is characterized by huge reserves that are likely to supply industry effectively for a long time, or 70–100 years or more (Table 3.3). These are ferrous metals, coal and nonmetallic ores. The second group comprises the major nonferrous metals, whose reserves will be able to provide supplies over the long term (50–100 years), although a considerable share of these have been inefficiently developed. As such, much of the reserves cannot be put into production in the near future.

The third group includes minerals with insufficient reserves to ensure further industrial development after the year 2000. These minerals include tin, antimony, mica, muscovite, and fluorite, as well as nickel-rich ore, oxidized manganese ore, and others. Reserves of these minerals contain supplies sufficient for less than 50–60 years.

In general, the present day mining industry is based on reserves sufficient for 15–40 years, excluding some limited types of raw materials and certain mines.

Table 3.3. Mineral Reserves of the Former Soviet Union in 1991 and 2006[a]

Commodity	January 1991 Mineral Reserves as % of January 1986 Reserves	Years of Remaining Reserves			
		January 1991		January 2006	
		Total	Developed	Total	Developed
Copper	102	85	43	73	42
Lead	106	104	31	75	33
Zinc	106	103	37	74	34
Nickel	103	59	43	60	39
Cobalt	100	62	46	60	42
Bauxite	102	178	68	88	44
Nepheline	94	74	39	75	28
Tin	107	57	22	34	11
Tungsten	107	106	32	59	24
Molybdenum	106	163	61	79	38
Antimony	76	23	16	34	13
Mercury	102	57	46	74	40
Titanium	100	535	37	135	29
Zirconium	133	127	33	100	20

a. Preliminary forecasts.

However, most gold and diamond-processing plants in northeastern Russia and the Far East are not adequately supplied with reserves. Limited reserves are also a problem for tin-processing plants in the Magadan region, copper plants in Kazakhstan and in the Urals, lead and zinc plants in the north Caucasus, Primoriye, and Baykal regions and Kazakhstan, tungsten mines in Primoriye and Buryatiya, and open-pit chromium ore mines in western Kazakhstan.

One of the troubling characteristics of mineral resource development in the Commonwealth is the inconvenient geographical location of mineral resources in relation to the already existing mining and processing facilities. The majority (75–90%) of manganese and iron ore, phosphorite, apatite and bauxite deposits were, for example, discovered in the western part of the former Soviet Union and in Kazakhstan. At present, there are relatively few mineral deposits being developed in Siberia and the Russian Far East, where abundant sources of energy are available for energy-intensive production. The bulk (30 million tonnes) of iron ore is shipped to the Urals from Kursk, Kazakhstan and Karelia which generates added transportation costs of 200 million rubles. About 80 percent of the coal reserves occur in the Asian part of the former Soviet Union, while 80 percent of the industrial demand is in the European part.

DEMAND FOR MINERAL RAW MATERIALS IN 2005

The former Soviet Union is the world's largest consumer of minerals and metals. Table 3.4 lists consumption levels of selected minerals in the country for the years 1985–1991. The table suggests that Soviet consumption of minerals declined in the past several years due, in part, to a deteriorating economy.

An analysis of recent world trends of consumption and utilization of raw minerals and their endproducts shows that despite deceleration in the growth rate of consumption to date, there is no basis to forecast a decline in demand for energy, metallic and nonmetallic mineral resources. Many recent forecasts project that the reverse will occur. Even with extensive introduction of substitutes, an increase in the role of secondary resources, and the application of material- and energy-conserving technologies, the demand for primary ore mineral resources will grow through 2000, though the rate will be slower than that prior to the mid-1980s. In estimating future demands of industry for mineral raw materials, it should be pointed out that the level of per capita consumption for most types of mineral resources in the former Soviet Union is lower than in the developed foreign countries.

The demand for the most important types of mineral raw materials in 2005 will exceed that of 1985, with the exception of iron ore and manganese ore, which will be consumed at approximately the same level as in 1985. The highest growth rates in demand will be for titanium (2.9 times) and sulfur (2.4 times), while demand for aluminum, cobalt, tungsten, molybdenum, chrysotile asbestos, barite and coal will be 1.5 to 2.3 times higher. The demand for chromium, copper, nickel, zinc, lead, mercury, antimony, phosphate raw materials, potassium salts, mica

Table 3.4. Minerals Consumption in the Soviet Union, 1985–1990

Commodity[a]	Year						% of world[b]
	1985	1986	1987	1988	1989	1990	
Aluminum	1,750,000	1,750,000	1,800,000	1,800,000	1,750,000	1,650,000	9.5
Cadmium	2,900	2,700	2,500	2,550	2,300	2,000	10.2
Copper	1,305,000	1,300,000	1,270,000	1,210,000	1,140,000	1,000,000	9.3
Lead	790,000	760,000	775,000	790,000	700,000	650,000	11.8
Magnesium	76,000	78,000	80,000	80,000	80,000	77,000	22.9
Nickel	138,000	137,000	135,000	130,000	120,000	115,000	13.6
Tin	31,500	31,500	29,000	28,000	24,000	20,000	13.6
Zinc	1,000,000	1,010,000	1,030,000	1,080,000	1,020,000	920,000	13.2

Note: figures are rounded.

NA Not available.

a. Refined

b. For latest year available

and fluorite is projected to rise by factors of 1.1 to 1.4. Per capita minerals consumption and production in the Commonwealth lags behind developed countries in the West; the gap will not be narrowed until the beginning of next century.

Mineral deposits that have already been explored and prepared for development will support planned increases in mining activities of 10–50 percent to the year 2005 for chromium ore, copper, antimony, mercury, lead, zinc, potassium salts, crysotile asbestos, coal and sulfur; and 60–100 percent for vanadium, tin, tungsten, molybdenum, fluorite, bauxite, and phosphorite. The level of mining of apatite and manganese ore will remain stable.

Mineral raw materials in the former Soviet Union will continue to meet the demands of industry. The share of secondary resources in total mineral raw materials production will comprise 10–30 percent, sometimes more. By 2005 the shares of secondary resources in the production of various metals are estimated to be the following: for aluminum production, 23 percent; copper, 14.6 percent; nickel, 8.5 percent; cobalt, 9.4 percent; lead, 42.4 percent; tin, 25.9 percent; tungsten, 28.6 percent; molybdenum, 12.8 percent; antimony, 25.9 percent; and mercury, 10 percent.

The growth in demand for mineral resources and the resulting increased mining activities will lead to the depletion of many mineral deposits and fields. Between 1991 and 2005, 25.4 percent of explored and developed coal reserves, 19 percent of iron ore reserves, 12.5 percent of manganese ore reserves and 24 percent of chromium ore reserves will be exhausted. The increase in the mining output of nonferrous, scarce and alloying metals will be even greater, causing within the span of fifteen years the depletion of 37 percent of all explored resources of copper, nickel, and molybdenum; 53 percent of all zinc reserves; 63 percent of all lead reserves; and 65 percent of all tungsten and zirconium reserves. Antimony, tin (97.7%) and titanium reserves will be completely or almost completely depleted. Considerable exhaustion of nonmetallic mineral deposits also is anticipated: apatite, 34 percent; mica, 33 percent; chrysotile asbestos, 25 percent; and phosphorite, 19 percent. The fluorite reserves will be completely depleted.

The considerable depletion of reserves and cessation of the industrial exploitation of mineral deposits will inevitably affect the supplies of some mineral raw materials for a number of industries and regions. It is anticipated that by the end of 2005 (if there are no increases in reserve amounts) supplies of mined nickel, bauxite, copper, zinc, lead, tungsten, mica, and apatite will no longer satisfy demand. Explored reserves of tin, antimony and fluorite will be used up almost completely. Only the mined supplies of coal, iron and manganese ores, molybdenum, potassium salt and raw cement will exceed demand for the next 50 years.

At the abovementioned rates of development of the mining industry, the share of old mining regions in the total output of mineral resources will decrease dramatically. Depleted capacities will be compensated for by the establishment of new mining enterprises based on already explored mineral deposits, as well as mineral deposits discovered during the 13th through 15th Five-Year Plans. In order to compensate for depleted resources, the rate of geological surveying in 1991–2005 should exceed the rate of development of the mining industry in order to provide the industry with explored resources and focus on sites with the best technical and economic characteristics.

CHARACTERISTICS OF PROJECTED MINERAL RAW MATERIAL RESOURCES

An evaluation of projected or forecasted mineral raw material resources, carried out at the beginning of 1988, assessed the prospects for development of the Soviet raw minerals base and the changes resulting from various geologic surveys in recent years. Projected resources of basins, areas, and fields are estimated by analogy with similar and explored resources elsewhere. The majority of forecasted mineral resources is 1.1–2.9 times greater than their balance or economically mineable resources. However, the raw materials base of bauxite and manganese is unsatisfactory, as forecasted resources are limited.

In terms of ore quality, projected resources are uneven. Average content of ore in the forecasted resources of manganese, copper, lead, zinc, tin, tungsten, molybdenum, titanium and phosphorite is slightly higher than in the balance ores, which indicates that improvement is possible in the quality of the raw materials base of these mineral resources. The quality of forecasted mercury and potassium salt reserves remains the same as that of the balance reserves, while for other mineral resources (iron and chromium ore, nickel, aluminum, antimony, apatite, etc.), the average content of ore in forecasted resources declines in comparison with the balance reserves. Thus, the most important task for prospecting is the discovery of newer, higher grade deposits which can considerably improve the qualitative characteristics of the minerals being mined.

About 30 to 50 percent of all forecasted resources fall into the category of extrapolated reserves, the basis on which geologic surveys at a scale of 1:5,000 are planned. Sufficient opportunities for identifying additional extrapolated reserves exist with further geologic surveys. Nevertheless, for a number of minerals the reserve supply is very short; for iron ore, bauxite, titanium and potassium salts the shares of extrapolated reserves are less than 15–20 percent, and for mica the share comprises only one percent.

As of 1988, for selected minerals, 10 to 20 percent of the total resources are categorized as explored and blocked-out reserves, and the quality of ore tends to be poor or their geographic, economic, mining and technical conditions are not favorable. The increase in reserves for a majority of Commonwealth mineral resources (with the exception of crude oil, natural gas, gold, tin and some other scarce resources) will be influenced by the degree of the resource depletion and future demands for the raw material. The projected increase of reserves in 1991–2005, compared with the previous 15-year period, will be 10–30 percent for iron ore, vanadium, bauxite, nepheline, coal, and potassium salts; 40–60 percent for copper, lead, zinc, nickel, cobalt, mercury, antimony, titanium, manganese ore, apatite, barite, and phosphorite; and 70–90 percent for gold, tin, tungsten, sulfur, and fluorite.

One of the main tasks for future geological surveys is the improvement of efficiency in exploration and reliability of forecasts. In the immediate future, it will be necessary to develop and adopt radically upgraded scientific and methodological foundations for geological surveys in order to establish reliable bases for exploration.

Improvement of geological surveying programs and expansion of the mineral raw materials resource base will require considerable funding. Total expenditures for geological surveys between 1991–2005 are estimated to approach 140–150 billion rubles. An estimated 105–110 billion rubles, or 70 percent of all expenditures (including 30–32 billion rubles in 1991–1995), will be allotted for the prospecting and exploration of oil and gas deposits, which is considered to be the most important and labor-consuming task. The number of geological surveys of solid mineral resources to 2005 will remain at the level of the last five years, and nearly 70 percent of all funds will be allotted for the improvement and increase of the raw material bases of the operating enterprises of ferrous and nonferrous metallurgy, gold mining, diamond mining, coal mining, and the chemical and construction industries.

Due to the high rate of reserves depletion and the necessity of finding new mineral deposits and building new operations, expenditures will rise. The share of funds for reserve studies in the total amount of allocations for geological surveys will grow from 60 to 75 percent. At the same time, the decline in the share of expenditures for prospecting and exploration will continue.

Raw materials for ferrous metallurgy

Though the amount of explored reserves for ferrous metallurgy is comparatively high (with reserves sufficient for 75–150 years), there are significant problems associated with the locations and industrial exploitation of the raw material resources. Explored reserves encompass categories A, B, C_1 and C_2 of the Soviet classification system, indicating that the mineral deposits have been thoroughly investigated and the mode of occurence, shape and structure of the ore bodies are known. There is an iron-ore deficiency in the Urals, where the metallurgical plants receive marketable ore transported from the Kursk magnetic anomaly, Karelia and the Kola Peninsula. Every year, the metallurgical works of Novokuznetsk in Western Siberia receive nearly two million tonnes of marketable iron ore from the Kursk magnetic anomaly region, although there are explored reserves nearby. The marketable ore is transported over a distance of 4,200 kilometers (km).

Because of the considerable deepening of strip mines (deeper than 300 meters) within the former Soviet Union's largest integrated mining and dressing works in the Krivoy Rog coal basin, problems in maintaining productive capacity have worsened. Of the former Soviet republics, only Kazakhstan has commercial deposits of chromium ore, while more than 90 percent of the metallurgical potential of the new Commonwealth is concentrated in Russia and the Ukraine. The major problem with the chromium ore base of western Kazakhstan is a depletion of reserves suitable for open-cast mining; the remaining reserves total 24.1 million tonnes, or seven percent of all reserves of the republic. Open-cast mining output in 1989 reached 2.4 million tonnes, leaving reserves that will last not more than ten years. Underground mining in the deposits of western Kazakhstan requires considerable expenditures due to rigorous mining and geological conditions. Another problem of the chromium ore base is an almost complete lack of ore deposits with a high content of alumina; instead, the industry uses the scarce metallurgical ores of western Kazakhstan.

The main problem of the raw materials base of manganese ore is the disproportion between the amount of the reserves and mining output of easily beneficiated oxide ore. The reserves of manganese oxide ore amount to only 17.6 percent of total manganese ore, while mining output represents 63.2 percent of all mining output for the oxide in the former Soviet Union. The situation with the reserves of scarce peroxide ore (0.22% of all the former Soviet Union reserves) and low phosphorite oxide ore (3% of all the former Soviet Union reserves) is even more critical.

The geographical locations of the explored reserves of manganese ore also do not correspond with the locations of their users. Reserves are concentrated in Georgia, the Ukraine and Kazakhstan (but actual mining is conducted mainly the Ukraine and Georgia), while 50 percent of all users are located in Russia, which lacks its own raw materials base of this mineral.

Raw materials for nonferrous metallurgy

Demand in the former Soviet Union for raw materials required in the nonferrous metals industry is satisfied by its own reserves. Through 2005, its self-sufficiency in almost all types of raw materials is expected, as a result of a 1.2–1.6-fold (for tin, 3-fold) increase of mining output at operating deposits and deposits to be put into production, excluding aluminum, zirconium and a number of scarce metals. The planned growth rates of mining output will considerably reduce the existing raw minerals reserves, and for such mineral resources as tin and antimony, reserves will become insufficient for continued production at the planned rates after 2005.

A considerable portion of the explored reserves of nonferrous metals consists of resources that are difficult to beneficiate. This is especially the case for stratiform complex deposits, copper-pyrite deposits, cassiterite and sulfide deposits, as well as for nepheline and alunite ores that are processed in order to obtain alumina.

There are some cases where reserves are located at great distances from processing capacities. Due to a lack of sufficient bauxite resources in Siberia, considerable amounts of alumina are transported to the processing facilities located in this region from other parts of the Commonwealth. Extensive transportation operations are needed to provide raw materials for the Severonickel integrated works, and the Almalik, Balkhast and Sredneuralsk plants.

A peculiarity of nonferrous metals reserves, and especially tin and antimony, is their concentration in unique and very large deposits that comprise a significant part of the raw mineral reserves in the extreme northern regions of Russia. Due to a lack of investment, these deposits have not been put into operation.

Precious metals and stones

Gold. A characteristic of gold mining in the former Soviet Union which distinguishes it from other precious metal-producing countries is the considerable share of gold that is extracted from placers (45.5%) and the comparatively small amount of gold in explored (11%) and forecasted reserves (8.8%). There are considerable gold ore reserves in various unexploited deposits in the former Soviet Union which, when put into operation, would make it possible to increase total gold output by a factor of 1.5. However, 40 percent of all gold reserves occur in large deposits containing poor quality ores (for example, Sukhoi Log, Kumtor, and Vasilkovskoye) that are not developed due to lack of investment. Thirty percent of the reserves (Maiskoye, Nezhdaninskoye, and Kokpatas) have ores containing arsenic, which makes beneficiation difficult, thus complicating their industrial development.

Notwithstanding, the explored and forecasted or projected reserves of placer gold in the former Soviet Union are still very large, while the task of maintaining and increasing extraction remains very difficult. Due to the depletion of the most important gold placers in Magadan Province and Yakutiya, the former Soviet Union has experienced a permanent decline in the level of growth of the reserves and output of the metal in these regions. In order to maintain the Commonwealth's gold production, it is necessary to develop other reserves and increase mining output in other regions, including Amur Province of the Russian Far East.

The Navoiiskii mining and metallurgical integrated works in Uzbekistan is a primary installation for gold mining in the Commonwealth. Its endowment of explored reserves is sufficient for supporting 16 more years of production.

Diamonds. For gem quality diamonds, explored reserves in kimberlite deposits account for 97 percent of reserves, while only three percent of all reserves occur in placer deposits. Based on current levels of diamond mining output, enterprises will have reserves sufficient for 20 to 40 years of continued production. However, the reserves for open-cast mining will last only 15 to 17 years. A critical situation involving diamond raw materials has developed in Yakutiya, which is the main diamond mining region of the country. As a result of intensive exploitation, reserves for open-cast mining are almost completely depleted in the three largest diamond deposits (Aikhal, Mir, and Internatsionalnaya). Preparatory operations for underground mining are being carried out at the Internatsionalnaya diamond pipe.

Recently, nearly 90 percent of the former Soviet Union's total diamond mining output occurred in the Udachnaya diamond pipe. Explored reserves for opencast mining will ensure the production capacity of this deposit for only 10 to 12 more years.

A decision as to whether to begin construction of a mining and beneficiation integrated works at the Lomonosov diamond field in Arkhangelsk Province has not been reached yet. The amount of reserves in this region suitable for mining will not compensate for the probable decline of diamond mining output in Yakutiya.

Other minerals

Raw phosphates. Despite large explored reserves, there is a considerable shortage of phosphorus fertilizers in the former Soviet Union, resulting in a level of fertilization that is 2–2.5 times lower than in the developed countries. The main problems with the country's phosphate raw materials reserves which limit growth in mining output and prevent the industry from meeting the demands of the economy include: low quality ores in reserve deposits; the occurrence of very deep-seated phosphorite-bearing horizons at great depths; and a lack of deposits suitable for industrial development in Siberia and the Far East.

Potassium salts. The bulk of explored reserves of salt in the Commonwealth consists of chloride salts; five percent of the reserves contain scarce phosphate salts. New deposits of sulfate salts are not likely to be located as they make up only about one percent of all forecasted reserves.

Raw materials for agriculture. A broad variety of nontraditional mineral raw materials, such as peat, bentonite, ceolite, and glaukonite, are used to improve the quality and richness of soils in the Commonwealth. These minerals can be used for plant growing, as well as poultry breeding, and livestock and fish farming.

The total amount of explored reserves of agricultural raw materials in the former Soviet Union is 3.6 billion tonnes, including ceolite, 250 million tonnes; bentonite, 627 million tonnes; diatomite and tripoli earth, 1.9 million tonnes; vermiculite, 47 billion tonnes; and palygorskite, 23.0 million tonnes. Preliminary estimates indicate that consumption of nontraditional raw materials and natural sorbents by the various branches of agriculture and industry in the country will probably total 1.8 million tonnes in 1995 and 3.8 million tonnes by the year 2005. Use of these raw materials is limited due to insufficient technologies for utilization.

Energy raw materials

Oil and natural gas. By 2005, oil production in the former Soviet Union will stabilize at the 1985 level, while natural gas production will be 1.7 times higher. The projected growth rate of oil and gas reserves will exceed the rates of expansion recorded between 1966 and 1985 by 1.2 and 1.23 times, respectively. The increase in reserves will be primarily concentrated in Western Siberia, in Timano-Pechora Province, and in the Caspian depression.

In order to initiate detailed geological surveying activities for oil and natural gas, there are plans to increase Russia's state budget and capital investments 25–30 percent for each pending five-year period which will total more than 75 percent of all investments allocated by the Russian Ministry of Geology for the period until 2000. The largest share of the funding will go to Western Siberia; its share will exceed 50 percent of all appropriations.

The planned increase in natural gas output in Western Siberia will be facilitated by abundant explored reserves. The oil reserve base is, however, more limited. It will require a considerable increase in reserves and the opening up of new oil fields to maintain the current level of oil production.

Explored reserves and forecasted resources of hydrocarbons are concentrated in the Komi ASSR, Timano-Pechora Province, and the Ninets Autonomous District. There are also good prospects for finding new oil and gas fields on the continental shelf of the Barents Sea, but the bulk of the reserves is thought to be located in waters deeper than 3,000 meters.

The hydrocarbons presence in the Caspian depression, which is considered to be one of the most promising areas for new reserves, is located across a wide stratigraphic sequence. The bulk of the forecasted resources is associated with complex rock formations, although knowledge down to depths of 4–4.5 km is very limited. Further geological surveys are recommended to depths of between five and seven kilometers. The surveys will focus on the southern Mangishlak and northern Ustyurt gas and oil-bearing regions.

The opportunities for finding new hydrocarbon reserves in the Urals-Povolzhlskii region to compensate for historical depletion are diminishing. The geology and subsoil of the region have already been explored in detail. Possibilities for increases in reserves are limited, and a gradual decrease in reserve quantities and production amounts is anticipated.

The prospects for developing oil and gas in eastern Siberia are dependent on the forecasted resources of the region. Local mining operations would have to be expanded. In the Far East, there is a high probability of finding hydrocarbons on Sakhalin Island and its continental shelf, as well as in western Yakutiya. Intensification of local gas and oil surveys will make it possible to define the most promising areas for future exploration.

As for the historical oil-producing regions of the Commonwealth (Azerbaijan, the Ukraine, Uzbekistan, Turkmenistan, and Georgia), the growth of reserves will only be sufficient to maintain oil production at current rates. At the same time, there are good prospects for finding new gas fields in areas of present-day operations. Prospects for discoveries are better in Central Asia than in the Ukraine and Azerbaijan.

Recent surveys for oil and gas occurrences on Soviet continental shelves yielded positive results. Surveying of the continental shelves is planned to increase considerably in the future, ensuring a steady growth of explored reserves. Total expenditures for oil surveys on continental shelves will grow 3.6 times.

Coal. Coal mining output in the year 2000 will exceed the 1985 level by 63 percent. However, the expansion of coal reserves between 1986 and 2005 will amount to only 32 percent of their growth in 1966–1985. The main focus of geological surveys will continue to be on coking coals, as well as bituminous coals suitable for open-cast mining.

The primary reserves at the Donetsk coal fields are concentrated in deep horizons, leading to rigorous mining conditions. In the future, exploration activity will decrease and concentrate on areas with probable occurrences of scarce coking coal and anthracite. Surveys will continue in the northern and eastern parts of the Donetsk coal fields.

Surveying, exploration and exploitation of coking coal reserves will be continued at the Kuznetsk coal field. These reserves will be reevaluated for possible open-cast mining. The geological surveys in the Karaganda and Pechora coal fields will also serve to reevaluate reserves, particularly the vast reserves of steaming coal, despite their high ash content. There is no need for further increases of reserves in the Kansk-Achinsk coal field, though it is necessary to conduct surveys to better define coal seams found in the southern part of the field. Surveys of new areas for coking coals suitable for open-cast mining will continue in the southern Yakutiya coal field. Coal surveys will concentrate on coal-bearing areas adjacent to the abovementioned fields. Exploration in the Urals will aim to put into production forecasted coal resources located in the area.

CONCLUDING REMARKS

In the past, the development of the raw materials resource base of the Soviet Union took place within a system of command and control, under which the country was almost completely isolated from the world economy. Under such conditions, the Soviet Union relied on self-development of its resources. Policies in the raw material sectors caused imbalances and distortions in the pricing of the minerals, in the organizational structure of the country's mining industry, the determination of exploration and output targets, and in the allocation of investments and use of technologies. Clearly, the mining industry adopted by the new Commonwealth is plagued by severe problems of inefficiency and waste which will take years to overcome.

The mining industry of the 1990s will be characterized by the following conditions:

- an integration of the Commonwealth into the world economic community by permitting free exchange of raw minerals in domestic as well as international markets at world prices;
- the attraction of new domestic and foreign investment for the mining of mineral deposits;
- more rational exploration, mining and utilization of the mineral resources;
- abandonment of the exclusive right of Commonwealth mining enterprises to the subsoil and all types of mineral resources (including diamonds and precious metals);
- a shift toward a balance between the demand for mineral resources and their supply on the world market;
- thorough consideration of environmental consequences during the mining, processing and use of mineral resources; and
- a greater use of economic criteria in the implementation of the mineral raw materials minerals policy of the Commonwealth.

Chapter 4

JOINT MINING VENTURES IN THE FORMER SOVIET UNION: A FOCUS ON THE RUSSIAN FEDERATION

James P. Dorian and Pavel V. Karaulov

BACKGROUND AND INTRODUCTION

Despite the historical achievements of mining in the former Soviet Union, the industry inherited by the new Commonwealth of Independent States (CIS) is today encountering problems in performance, efficiency, transportation and utilization. The industry operates below capacity and is plagued with outdated technology and equipment, rising production costs, capital shortages and labor strikes, in addition to general economic deterioration and regional civil unrest. Attempts to boost domestic minerals supplies in the republics have failed so far because of economic difficulties. And the former Soviet republics lack the financing required to initiate new, large-scale development projects.

In an effort to reinvigorate the mining industry, the former Soviet republics are now encouraging foreign participation in minerals and energy development through joint ventures. Having been effectively closed to foreign investors for decades, the newly independent states are now trying to attract overseas investment. State leaders desire to expand their nations' mineral and energy sectors and improve the efficiency of raw material utilization in their countries by introducing capital, technology, machinery and science from abroad. This first massive effort by the leadership of the new states to seek joint ventures in mining represents a departure from traditional self-sufficiency principles.

This chapter presents an examination of joint mining ventures in the former Soviet Union, including the major spheres of activity pursued, problems and concerns of the international investment community, and recent developments in investment legislation. The text also includes a detailed analysis of joint venture opportunities in the Russian Federation following the dissolution of the

J.P. Dorian et al. (eds.), CIS Energy and Minerals Development, 49–69.

Soviet Union in late 1991. Russia's foreign investment climate is closely examined in this chapter, including the republic's privatization law and taxation legislation. Of the fifteen former Soviet republics, Russia is in the forefront in establishing investment policies. Most importantly, Russia possesses the largest mining industry of the new independent states, accounting for a significant share of the former Soviet Union's total minerals and energy output and trade. The chapter concludes with an assessment of the long-term outlook for joint mining ventures in the Russian Federation.

SOVIET EFFORTS TO ATTRACT INVESTMENT

In an effort to foster economic cooperation with foreign nations, Soviet government officials restructured legislation in the late 1980s pertaining to the establishment of joint ventures in the country. Initial legislation was established on January 13, 1987 with two decrees of the Presidium of the USSR Supreme Soviet and the USSR Council of Ministers. On November 18, 1989, the USSR Supreme Soviet adopted a resolution to produce and finalize a law on ownership rights in the Soviet Union which, remarkably, represented the first time in seven decades that private ownership was actually considered. With the creation of a legal environment guiding joint ventures in the Soviet Union, hundreds of firms from around the world have established joint ventures in a variety of fields, although a disproportionately high number of joint enterprises are in the services sector. Table 4.1 presents a breakdown of the distribution of foreign investment in Soviet joint ventures by branch of activity, as of November 1991.

In order to open up the country's vast minerals wealth to the international mining community, in December 1989 the Soviet government issued the decree "On Reconstruction of Foreign Economic Activity of the USSR Ministry of Geology," which delegated the authority to the Vniizarubezhgeologia (VZG) or All-Union Research Institute of Geology of the USSR Ministry of Geology to furnish overseas companies with geological data with the objective of creating mining joint ventures in the Soviet Union. To further this goal, the Soviet government hired the Robertson Group of the United Kingdom under an exclusive agreement to serve as a marketing agent for the country's resources. Working with the VZG, the private consulting firm provided international mining companies with data necessary to assess project potential and guidelines for establishing joint ventures. Information supplied by the Robertson Group in cooperation with the VZG included ore deposit data summary sheets, mineral data passports, and full data packages.

Robertson's role as a marketing broker of Soviet mineral deposit opportunities eventually disappeared as foreign mining companies chose to negotiate directly with Soviet authorities rather than through a third party.

Despite the many efforts to improve the investment environment, international mining companies were slow to respond to the Soviet offers, though interest remained extensive. Clearly, most companies adopted a cautious approach in seek-

Table 4.1. Foreign Investment in Soviet Joint Ventures by Branch of Activity[a] (thousand transferable rubles)

Country	Branch of Activity				
	Building materials	Producer goods	Con-struction	Services	Total
Sweden	12,834	2,541	—	59,843	75,218
Germany	—	1,000	537	—	1,537
Bulgaria	—	3,510	—	—	3,510
Finland	400	47	—	160	607
Cyprus	—	—	32	835	867
Italy	—	323	175	—	498
Poland	50	1,115	500	5	1,670
Switzerland	16	1,476	—	44	1,536
United States	165	624	—	150	939
United Kingdom	—	—	459	—	459

a. Through November 1991

ing joint venture opportunities in the Soviet Union. The lackluster response by the overseas mining community was evidenced by the disappointingly small number of joint ventures that were registered involving geology, metallurgy or mining (Table 4.2). Although nearly 3,000 joint ventures were registered with the Soviet Ministry of Finance through 1990, less than four percent were in the minerals and energy fields. Twenty-eight countries were represented in the agreements, with thirteen joint ventures involving firms from the United States. A few large foreign commitments to oil and gas development during the year brightened an otherwise gloomy picture for the Soviet government. Oil-related joint ventures contributed a modest one percent of Soviet national product by the end of 1990.

In 1991, several additional projects were negotiated or finalized, suggesting continued interest and some promise for the long term. However, the failed coup in August and expanded political disintegration did little to comfort foreign investors seeking opportunities for participation in the country's domestic mining industry. By year-end, the Soviet Ministries of Geology and Metallurgy were negotiating with a handful of overseas companies for additional joint ventures to process secondary metals, modernize existing metallurgical works, exploit ore deposits, rework previously discarded tailings and dumps, renovate environmental pollution-monitoring devices, and set up enterprises in partnership with foreign firms in developing countries. Negotiations for an additional forty joint ventures in oil extraction were held during 1991, with more than half of the projects eventually registered. Through October 1991, around fifty joint ventures in the petroleum sector were registered, of which sixteen are functioning (Interfax News Agency, October 24, 1991).

Many joint ventures being discussed today as well as those already finalized involve transfers of mining technology to Commonwealth partners. For exam-

Table 4.2. Soviet Joint Ventures in Minerals and Energy, 1987-1990

Characteristic	Number/description
Total registered joint ventures in minerals and energy, 1987-1990[a]	118[b]
Minerals	91
Energy	27
Major spheres of activity	
Minerals	Nonferrous and ferrous metals processing; production and processing of building and industrial minerals; metallurgy technology, development and licensing; equipment manufacture; and construction/renovation of processing facilities
Energy	Petrochemical development, processing, and research; petroleum engineering, supply, and management; peat processing; marketing services for gas; oil industry processing and technology
Breakdown by country[c]	Australia (3), Austria (9), Belgium (1), Bulgaria (3), Canada (4), China (1), Cyprus (6), Czechoslovakia (1), Finland (10), France (5), Germany (15), Great Britain (11), Greece (1), Hungary (2), India (2), Ireland (1), Italy (8), Liechtenstein (2), Luxembourg (4), Netherlands (3), New Zealand (2), Poland (5), Singapore (1), Spain (3), Sweden (3), Switzerland (5), United States (13), Yugoslavia (1)
Mineral deposits offered through Robertson Group-VZG cooperation[d]	121
Breakdown by mineral category[e]	Industrial materials and construction stones (61), iron and ferroalloys (31), base metals (15), mineral fuels (6), rare metals (5), precious metals (2), and mineral waters (1)
Operational joint ventures[f]	1 in 3

a. Registration with the USSR Ministry of Finance. Not necessarily a complete list. Includes all registered joint ventures number 1-1,792, the first 100 U.S.-Soviet joint ventures registered in the USSR (numbers 13-913), and all other joint ventures announced in published sources through 1990.

b. Generally related to minerals and energy development, utilization, and processing, geology, research, management, and other related areas.

c. Through June 1990. Not necessarily a complete list.

ple, in 1990 the huge Soviet Far East gold mining association, Severovostokzoloto, launched a joint venture with an Alaskan-based trading company to market mining technology and develop mineral deposits on both sides of the Bering Straits. The Far East partners are expected to provide advanced permafrost technology as well as specialized techniques and equipment for large-scale placer operations, while the U.S. firm will make available extensive automation technology.

Factors prompting overseas mining companies to be cautious before investing in the Soviet Union were numerous. Social disintegration threatened the stability of the country, while economic conditions deteriorated at an alarmingly fast pace. Other factors included growing uncertainty over ownership rights to mineral resources; difficulty in obtaining detailed geological data despite the services of the Robertson Group; an unclear decision-making hierarchy; the nonconvertibility of the Soviet ruble; a failing banking and currency system; an untested legal and taxation environment; poor infrastructure; and the availability of predominantly small, low-grade, and remote ore deposits to the overseas investor community (Dorian, June 1991). The uncertainties created by perestroika clearly tarnished the investment climate of the Soviet Union, in addition to raising the levels of risk of investing in the nation.

At its first major conference held to educate international mining firms on business opportunities in the Soviet Union in October 1990, the Robertson Group and VZG jointly released a list of 121 mineral deposits considered suitable for foreign investors. While the deposits were of many types, including industrial minerals, construction stones, iron and ferroalloys, base metals, rare earths, and precious metals, a majority were of low unit-value commodities and located in remote parts of the country. In general, the Soviet government targeted mineral deposits that were partially or fully explored, and deemed mineable by the Ministry of Geology. To the dismay of the overseas mining community, however, relatively few deposits of base and precious metals were offered to foreign investors. Soviet gold resources remained off limits to overseas companies as they are were considered strategic materials.

---

d. The Robertson Group's role as a marketing agent for Soviet mineral deposits formally ended in June 1991, though cooperation continues between Robertson and VZG. Vniizarubezhgeologia has assumed Robertson's marketing role in Russia, offering the 121 preselected mineral deposits to foreign investors plus a host of other deposits considered suitable to foreign investors.

e. Some deposits contain multiple primary and secondary commodities.

f. Includes nearly 3,000 joint ventures registered in the Soviet Union as of January 1, 1991. On that date, 1,027 joint ventures were actually operating. By April 1, 1991, 1,188 joint ventures were active, of which 481 were industrial based. The number of enterprises producing output (jobs, services) as of April 1, 1991 was 948, worth an estimated 2.258 billion rubles. Most joint ventures are trivial in size and based in Moscow or St. Petersburg. Nearly two-thirds of the projects have an initial capitalization of under one million rubles.

Note: On January 1, 1991, the USSR Ministry of Finance turned over its registration duties to the individual Soviet republics. The Ministry stopped registration of joint ventures in the Soviet Baltic republics and Armenia in 1990 at their request.

Sources: Dorian, June 1991 and March 1991, and *Business Eastern Europe,* 1987–1990.

NEW OPPORTUNITIES FOR INVESTMENT IN THE COMMONWEALTH

The breakup of the Soviet Union and the transfer of ownership of mineral and energy resources to the former Soviet republics have broadened investment opportunities for Western companies, but also increased confusion and uncertainty. The bureaucratic layer of the former Soviet government is no longer a stumbling block to the establishment of joint ventures, and the new independent states have ample incentives to create favorable legislation conducive to foreign investment. At the same time, many of the former Soviet republics may begin charging companies transit fees for resources shipped through their territory which could reduce potential profits to foreign firms. Interest in establishing joint ventures in the new Commonwealth will tend to reflect economic climates in the individual states and will depend on the laws and regulations enacted in the republics.

Of the fifteen former Soviet republics, the Baltic states will likely embrace market reforms first, as their experience with capitalism is most recent and their economies are comparatively robust. Though small in size, they are already attracting a substantial share of overseas investment dollars, owing to their well-established economic ties with neighboring Scandinavia and long-standing dissatisfaction with socialist central planning. Through 1990, Estonia had more joint ventures registered than any other Soviet republic, except Russia, while Latvia and Lithuania were behind both Russia and the Ukraine.

The Russian Federation, which is clearly attracting the greatest foreign investor interest to date, had approximately 2,300 joint ventures registered at year-end 1991, with only 971, or less than half, in operation (Borisovich, January 1992). An estimated 4,500 joint ventures were registered during the same period in the Soviet Union, while more than 10,000 foreign companies had signed protocols of intention with Soviet counterparts to form joint enterprises.

Worldwide, business analysts tend to agree that the most attractive opportunities for investment in the former Soviet Union will be ventures capable of generating hard-currency earnings in the many resource-rich republics. States with resources will have more flexibility in setting up jointly funded projects and the hard currency companies need to repatriate profits. Many officials in the former Soviet republics seek foreign investment to help exploit natural resources and provide the critical hard currency needed to industrialize.

RUSSIA'S TREMENDOUS MINERALS WEALTH

While the Ukraine and many Central Asian republics are abundantly endowed with mineral resources, much of the new Commonwealth's energy and minerals wealth, both developed and undeveloped, lies within the borders of the Russian Federation. It is by far the wealthiest new independent state, accounting for 90 percent of the former Soviet Union's oil output and a majority share of gold, diamonds, platinum, and base metals production (Dorian and Borisovich, September 1992). Russia's minerals wealth extends from the major Tyumen petroleum fields of Western Siberia, the center of the former Soviet oil industry, to gold deposits in Magadan, the Yakutiya Autonomous Region, and the Amur area. Of the 7.9 million meters of oil exploratory holes drilled in the Soviet Union in 1989, nearly three-quarters were drilled in Russia, ensuring its continued leading role in the former Soviet Union's oil and gas industry. In the Western Siberian oil province, exploration will focus on Lower Cretaceous and Triassic prospects in the northern areas of the Tyumen region.

Of particular importance to the mining future of Russia is the Far East region, which was the Soviet Union's leading producer of gold and diamonds, as well as tin, antimony, fluorite and boron. The Far East region, considered by many to be a new frontier for investment, is richly endowed with natural resources and is advantageously positioned geographically along the northwest Pacific rim. It borders on China, Korea, and Japan (already its biggest trading partner), and is also accessible to the North American market. Geologically complex and diversified, the Far East contains more than 70 types of minerals, including gold, diamonds, tin, zinc, iron ore, oil, natural gas, and coal. The region possesses 95 percent of Russia's tin reserves and contributes nearly the same percentage of tin production. The Far East is also a significant supplier of tungsten (24% reserves; 37% production), lead (8% reserves; 49% production), zinc (4% reserves; 14% production), fluorspar (41% reserves; 91% production), and coal (9% reserves; 14% production).

The Far East can be conveniently divided into four mining zones: the South Zone incorporating parts of Khabarovsk Territory and the Amur region (tin, gold and coal); the Pacific Zone including the Maritime Territory and the Sakhalin and Kamchatka regions (polymetallic ores and tungsten); the Northern Central Zone including Yakutiya, Magadan District and part of Khabarovsk (precious and nonferrous metals, diamonds, coal, iron ore and natural gas); and the Far North Zone which includes the Chukchi National Area (nonferrous metals, gold and diamonds).

Gold and diamonds are mined principally in the basins of the Kolyma and Indigirka Rivers in Magadan and Yakutiya and, in the south, in the Upper Amur River basin. As an example of the scale of gold mining operations in the Far East, at Susuman in the Kolyma mining district, approximately 150 km north of the town of Magadan, about 13 dredges and up to 200 open-cut placer mines are operating. Lode mining also takes place, based mainly on stockworks in diorites and sheeted shear zones, and the mineralization is very similar to that which occurs in the Juneau gold belt and Valdez Creek mining district of Alaska. Total production from the Susuman area is some 350,000 troy ounces per year (tr. oz/y) or about 35 percent of the total annual output from the Kolyma region. Reportedly, less than 10 percent of the gold resources in the Magadan region are currently being worked.

Exploration in the Far East region over the next few years will focus on gold, oil and natural gas, coal, tin, tungsten, antimony and phosphate. Deposits al-

Table 4.3. Distribution of Mineral Resources in Russia by Economic Planning Region (%)

Economic district	Value Share of Russian Resources/Value Share of Region's Resources[a]		
	Total reserves	Explored reserves[b]	Projected resources[c]
Far East	34.4/100	5.2/1.6	40.4/97.4
West Siberia	26.2/100	45.9/18.3	23.5/74.4
East Siberia	26.1/100	12.3/4.9	28.5/90.7
North	5.4/100	18.5/35.9	3.0/56.4
Urals	3.1/100	8.7/28.9	0.9/24.8
Blackearth Center	2.3/100	3.1/14.2	2.0/71.6
Volga	1.4/100	2.9/21.9	1.1/66.1
North Caucasus	0.8/100	2.4/30.6	0.5/44.4
Center	0.1/100	0.4/32.4	—/33.4
Northwest	0.1/100	0.4/39.0	0.1/31.0
Volga-Vyatka	0.1/100	0.2/28.6	—/23.6
Russian Federation	100/100	100/10.4	100/82.9

a. Percentage share of resources in individual economic regions limited to explored reserves and projected resources. Figures indicate the level and extent of exploration in each specific region.

b. Explored or proven reserves encompass categories A, B, and C_1 of the Soviet mineral resource classification scheme. Explored resources consist of the most highly proved part of a minerals base which has been thoroughly investigated (A category); deposits whose characteristics have not been evaluated so thorougly (B category); and resources which have not been well defined or delineated (C_1 category).

c. Projected or undiscovered resources of basins, areas, and fields are estimated by analogy with similar and explored resources elsewhere. They are subdivided into three categories: projected resources of explored deposits or those under exploration (P_1); projected resources or undiscovered deposits thought to exist on the basis of evidence from geologic surveys, prospecting, and geophysical and geochemical tests (P_2); and projected resources of potentially promising areas where new deposits may be discovered (P_3).

Source: Goskomgeologii RSFSR, 1991.

ready made available for joint development with foreign investors include the stratiform copper-sandstone Udokan ore body in Yakutiya (700 million tonnes of ore grading 1.5 percent Cu); a tungsten-molybdenum stockwork at Bugdainskoe; tantalum/niobium at Katuginskoe; and titanium at Bolshoi/Seyiim. Because of infrastructural problems, however, foreign investors are unlikely to be attracted at this stage to the development of bulk commodities and would probably prefer to seek opportunities in high unit-value commodities such as gold and diamonds.

The distribution of mineral resources of the Russian Federation by region is outlined in Table 4.3. By value, the greatest resource potential occurs in the Far East economic region, although the area remains relatively lightly explored. Only 1.6 percent of the resources in the Far East have been identified, whereas 97 per-

Table 4.4. Distribution of Value by Mineral Group in Russia[a]

	Value Share of Mineral Resources		
Mineral category	Total	Explored reserves[b]	Projected resources[c]
Energy fuels (includes coal)	86.4	9.9	76.5
Coal	75.4	1.3	74.1
Ferrous metals	3.0	0.8	2.2
Nonferrous metals	6.8	3.8	3.0
Nonmetallic metals	3.8	2.6	1.2
Total	100.0	17.1	82.9

a. Relative comparison between explored reserves and projected resources indicates level of exploration of various mineral groupings.

b. Explored or proven reserves encompass categories A, B, and C_1 of the Soviet mineral resource classification scheme. Explored resources consist of the most highly proved part of a minerals base which has been thoroughly investigated (A category); deposits whose characteristics have not been evaluated so thoroughly (B category); and resources which have not been well defined or delineated (C_1 category).

c. Projected or undiscovered resources of basins, areas, and fields are estimated by analogy with similar and explored resources elsewhere. They are subdivided into three categories: projected resources of explored deposits or those under exploration (P_1); projected resources of undiscovered deposits thought to exist on the basis of evidence from geologic surveys, prospecting, and geophysical and geochemical tests (P_2); and projected resources of potentially promising areas where new deposits may be discovered (P_3).

Source: Goskomgeologii RSFSR, 1991.

cent are projected based on geological evidence. Western and eastern Siberia are also relatively wealthy in mineral resources, containing more than half of the explored reserves of Russia. The North and Ural regions of Russia are well-endowed with minerals and moderately explored. In the North region, more than one-third of the reserves are explored, accounting for 18.5 percent of total explored reserves in Russia. The Central, Northwest, and Volga-Vyatka regions of Russia possess the fewest resources throughout Russia.

Russia's minerals wealth has been roughly valued at 117 trillion rubles, which includes the sum of value of the explored reserves (Soviet categories A, B, and C_1), preliminary assessed reserves (C_2), and projected or probable resources (P_1, P_2, and P_3) (Goskomgeologii RSFSR, 1991). Resources in the Russian Federation represent nearly 90 percent of the value of minerals and energy in the former Soviet Union. Table 4.4 presents a breakdown of the value of mineral categories in Russia, including energy fuels, ferrous metals, nonferrous metals, and nonmetals. As shown, energy fuels, most notably coal, account for the overwhelming value of mineral resources in Russia. However, the principal value of coal occurs in the projected resources category, which is based in part on the presence of a large quantity of unexplored fields in Russia and the exaggeration of coal's industrial importance. Projected coal resources are largely undiscovered and located in remote regions with poor infrastructure, suggesting that little or no exploitation is likely to take place in the foreseeable future. Projected resources of coal in the former Soviet Union frequently exceed the value of explored reserves, while in other mineral groupings the values are often similar.

Russia's explored or proven reserves—a more accurate indicator of economic viability and present-day market considerations—reflect the relevant characteristics of Russia's underground minerals wealth. A large part of this reserve inventory is comprised of oil, natural gas, and condensates, whose value is more than six times that of coal resources, oil shale, and peat. In terms of explored reserves, the energy fuels are first in value, followed by nonferrous and rare metals, nonmetals, and ferrous metals. Values of explored reserves in Russia may be based, in part, on the importance of the particular mineral group to the national economy, technological efficiencies, and profitability.

INVESTMENT NEEDS IN RUSSIAN MINING

The Russian economy accounted for nearly two-thirds of the net material product (NMP) of the Soviet Union. With a population of 146 million, Russia today serves as the economic and political hub of the new Commonwealth of Independent States. Although the current difficult economic environment in Russia is characterized by an unstable political system, the lack of a functioning infrastructure, a nonconvertible currency, high inflation in the range of 200–300 percent, and declining exports, the long-term prospects for economic growth are promising given Russia's skilled labor force, an educational standard on par with the industrialized world, an extensive minerals and agricultural resource base, and tremendous land area. As Russia's economy rebounds in the years ahead, there will be without question substantial reliance on the expansion of the state's mineral and energy industries. This, in turn, suggests a greater need for foreign financial and technical assistance in the 1990s.

Though Russia's mining industry is the largest in the CIS, it is plagued by numerous problems which may threaten future expansion, including low efficiencies in recovery, production, and utilization, outdated mining equipment and technology, and severe transportation bottlenecks. Billions of dollars of investment will be required to renovate the industry during the 1990s. The availability of capital for priority investment areas will be critical in order to ensure expansion of the industry into the twenty-first century. Priority will be given to the development of mines with large volumes and high-grade deposits, low production costs and economically viable operations. Emphasis will be placed on comprehensive prospecting, mining, recovery and utilization, and the intensive processing of manufactures for the purpose of absorbing foreign investment. The primary investment areas in Russia's mining industry are listed in Table 4.5.

The strategy to tap enterprise potential and promote production, economic growth, and technical development in Russia will rely on the use of imported technology and equipment to renovate existing enterprises. During the 1990s and early next century, technology transfers could play a pivotal role in promoting

Table 4.5. Priority Investment Requirements in Russia's Mineral and Energy Industries

- Modernization of existing smelting and refining facilities, including those that process nickel, copper, aluminum, and gold
- Technologically advanced concentrator facilities to curtail discharge of near ore grade tailings
- Pollution monitoring and controlling devices aimed at curtailing sulfur dioxide and fluorine particle emissions
- Geological, geophysical, exploration, and remote-sensing technologies
- Physical infrastructure and telecommunications network, transport machinery (notably haulage trucks)
- Surveying equipment and technology for continental shelf studies
- Advanced gold recovery technologies
- Economic evaluation of mineral/energy deposits; management training and skills
- Advanced hydrocarbon seismic reflection/refraction methodologies
- Improvement of petroleum gas processing quality through the introduction of small-size gas processing devices
- Development of an integrated system for gas collection and utilization, including compressors for gas separation
- Industrial gas drying and purification units and gas processing factories
- Recovery of fields previously considered spent that still contain hydrocarbon resources subject to extraction with advanced technologies

domestic economic growth as well as cooperation with foreign nations. Although economic restructuring is under way in Russia, technological advancement of industry is needed to facilitate economic growth in the long term. It has been estimated that as much as 75 percent of all equipment used in Soviet industry must be renovated or replaced to improve efficiencies in operations (Kollontai, 1990). Such an accomplishment would require massive capital investments as well as technology transfers.

Potential growth areas for imports of technology and equipment are in priority development projects, which include minerals and energy extraction, processing, mining and safety, and pollution abatement equipment. Russia intends to expand its minerals and metals production in the long term in order to manufacture value-added goods to meet consumer needs. Production of minerals, metals and their products also requires addressing the environmental problems associated with these industries.

RUSSIAN LEGISLATION ON FOREIGN INVESTMENT

Following the August 19, 1991 military coup attempt, Soviet Union structures no longer functioned, and each republic began to independently solve its own problems by issuing business laws and regulations, including foreign investment legislation. There is a dire need for investment in the mining sectors of the many former Soviet republics, including Russia, as the transformation of the republics into sovereign states has had adverse effects on some aspects of mining. Some mines in the Donbass coal field in the Ukraine, for example, are reported to be idle today because of a lack of timber roof supports that used to be shipped from Russia.

While opportunities for investing in the new Commonwealth are becoming plentiful, the legal regimes within the new states have not kept pace with the rapid development of economic and social reform. In most former Soviet republics, rules and regulations that protect investors' interests are largely absent or untested. Clearly, the former Soviet republics will require sufficient time to develop legal foundations capable of guiding international business activities in their economies. Until such environments are firmly established, overseas enterprises will need to comprehend fully the terms under which they are investing and the details of any transaction.

In the minerals-rich Russian Federation, a law on foreign investment was approved in August 1991 (Table 4.6). It concerns all types of property and investment in the form of both joint ventures and wholly foreign-owned enterprises, including 100 percent-owned affiliates established in the form of joint stock companies. The primary aim of the law is to establish a necessary legal basis for foreign investment primary activity in Russia. With the new law, the Russian government proceeds from the notion that foreign investment in the economy must not be temporary, but rather a powerful strategic force which would facilitate the process of privatization and the integration of Russia into the world economy. Practical areas governed or regulated by the law "On Foreign Investments in the RSFSR" include:

- conditions which must be observed by foreign investors;
- procedures for establishing enterprises with foreign investments;
- conditions for carrying out special studies;
- requirements pertaining to foundation documents (statutes);
- establishment of the values of investment;
- conditions of the state registration of enterprises;
- specifications of the documents to be submitted upon registration;
- issuance of licenses for banking, insurance, and trading activities connected with the purchase and sale of securities;
- tax and customs incentives;
- conditions for the export, import, and sale of goods in the Russian market;
- conditions for the protection of intellectual property;
- conditions for labor relations; and
- terms and conditions of concession agreements.

Table 4.6. Major Provisions of the Russian Federation Law on Foreign Investments in Russia

Characteristics

1. Signed July 4, 1991 by Russian Supreme Soviet Chairman B.N. Yeltsin.
2. Consists of seven chapters and 42 articles.

Chapter I: General Provisions

1. Foreign investments are all kinds of property and intellectual values invested by foreign investors into entrepreneurial projects and other kinds of activity in order to derive profit (revenue).
2. Foreign investors have the right to make investments in the territory of Russia through shared participation in enterprises set up jointly with legal entities and citizens of Russia and other Union republics; to create enterprises wholly owned by foreign investors; to acquire enterprises; to acquire rights to use land and other natural resources; and to engage in other investment activity not prohibited by the laws in force in the territory of Russia, including the granting of loans, credits, property, and property rights.

Chapter II: State Guarantees of Protection for Foreign Investments

1. Foreign investments in the territory of Russia enjoy full and unconditional legal protection, which is insured by this law and by other legislative enactments and international treaties in force in the territory of Russia. Legal treatment of foreign investments may not be less favorable than the treatment of property, property rights, and investment activity of legal entities and citizens of Russia.
2. Decisions on nationalization are made by the Russian government Supreme Soviet. Decisions on requisition and confiscation are made under procedures set forth in laws operating in the territory of Russia.
3. Compensation paid to a foreign investor should be in line with the actual value of nationalized or requisitioned investments immediately before it becomes officially known that nationalization or requisition will actually occur.

Chapter III: Creation and Liquidation of Enterprises with Foreign Investments

1. The following may be set up and operate in the territory of Russia: (1) enterprises with shared participation in the form of foreign investments (joint venture enterprises), and also their subsidiaries and branches; (2) enterprises wholly owned by foreign investors, and also their subsidiaries and branches; and (3) branches of foreign legal entities.

Chapter IV: Conditions for and Types of Activity of Enterprises with Foreign Investments

1. An enterprise with foreign investments may engage in any form of activity that meets the aims envisaged in the regulations of the enterprise except for those prohibited by the laws in force in the territory of Russia.

Table 4.6. *(continued)*

2. An enterprise with foreign investments may on a contractual basis determine the conditions for marketing of the output (or work or services) that it produces in the Russian market, including the price, and also the conditions for the delivery of goods and services from that market.
3. Enterprises with foreign investments, and also foreign investors, pay taxes as established by the laws in force in the territory of Russia.
4. For enterprises with foreign investments operating in priority sectors of the national economy and in particular regions, preferential taxation may be established.

Chapter V: Acquisition by Foreign Investors of Shared Participation in Enterprises, and of Stock and Other Securities

1. Foreign investors may participate in the privatization of state and municipal enterprises and also uncompleted capital construction projects in the territory of Russia. Conditions for participation in competitions and auctions to privatize state and municipal enterprises are established by laws in force in the territory of Russia.

Chapter VI: Acquisition by Foreign Investors and Enterprises with Foreign Investments of Rights to Use Land and Other Property Rights

1. The offer to foreign investors and enterprises with foreign investments of rights to use land, including its lease, and other natural resources is regulated by the Russian Land Code and other legislative enactments in force in the territory of Russia.
2. Offering foreign investors rights to work and exploit renewable and nonrenewable natural resources and engage in economic activity connected with the use of objects owned by the state, but not transferred to enterprises, establishments, or organizations for full economic management or operational control, is done on the basis of concession contracts concluded with foreign investors by the Russian Council of Ministers or other suitably empowered state bodies using the procedure established by the legislation of Russia on concessions.
3. The period of concession contract is determined depending on the nature and terms of the concession, but cannot be more than 50 years.

Chapter VII: Foreign Investments in Free Economic Zones

1. For the purpose of attracting foreign capital, advanced foreign techniques and technology, and management experience, and to develop the export potential of Russia, free economic zones are created in its territory. Within free economic zones, compared to the general regime, a preferential regime is established for economic activity by foreign investors and enterprises with foreign investments.
2. Foreign investors and enterprises with foreign investments engaged in economic activity in free economic zones are offered the following additional privileges: (1) a simplified procedure for registration of enterprises with foreign invest-

Table 4.6. *(continued)*

ments; enterprises with a volume of foreign investments up to R75 million must be registered by the suitably empowered bodies within the free economic zone; (2) a preferential tax regime: foreign investors and enterprises with foreign investments are taxed at lower rates, including the tax on profit transferred abroad (here, the rate of taxation may be less than 50 percent of the tax rates in force in the territory of Russia for foreign investors and enterprises with foreign investments); (3) lower rates of payment for the use of land and other natural resources, and rights for long-term leasing for a period of up to 70 years with right of sublease; (4) a special customs regime, including lower customs duty on imported and exported goods and a simplified procedure for crossing the border; and (5) a simplified procedure for the entry and exit of foreign citizens, including without visas.

Sources: *Interflo,* September 1991.

Under the law, foreign enterprises will be permitted to set up joint ventures, establish subsidiaries, purchase shares in Russian organizations, and be awarded concessions for the exploitation of natural resources. The law guarantees that foreign investors will receive any rights and privileges available to Russian organizations and will be free of discriminatory restrictions, except in sensitive activities such as banking. Foreign investments in Russia cannot be nationalized and cannot be subject to requisition or confiscation, except in exceptional cases as provided for by legislative enactments. Priority areas for investment in the Russian Federation, in addition to mineral and energy resources, include electronics; automotive parts and details; machine building; machines and equipment for agriculture; parts, details and mechanisms for precision metalcasting; packaging materials; new pharmaceutical materials; goods suitable for export; new plant strains and animal breeds; energy-saving materials; communications; construction of resort and tourism facilities; and biotechnology and medical equipment. For foreign investments of at least 20 percent or US$500,000 in these fields of activity, tax breaks will apply.

ESTABLISHMENT AND LIQUIDATION OF ENTERPRISES WITH FOREIGN INVESTMENT

Creation and liquidation of enterprises with foreign investment in Russia are exercised in accordance with the state's new investment legislation. Newly established joint ventures are to be registered with the Ministry of Finance of the Russian Federation or another authorized state body. If foreign investment exceeds 100 million rubles, joint enterprises are to be registered with the Ministry of Finance of the Russian Federation in addition to receiving the approval of the Council of Ministers of the Russian Federation.

State registration requirements for joint ventures include:

- an application for registration;
- two notarial witnessed copies of foundation documents (statute and the agreement);
- expert findings (in some specific cases);
- papers on solvency of the foreign investor from a bank (with witnessed translation into Russian);
- an extract from the trade register or other evidence of the foreign partner's juridical status (with witnessed translation).

The Russian Federation Ministry of Finance shall register the new enterprise within 21 days of the day of application.

INVESTMENT IN FREE ECONOMIC ZONES

Another feature of Russia's investment legislation is that it contains a definition of the status of free economic zones and provides for additional incentives for investments in such zones. In search of new ways and forms to promote foreign economic relations, Russia considers the establishment of special economic zones as a means of enhancing the competitiveness of Russian goods and stimulating production and trade. Styled after Chinese free economic zones such as Shenzhen near Hong Kong, the objective is to take advantage of a region's nearness to foreign investors by attracting them with tax privileges, special joint venture provisions, infrastructural improvements, and low-cost labor.

Though the intent to establish special economic zones is noteworthy, Russian officials realize that such zones will not serve as substitutes for economic reform, and will only work if general economic reform is implemented on a successful, continuing basis. To enhance the viability of free economic zones in Russia, comprehensive legislation is needed before any economic activities can begin. Such laws must address issues of labor regulations in the zone, nationalization, taxation, currency usage and payment schemes, import tariffs, and legal rights of companies operating in the zones. With a favorable tariff schedule on goods originating in a free economic zone, companies could have suitable access to the domestic Russian market, while Russia could have access to materials not widely produced in the country.

Foreign investors in Russia's free economic zones may be accorded the following additional incentives:

- a simplified procedure for registering enterprises with foreign capital of up to 75 million rubles;
- a preferential tax regime (lower tax rates; but in any event not less than 50% of the existing rates for foreign investors in Russia);
- reduced rates for payments for the use of land and other natural resources;
- an extension of the right to long-term leases of land of up to 70 years with the right of sublease;

- customs benefits and a special customs regime; and
- a simplified procedure for entry and exit of foreign citizens.

Taxation

On December 27, 1991, the Russian Parliament passed the law "On Tax on the Profits of Enterprises and Organizations." It retained the basic tax rate of 32 percent, the same as in 1991. The law became effective as of January 1, 1992. By parliamentary decision, it was expected to be in force during the first quarter of 1992, after which it was to be replaced by the law "On the Taxation of Incomes of Enterprises," adopted December 20, 1991. Final rulings on both laws were expected sometime in 1992.

The final draft of the December 27 taxation law cut the profits tax rate from 35 percent proposed by the government to 32 percent—that is, it remained at the 1991 level (*Commersant,* December 30, 1991). This was done to facilitate the stabilization of Russian industry by imposing favorable taxation levels, even though officials had suggested that a 35 percent rate was necessary for the Russian budget to break even. To complete the Russian tax legislation package on December 27, 1991, the law "On the Fundamentals of the Tax System in Russia" was also passed the same day.

The law classifies the numerous taxes endorsed by Parliament on the basis of which government body imposes them and into what budget category they will go. For instance, the value-added tax, excises and income taxes (taxes on profits) on enterprises, and personal income taxes are national taxes established directly by the Russian Parliament. But value-added taxes and excises go into the federal Russian budget, while the income taxes on enterprises and individuals are divided between the budgets of the republics and territories.

Taxes of republics within Russia include taxes on the assets of enterprises and taxes for the utilization of minerals. Local taxes include the tax on the property of individuals, land tax, registration charges on individuals engaged in business activities, and license fees for retailing of alcohol. In the main body of the law, the latter is stipulated not in rubles, but in the recently popular unit of monthly wages—fifty minimum wages a year on legal entities and twenty-five minimum wages a year on individuals.

Joint ventures, as legal entities with foreign capital in Russia, are subject to a profits tax, turnover tax, import/export tax, income tax, and a tax on an increase in funds allocated for consumption. Taxable profit shall be determined on the basis of gross profit, which is the sum of proceeds or losses from the marketing of goods or services, basic assets and other inventory of the enterprise, and income from nonmarketing operations less the cost of these operations. In determining taxable profits, the gross profits shall be reduced by the sum of deductions to the reserve or similar funds for enterprises which are required to have such funds under Russian legislation after these funds reach the size stipulated in the founding instruments, but not more than 25 percent of the authorized fund. The sum of deductions to the said funds shall not exceed 50 percent of the taxable profits of enterprises.

Profit tax rates are 30 percent if the foreign investor's share exceeds 30 percent of registered equity and 32 percent if the foreign investor's share in registered equity is equal to or below 30 percent. Joint ventures in the Russian Far East are taxed at 10 percent. Profits of enterprises wholly owned by foreign investors shall be taxed at the rate of 32 percent, while profits of brokerage offices and investment institutions are taxed at 45 percent.

Joint ventures with foreign investor shares exceeding 30 percent are subject to certain tax rebates or preferences. Projects with 30 percent or more foreign investment receive tax holidays for two years (or three years for joint ventures in the Far East). Part of the profit is exempted from taxation for five years if losses recorded in balance sheets exceed reserve funds. Taxable profit may also be reduced by the sum of donations to charities, ecological and health funds, religious associations and the sums remitted to enterprises and institutions of culture, education, health, social insurance, physical culture and sports, but not more than two percent of taxable profit. The total sum of tax rebates granted to enterprises may not decrease the sum of tax on profit by more than 50 percent.

Turnover taxes are charged for the production and sales of goods subject to a profits tax. Tax rates depend heavily on the category of goods being produced and sold; for example, alcoholic drinks have a turnover tax of 90 percent; jewelry, 50–70 percent; and refrigerators, watches, and perfumes, 30 percent. The turnover tax rate for the majority of industrial goods and semifinished products in Russia is 10 percent.

Import/export taxes are charged to all joint ventures conducting foreign trade operations. Maximum rates are imposed on the export of scrap metal and ferrous metals, 50 percent; nonferrous metals, 45 percent; oil and gas, 40 percent; kerosene, diesel, fuel and engine oils, 35 percent; and ores of ferrous and nonferrous metals, coke, nonmetallic minerals, and building materials, 30 percent. Maximum import tax rates are imposed on consumer goods, including 1,000 percent for cigarettes, 800 percent for alcohol, and 600 percent for calculators, coffee, umbrellas, and other similar items.

Income taxes are also charged on incomes derived from shares, bonds, and other securities, including shareholding in joint ventures. Incomes of foreign investors are taxable at a rate of 15 percent if no other provisions are stipulated in international agreements on taxation. Interest from state bonds and other securities are not taxable. A six percent tax rate is imposed on revenues from freight shipped via international cargo transportation by foreign legal entities.

Royalties and use of profits

In the near future, companies and enterprises, regardless of the form of ownership they take, will be required to pay for the right to explore and exploit mineral deposits in Russia, as is the case in other countries. As recipients of such tax revenues, regions and districts may spend the funds locally. A part of the revenues is to be spent on social and economic development programs for native populations whose historical lands are subject to mining.

In Russia, the state will regulate the payment rate for the deposit exploitation rights or the royalty. It will average five percent of domestic prices for solid minerals (except gold) and up to 10–15 percent of domestic prices for oil and gas. According to the agreement between the subsurface user and the recipient of payments, the former can pay in kind, including a portion of the extracted minerals. Local authorities receiving royalty payments can grant privileges to the user.

Russian investment legislation not only permits a foreign investor to repatriate income received in foreign currency, but also to use profits received in Russian currency in the territory of Russia. Thus, in accordance with Article 11 of the investment law, foreign investors may open current and deposit accounts in Russian currency in banks in Russia without, however, the right to transfer amounts from these accounts abroad. They may also use these ruble accounts to acquire foreign currency on the domestic currency market. Under certain circumstances, enterprises engaged in the production of import-substitution goods may exchange profits in Russian currency for hard currency.

RUSSIAN MINING CODES TO ENCOURAGE DEVELOPMENT

While Russia and the other independent states are creating investment laws and provisions, many are also considering establishing policies to guide mineral exploration and development activities within their boundaries. The lack of specialized mining legislation has been one of the principal deterrents to foreign investment. For some states, there is an emphasis on devising and implementing mineral strategies that will meet economic development objectives and improve social welfare by using natural resources as an impetus to growth. Mineral policies in these former Soviet republics will most likely resemble those in market economies, since in both the former Soviet republics and market economies resources are viewed as national goods which should be developed for the well-being of the entire society. The means by which the policies achieve their objectives will not, however, be similar.

In November 1991, Russia set a precedent by issuing a formal decree on precious metal and diamond exploration, development, and sales in the former Soviet republic, in part to boost production of the valuable commodities. The decree stipulated that all activities related to precious metals and stones must be in accordance with the October 1990 law "On Guaranteeing the Economic Foundation of Russia's Sovereignty." The decree established a Committee for Precious Metals and Stones, which will monitor all operations involving precious metals and stones in Russia and oversee a state depository for valuables. The new committee will also work out new principles for setting prices for precious metals, precious stones, and items made of them by following world market prices. The decree also established that payment for 25 percent of gold mined and 25 percent of diamonds and platinum group metals sold in the external market be made to enterprises mining precious metals and diamonds in freely convertible currency.

Another precedent was set by Russia in February 1992 when the Federation passed a comprehensive law on minerals exploration and development. Under the law, subjects of entrepreneurial activity regardless of the form of ownership, including legal persons and citizens of any country, may become mineral users in Russia, provided they have obtained a license. Bodies of the state geological inspection system are to monitor the rational use and protection of Russia's mineral wealth. Money paid for the use of minerals onshore, in the offshore economic zone, on the continental shelf, and within inland waters of the Russian Federation will be distributed as follows: 30 percent will go to replenish local budgets; 30 percent to the budgets of republics, territories and regions within the Federation; and 40 percent will go to the federal budget.

The offer to foreign investors and enterprises with foreign investments in Russia of rights to use land, including its lease, and other natural resources is regulated by the Russian Land Code and other legislation in force in the territory of the Russian Federation. Foreign investors obtain the rights to work and exploit renewable and nonrenewable natural resources through concession contracts concluded with the Russian Council of Ministers. The period of a concession contract is determined depending on the nature and terms of the concession, but cannot exceed 50 years.

CONCLUSIONS

Overseas interest in Russian joint mineral ventures will remain subdued until the nation's economic and political situation stabilizes. The Russian Federation is in the midst of a particularly harsh transitional phase in moving from one type of economic system to another. The uncertainties that arise during such a transition have tarnished the investment climate of the country, despite its abundance of natural resources and large domestic markets. In the long term, however, the Russian mining industry will likely be attractive to foreign companies as the joint venture potential could be enormous. As Russia's economy begins to expand once again, its minerals industry will require increasing amounts of financial and technical assistance in exploration, mining, and processing. And as foreign participation in Russian minerals development increases, newer technologies and sound management techniques will likely be introduced, leading to an overall increase in production performance and a reduction in materials wastage.

REFERENCES

Borisovich, Vitaly T., January 1992, "Russian Coal in Eastern Europe's Energy Balance," unpublished manuscript, Moscow Geological and Prospecting Institute, Moscow, 6 p.

Business Eastern Europe, 1987–1990, various issues.

Commersant, December 30, 1991, "Russian Law on Taxation of Profits of Enterprises and Organizations," Moscow, pp. 14–15.

Dorian, James P., June 1991, *Mining in the USSR: Investment, Trade, and Cooperation in a Changing Environment,* Special Report No. 2133, Economist Intelligence Unit, London, 98 p.

______, March 1991, "USSR-Mongolia: A Minerals Association About to End," *Resources Policy,* Vol. 17, No. 1, Butterworth-Heinemann Publishers, London, pp. 42–53.

Dorian, James P., and Vitaly T. Borisovich, September 1992, "Energy and Minerals Resources in the Former Soviet Republics: Distribution, Development Potential, and Policy Issues," *Resources Policy,* Vol. 18, No. 3, Butterworth-Heinemann Publishers, London, pp. 205–229.

Goskomgeologii RSFSR, 1991, "Mineralinie Resursi Rossii: Economica i Upravlenie," Issue O, Moskva (State Committee of RSFSR for Geology and Production of Mineral Resources and Energy Fuels, 1991, "Mineral Resources of Russia: Economics and Management," Issue O, Moscow.)

Interfax News Agency, October 24, 1991, "A Lack of Appropriate Legislation Holds Back Investment in Oil Extraction," *Oil, Gas and Coal Report,* No. O, Moscow.

Interflo, September 1991, "Russian Federation Law on Foreign Investments in Russia," Vol. 10, No. 11, New Jersey, pp. 7–12 (translation of *Sovetskaya Rossiya,* July 25, 1991, by Foreign Broadcast Information Service.)

Kollontai, Alexander, 1990, "Technology Transfer and Economic Relations with the Soviet Union," speech presented at the USSR: Markets of the Future seminar held at the Ilikai Hotel, May 3, 1990, Honolulu, Hawaii.

Chapter 5

NEW PERSPECTIVES ON AN OLD GOLD MINING COUNTRY*

Stewart Murray

INTRODUCTION

Along with South Africa, the former Soviet Union was traditionally one of the biggest gold producers throughout the world. Historically, the bulk of the Soviet Union's gold reserves were considered as a strategic commodity to be used as a source of hard currency to pay for imported goods or finance balance of payments deficits. Western analysts were shocked in 1991 when the former Soviet Union government revealed that only 240 tonnes of gold remained in government vaults. Previous estimates ranged from 2,000 to 3,000 tonnes; the new figure represents a loss of billions and billions of dollars.

This chapter will focus on Russia: first, because it represented around two-thirds of total Soviet production and, second, because its output has been falling recently. By contrast, the chapter will refer only briefly to the other gold-producing republics of the new Commonwealth, namely Uzbekistan and Kazakhstan.

With independence, the Russian Federation has been struggling to prevent complete deterioration of its already fragile economy. Even a brief glance at the description of the Russian economy published by the International Monetary Fund in May 1992 showed the frightening magnitude of the problems. But progress is being made and the authorities deserve applause and encouragement.

In terms of economics, an increasing range of products and services is slowly becoming available for rubles, although at high prices. As with all revolutions, this one is reflected in the appearance of new street names (or rather the reappearance of old ones) together with the changing names of organizations. In the gold sector, the state-run Glavalmazzoloto organization changed first to Rosalmazzoloto and then simply Almazzoloto between June 1991 and June 1992.

---

*An earlier version of this chapter was presented as a paper at the *Financial Times* "World Gold Conference" held in Montreux, June 1992.

J.P. Dorian et al. (eds.), CIS Energy and Minerals Development, 71–82.

Printers of business stationery are indeed facing unprecedented demand in the former Soviet republics.

But, in terms of fundamental restructuring of their commercial and industrial activities, the new republics are only just setting out on a long and arduous road. Especially for a country in which centralized, czarist control was followed by 70 years of Soviet central planning, the process of setting up a privatized market economy is going to be infinitely difficult. At least the widespread nature of the mining industry will make it easier to privatize than some of the monolithic manufacturing enterprises, and, in fact, the gold mining industry has possibly the longest record of private enterprise (in the form of the alluvial gold cooperatives, or "artels," the first of which was established some 35 years ago). There has been an apparent paradox here in that, until recently, it was widely believed that the state would continue to preserve a monopoly on the formal mining of gold, but this view now appears to have changed.

GOLD MARKET ISSUES

For the world market, the main questions about the gold industries of Russia and the new republics are:

- first, how much will be produced (not only from primary and by-product mines, but also from scrap);
- second, how much will be consumed (not only by industry in the form of electronics, but also increasingly by consumers as jewelry);
- and finally, what will be the official policy on the reaccumulation or sale of gold reserves.

These questions defy simplistic analysis. They need to be examined from a number of angles. In order to ensure the continued output from the country's gold resources, which are geographically dispersed, geologically diverse and climatically challenging, the authorities have to use the available levers of power to promote the financing (both capital and current) of the factors of production.

GOLD MINE PRODUCTION

The Soviet Union was often considered a large producer of gold (Table 5.1). Production is certainly very widespread (Figure 5.1); but in relation to the country's huge area (somewhat larger than South Africa, the United States and Australia combined), it is revealed that its output per unit of area was actually quite modest, around half that of Canada, to make the most relevant comparison. And, for instance, New Zealand now produces almost three times as much gold per unit of area as did the Soviet Union in 1991.

However, the vast size of the country and its role as a supplier of gold to the world market has ensured Russia's special place in the world's gold market. Sim-

Table 5.1. Gold Production in 1991

	tonnes	tonnes/ million sq. km
Soviet Union	242	11
South Africa	601	492
United States	300	32
Australia	234	30
Canada	177	18

ply by flying over the gold-producing mountains of the Russian Far East and, even more so, the interminable swamps of Western Siberia is the clearest way to appreciate the size of the country and the challenge of exploiting the minerals wealth east of the Urals. The development of the gold industry followed similar lines to the expansion of the Russian and, later, Soviet empires. Initially based in the Urals, the gold industry developed in the late 19th and mid-20th centuries by expanding eastward into Siberia and then to the Far East. Later on, production was established in the Central Asian republics, most significantly with the development of the Muruntau deposit in Uzbekistan from 1967.

Of all the main producing countries today, Russia has the longest history as a gold producer. Although eclipsed at various times in the past century by the United States, Canada and Australia, the cumulative output of Russia (and later the Soviet Union) has been second only to that of South Africa during the past 150 years.

The question for both of these traditional gold producers is how future production can be sustained. Technologically, the gold mining industries in the former Soviet Union and South Africa could hardly be more different, but the problems are, in a sense, the same. On the one hand, the richest and most easily exploited deposits have been worked out, and on the other, although resources are enormous, so too are the capital and human costs of developing and extracting them.

ORGANIZATION OF GOLD PRODUCERS IN RUSSIA

It is not easy to understand the political administration of the Russian Federation, with its various tiers of autonomous republics and oblasts, okrugs and krais, some overlapping, some hierarchical. And with the breakup of the Union, the word "autonomous" has taken on a real significance in 1992. The gold industry is organized along similar if somewhat less precise geographic lines: a number of principal amalgamations represent the state-owned mining industry (Figure 5.2). These are known as zolotos, the Russian word for gold. The most important in terms of output are in Yakutiya, where Yakutzoloto is in turn divided into a number of local associations in the Lena River basin (Lenzoloto) and in Magadan (where Severovostokzoloto means Northeast Gold).

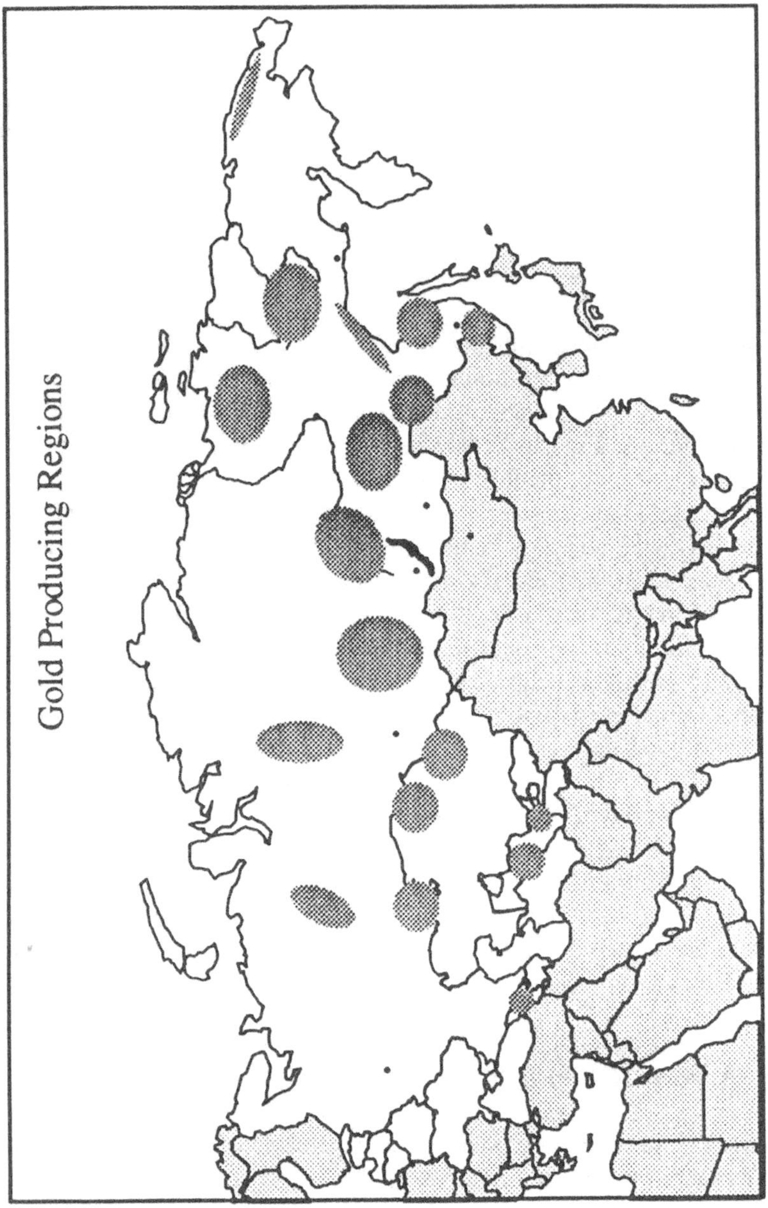

Figure 5.1. Gold Producing Regions of the Former Soviet Union

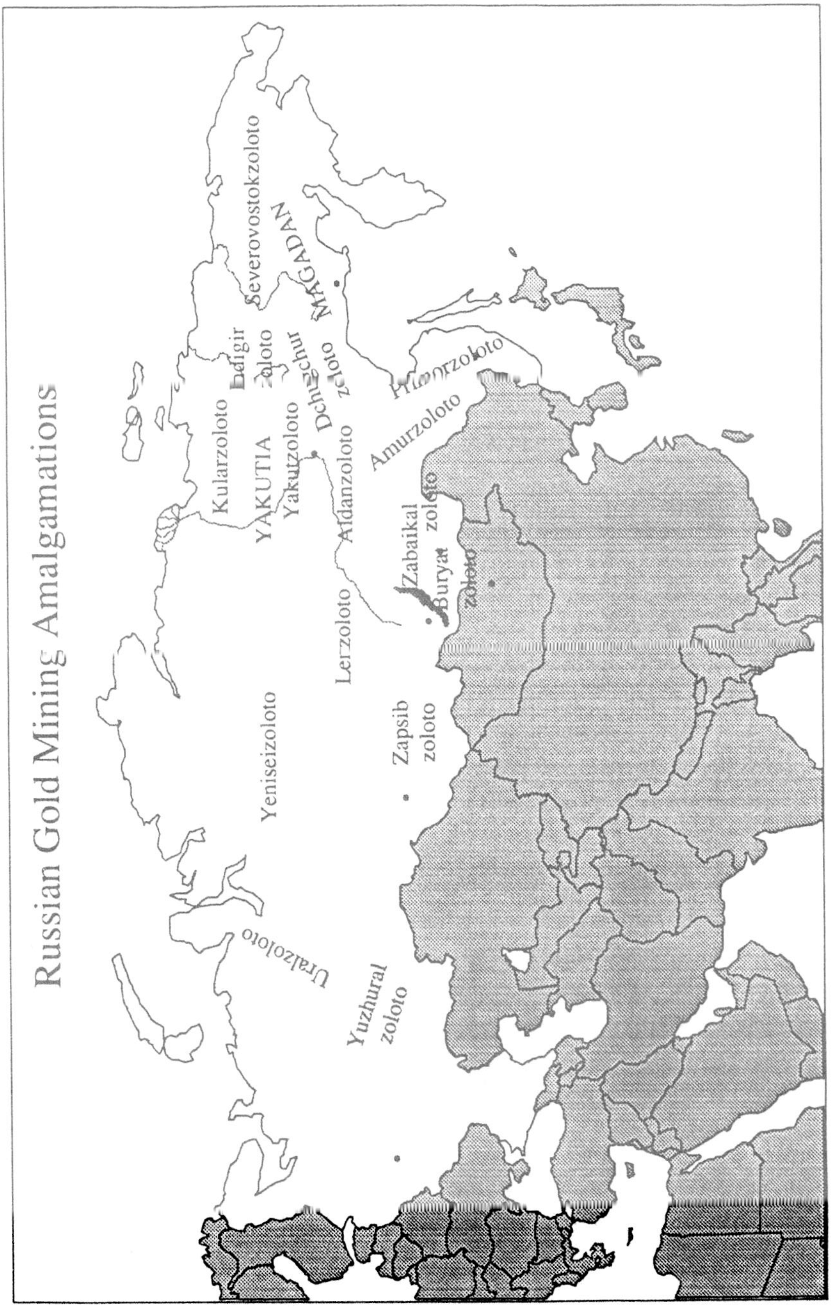

Figure 5.2. Russian Gold Mining Amalgamations

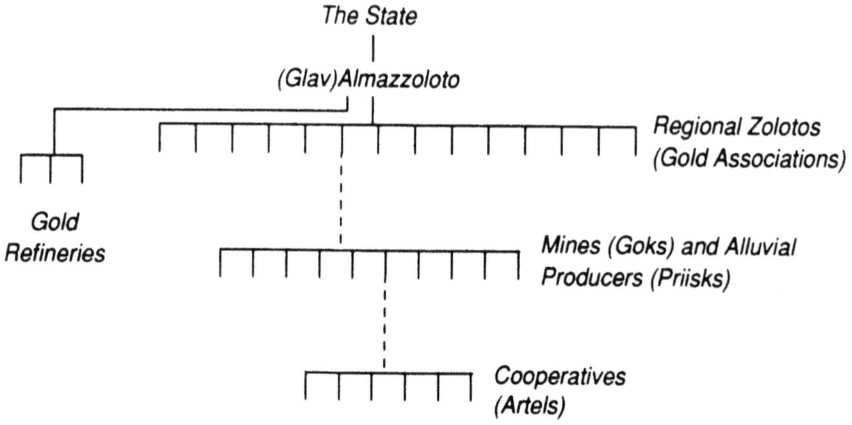

Figure 5.3. Organization of Gold Mining in the Soviet Union/Russia

The gold industry resembles a pyramid, a section of which is illustrated in Figure 5.3. At the top is Almazzoloto, the state-owned Russian corporation derived from the former Soviet Glavalmazzoloto, but shorn of the production units in the other Commonwealth republics. Almazzoloto owns a number of the refining plants, and it appears to own some of the assets of the zolotos, but it does not own the zolotos themselves which instead adhere voluntarily to Almazzoloto and depend on it for funding, equipment and information.

Within each zoloto amalgamation, there are a number of operations known as goks, when they represent mines, and as priisks, when they represent alluvial production.

Finally, adhering even more loosely to the operating companies, we have the artels, which manage their own affairs and pay their own workers, but in practice depend on the amalgamations for a number of supplies, including equipment, and they generally use the infrastructure of the state sector to sell their gold (which must by law be sold to the Ministry of Finance's Committee on Precious Metals).

This structure is by no means set in stone. Some of the enterprises in the state sector are trying to break away from it, either by bringing in domestic or foreign joint venture partners or even by transforming themselves, in effect, into artels.

The artel movement is widely recognized as being the most productive of all the structures shown here. It has spread throughout the Commonwealth of Independent States (CIS), though the majority are in Russia. Wherever there is alluvial gold, artels are likely to be found. Some 250 of these exist in the CIS, employing from 1000 down to 30 workers. They are manned by a tough, self-reliant and disciplined workforce that works in harsh conditions, often on inferior deposits and with limited availability of equipment. Their extraction methods are almost exclusively gravitational; they use no mercury, though one operates a carbon-in-pulp plant. Until recently, the artels were discriminated against by the system of gold purchasing, whereby they would have to sell their output to the associations at a much lower price than that at which these companies could sell it to the state. Nevertheless, some artels have achieved productivity and salaries three to four times higher than in the amalgamations; although the latter argue that they are faced with having to provide infrastructure such as roads, towns, and all the other facilities required to support gold mining in some very remote areas. The artels now reportedly produce more than half of Russia's gold output as a result not only of their increasing production but also because of the falling production in the state sector.

Two or three years ago, when Western companies began looking at the prospects for investing in the Soviet gold mining industry, it looked as if there was an official "no entry" sign on the boundaries of all gold deposits. However, during the past year, it has become clear that the Russian government would welcome alternative gold mining structures. These include, first of all, a number of Russian joint-stock companies, although it is too soon to say whether they will be successful. Privatization of state mining companies may also take place over the coming year. And finally, foreign mining companies are now looking again at the prospects of investing in gold mining joint ventures.

To date, ten major Western mining companies have seriously considered investment in the CIS states. Interest has ranged from brief visits to Moscow to Newmont's agreement with the Republic of Uzbekistan. Most of these companies have not yet developed their plans to a stage where it is appropriate to inform their shareholders; indeed, they almost universally describe the difficulties and frustrations which they have experienced in the past and express caution about the prospects for investment in this region.

The new openness of the Russian gold mining sector to foreign investment is, it is believed, a symptom of the general shortage of capital. Moreover, there is now a widespread feeling that the gold-producing sector is in crisis. Analysts suggest that Russian gold production in 1992 may be 30 percent below the previous year's level. What has gone wrong?

There has been a perception for some time that the rich alluvial deposits of the Russian Federation are running out; this is an oversimplification. There are plenty of unexploited alluvial deposits, but there is a desperate shortage of the financing to find, develop and operate them. There is even a shortage of simple drilling rigs and the facilities to evaluate alluvial deposits, which are by no means all the same and which require quite a range of recovery techniques if optimum yields are to be achieved.

For small alluvial deposits in the frozen Northeast, development, infrastructure and production costs can be minimized by the "shift" or "expedition" method of working (where the workforce is only actively processing the ore during the summer season), but there is still a need for efficient bulldozers, scrapers and other earth-moving equipment, which is costly to buy and maintain.

Cash payments to workers have already suffered from shortages of rubles in the past year, causing many to leave the gold fields and seek employment elsewhere. Then again, when and if energy prices are liberalized, the gold miners will face a substantial increase in costs.

In January, northeastern Russia is the coldest place on earth, indicating the special problems involved in this region. The climate necessitates either the use of vast amounts of energy to thaw ore deposits subject to permafrost or restriction of operations to a short season, which can last from as little as 30 days in the extreme north to around 200 days in many parts of the Far East. Workers also need to be paid much more to work in such a challenging environment. In Khabarovsk, a mechanical engineer from a coal mine in Yakutiya may have a salary of 40,000 rubles per month, or approximately ten times that of a university professor in Moscow.

Somewhat similar problems apply to the many undeveloped hard rock deposits, except that the infrastructure and capital requirements are much greater, as is the need for hard rock mining experience and the ability to process often mineralogically complex ore bodies, skills which are often lacking in the primarily alluvial gold-producing regions. An effective response to the developing crisis for the mining industry generally has been made more difficult by the vacuum in power and administration which has accompanied the breakup of the Union and the emasculation of the Communist Party. Instead of an all-powerful set of ministries in Moscow (whether Union or Russian), control has passed to a large extent to regional committees (Table 5.2).

Central planning, often carried out through the Party in the past, is slowly being replaced by devolved decision making at republic or local authority level. As mentioned earlier, the Union's Glavalmazzoloto has already lost some of its control and the process of fragmentation could intensify.

THE CHALLENGE FOR THE STATE

How then has the Russian government responded to the challenges inherent in this evolving situation? The government clearly realizes that it has a problem in maintaining gold output. It needs not only to stimulate a short-term recovery, which *may* be achievable by paying higher prices (that is, up to the world level) to all gold miners. In May 1992 it doubled the price it pays for gold production. It also needs to ensure an orderly longer-term development of the industry. Furthermore, given the traditional reliance on this sector as a hard currency generator, the state wants to prevent "leakage" or smuggling of production out of the country.

Finally, the government wishes to optimize its "take" or share of the revenue from gold production through corporate taxes, royalties and income taxes or simply by buying up production at a cheap price in domestic currency. As in every other mining country, this desire for a share of the revenue conflicts to some extent with the first three goals and requires a compromise in the form of a just and workable mining code.

The new Russian mining law, which was released in May 1992, provides a framework for the licensing of all mineral deposits. It is intended that all kinds of enterprises, whether local or foreign, should have a level playing field on which

Table 5.2. The Mining Environment Following the Breakup of the Soviet Union

Before	After
Ministry of Geology	Regional committees
Central planning	Republics and local authorities
Glavalmazzoloto	Fragmentation

to make their applications for a license to explore and develop mineral deposits. The mining law envisages settlement of disputes in Russian courts, but, because of the relatively undeveloped nature of the legal system as far as foreign companies are concerned, it remains to be seen whether this will be regarded as satisfactory.

The fundamental idea of the license is that mining companies will pay a premium for the benefit of the land having previously been explored and then will pay a royalty of around 10 percent or more once production starts. Existing operations will also have to apply to their local authorities (in which the mineral rights are vested); although it is possible that they will be given the license without payment.

What the mining law does not envisage is the system which exists in many other successful mining countries, where a junior company can obtain the mineral rights, carry out exploration and, if successful, sell these to a senior mining company. This is regarded by the legislators in Moscow as encouraging the churning of paper rather than the development of new mines.

FOREIGN INVESTMENT PROSPECTS

In stark contrast to past policy, Russia is pursuing foreign investment in the exploration and development of gold ore bodies. Gold is the favored target in Russia of most Western mining companies because there are no problems in getting the product to market—a major consideration when it comes to the possibility of mining nonferrous metals or bulk minerals. Most of those companies which are still interested appear to favor the possibility of investment in hard rock operations, where they feel their mining and extraction know-how would be of greatest advantage.

The geology of the main gold-producing regions in southeast Siberia and the Far East of the country is very complicated. This is reflected in the hard rock deposits having a complex mineralogy with deleterious elements, such as arsenic, often present in substantial amounts. But the geological complexities are nothing compared with the legislative complexities which companies have experienced up until now. The publication of the new mining law will certainly help, as will the forthcoming regulations on licensing; but doubts about the legal system and the continued uncertainty about the ownership of mineral rights seem likely to result in continuing caution on the part of Western mining companies. A few mining companies have had an ongoing commitment to exploration in parts of Russia for several years; but for now, the general situation is that many have looked, but few have returned with their checkbooks.

OTHER REPUBLICS

Compared with the rapid developments in legislation in Russia, such developments in the other former Soviet republics have lagged behind, perhaps waiting to see how the authorities in Moscow develop the package of laws on mining. Thus, in states such as Uzbekistan and Kazakhstan, the state geological committees are still all-powerful. This means that foreign mining companies have to negotiate directly with such committees rather than act under an umbrella mining law. They thus have a tailored package rather than an off-the-shelf deal. The other main difference is that the proportion of alluvial mining in the other republics is much lower than in Russia which may mean a relatively greater number of opportunities for investment by Western mining companies.

OUTLOOK FOR GOLD MINING

In summing up the outlook for gold mining, particularly in Russia for 1993 and beyond, the industry is going to face great difficulties. This applies not only to the primary producers in eastern Siberia and the Far East, but also to the by-product producers from copper and lead-zinc ores which have suffered from the general difficulties of inter-enterprise debts and shortages of equipment which are now endemic throughout the country.

In the medium term, and as long as the gold price paid to local producers is maintained at close to the international price (a much higher level in real terms than has been seen in recent years), it would seem likely that there will be a recovery in alluvial production; although the consensus view appears to be that, with the depletion of the richest deposits, it is unlikely that the past peak levels of production will be regained.

In the longer term, if the investment environment can be made attractive enough for foreign companies (and bearing in mind that the country will have to compete with the improved mining codes of Latin America and Southeast Asia), there is the possibility of a future expansion in hard rock mining investment, although there are few signs of this at present.

SCRAP

Turning now to the question of gold production from scrap, in the past jewelry was scrapped periodically when the domestic price was artificially low. (People bought jewelry merely to sell it as scrap for hard currency.) This ceased when the jewelry purchasers had to pay world prices. But now a certain amount of distress sales are evident, due to the impact of inflation on people with fixed incomes who had previously managed to accumulate some jewelry. This process seems likely to continue in the medium term.

On the electronics side, the defense industries benefited in the past from very low procurement prices for gold, which was therefore used to a very high intensity relative to Western norms. As this sector accounted for a large share of national GDP, gold consumption in it was extremely high. Now, however, in addition to the general decline in industrial production, the peace dividend is causing the scrapping of much of this type of equipment, and we are certain to see an increase in the production of secondary gold from this source.

CONSUMPTION

Looking briefly at gold consumption, it is emphasized first that, with a total consumption of around 90 tonnes, the majority of which had been in military electronics, the Soviet Union had been a small per capita consumer. Per capita jewelry consumption was estimated at about 0.13 grams in 1991 (a quarter of that of Portugal and a twentieth of Italy's consumption). But, until 1991, there tended to be excess jewelry demand due to low prices. Shops were cleared as soon as a delivery of jewelry arrived; but all this changed with the moves to international prices over the past year. In addition, there does seem to have been a generally higher availability of jewelry in the Central Asian republics, perhaps indicating that even in the past some of the local production leaked into the local market. As mentioned before, electronics gold consumption has been very important in the past, but this is expected to fall in the short term and then to settle at a much lower level until a competitive computer and telecommunications industry is developed. By contrast, a boost in decorative uses of gold has occurred recently, as churches are refurbished.

As for the dental use of gold, we have some visual evidence that indicates that this was very high in some regions in the past, but again, this was based on low gold prices.

Investment demand for gold is almost absent, but it is just possible that if a future government were to ban individuals from holding foreign currency, gold jewelry or other fabricated forms of gold might be held as a proxy for it. Discussions have recently taken place about the possibility of issuing coins to workers in place of paper money—this would represent a unique use of monetary gold in modern times.

OFFICIAL POLICY ON SALES AND RESERVES

Finally, what will be the role of the authorities in determining the country's sales policy? Will the new republics eventually follow the example of the Soviet Union in the period up to 1953 and accumulate substantial gold reserves? There is no doubt that these reserves placed the Soviet Union in good stead during recent years and, in less desperate economic circumstances, might have been enough to tide the country over a short-term liquidity crisis, as we saw in India in 1991. Alternatively, of course, it may be that the gold-producing republics, in particular, will decide that they do not need to build up extensive gold bullion reserves. In the immediate future, they may be unable to resist the pressure to obtain hard currency for their gold output.

From an analyst's point of view, it would seem prudent to assume that the gold-producing republics will sell the bulk of their production, at least for the medium term and until such a time as the country's hard currency earnings from other sources are great enough to allow a meaningful level of reserves to be established.

OUTLOOK

To sum up the outlook for gold in the new Commonwealth states, in the short term it seems likely that the exportable surplus of gold will be low because of the further fall in Russian gold mine output. Over the next few years, the gold available for export may increase due to some recovery in output which could occur as long as the price paid to the miners is maintained in line with the international level.

In the long term, however, there seems no reason why Russia and the other republics which form the CIS should forever be net exporters rather than importers of gold. If free-market reforms are successful, the development of an affluent middle class and eventually a prosperous working class should be able easily to absorb a much larger quantity of gold in the form of jewelry than that which the country currently produces. Indeed, as industrial output in the more populous regions increases and as tertiary industries are established, it will prove increasingly difficult to attract workers to the more desolate regions of Siberia and northeast Russia where much of the gold resources are located. So, even factoring in some new production based on the exploitation of Western capital and technology at existing and yet-to-be-discovered hard rock deposits, it is not impossible that we will see Russia and some of the Asian republics becoming net consumers of gold rather than suppliers.

The above may sound like a post-communist gold market utopia, but this argument seems internally consistent, even if it is quite impossible to forecast whether we are talking about a period of 10, 20 or 50 years in the making. But one thing is clear: the future development of the gold market in the former Soviet Union will be every bit as fascinating as the changing political, economic and social circumstances in which it occurs.

Chapter 6

OPPORTUNITIES FOR TECHNOLOGY TRANSFER BETWEEN THE UNITED STATES AND THE COMMONWEALTH OF INDEPENDENT STATES: A FOCUS ON THE MINING SECTOR

Chennat Gopalakrishnan and Valentin N. Tarasov

INTRODUCTION

East-West technology transfer has taken on a new urgency in the wake of recent dramatic changes in the former Soviet Union that have led to the emergence of a new Commonwealth of Independent States (CIS). The key element in this transformation is the decentralization of political power and the resulting primacy of the states in the process of technology transfer. This remarkable development has provided a singular opportunity to put technology to work to help achieve accelerated growth and structural change in the economies of the former Soviet republics. This chapter attempts to explore against this background the crucial role of technology transfer from the United States to the independent states of the Commonwealth in achieving these objectives. The focus of the discussion is on the mining sector.

TECHNOLOGY TRANSFER: THE RATIONALE

The role of technology in economic growth has been the subject of serious critical scrutiny in recent years. Most studies confirm technological advancement as the single most important factor contributing to rapid productivity gains. In large measure, this ability of technology can be attributed to its catalytic role in making more effective use of labor, capital and natural resources, the other production inputs. Estimates of such productivity growth range from 40 to 70 percent, as opposed to 15 percent from capital formation and 12 percent from

J.P. Dorian et al. (eds.), CIS Energy and Minerals Development, 83–95.

better education and training of the labor force. Technology thus plays a pre-eminent role in promoting economic growth through stimulating a surge in productivity.

In this chapter, we will use "a broad definition of technology that embraces both hardware (machinery, equipment, instruments, etc.) and knowledge and practices associated with the creation and use of this hardware, including organizational forms and methods of management" (Cooper, 1985). As Mansfield (1974) has pointed out, "production designs, plant blueprints, patents, managerial know-how, marketing techniques, distribution channels and any other sphere of business operation from new idea generation to ultimate product production and promotion" would all be important elements in this broad conception of technology.

The basic objective of technology transfer is to maximize the efficiency of the production and distribution processes. Ledin (1978) states it as follows: $Q = f(C,L,T,t)$, where Q = production volume; C = input of capital; L = input of labor; T = technology, and t = time (cited in Gopalakrishnan, 1989).

U.S.-SOVIET UNION TECHNOLOGY TRANSFER: FORMS, MODES AND SCALE

For analytical purposes, Soviet imports of Western technology can be classified into broad categories. Cooper (1985) suggests the following typology: "normal," "structural," "conjunctural," and "nondiscretional" imports of technology. "These categories cannot be rigidly defined and inevitably shade into one another, but they may be useful in analyzing the economic significance of Western technology" (Cooper, 1985).

Normal imports include items that enjoy a comparative advantage in production in the West. These include, for example, ships, railway rolling stock, agricultural machinery, timber, paper- and pulp-making machinery and metallurgical equipment. Such normal imports, which benefit from the international division of labor, account for a major share of current Western technology transfers. Structural imports refer to imports aimed at stimulating the lagging sectors of the Soviet economy, e.g., the motor, chemicals, and oil and gas industries. The domestic industries, in these cases, have diminished productive capacity largely due to a shortage of inputs. Conjunctural imports result from "favorable international conditions or other advantageous circumstances that permit the Soviet Union to increase Western technology imports above the normal trend" (Cooper, 1985). Principal beneficiaries of such imports have been the Soviet chemicals and petrochemicals industries. The fourth category, nondiscretional imports, refers to imports necessitated by the inadequacy of "domestic capability" to produce certain goods and services. "This inadequate capability could stem from a variety of factors, including a lack of technological knowledge and experience, structural weaknesses of the economy, deep-rooted problems of the system of planning and management, etc." (Cooper, 1985). The Soviet paper and pulp industry provides a case in point: in 1977, 80 percent of paper-making machines used in the Soviet Union were imported.

Another broad distinction in the form in which technology is transferred in is between "embodied" and "disembodied." The former refers to the transfer of technology as embodied in goods or machinery (hardware), while the latter involves transfer of technological knowledge *per se*. Examples include "sales of licenses, subscription to technical periodicals, intergovernmental technical cooperation agreements, word of mouth and industrial espionage" (Schaffer, 1985). Imports of actual products, especially capital goods, represent embodied technology transfer and account for the bulk of technology imports.

Another dichotomy is represented by "active" and "passive" channels of technology transfer. Active channels involve direct interaction between the seller and the buyer of technology, such as the training of personnel or active guidance in the running of a turnkey plant. Passive channels are basically "one-way channels, involving the simple transfer of technology with no active assistance from the owner in how to use it properly" (Schaffer, 1985).

Technology transfer occurs in a variety of ways. The modes of technology transfer include joint ventures, license purchases, commodity imports, turnkey projects, coproduction agreements, industrial cooperation, and a number of noncommercial channels.

To date, the joint venture has been the dominant form of foreign direct investment, and by inference, foreign technology transfer. The joint venture is generally an ideal channel for technology transfer. As Ruoho (1990) points out, ". . . as a channel of technology transfer, joint ventures will surpass other forms of technology transfer in efficiency." "The joint venture involves an equity ownership in a separate legal entity by the foreign partner(s) of up to 99 percent, reflecting a commensurate investment of capital" (International Monetary Fund et al., 1991). The scope and dimensions of joint venture activity in the former Soviet Union are presented in Table 6.1.

The distribution of joint ventures by economic sector or sphere of activity is not available because there is no official classification scheme for such activities. A large majority of joint ventures (1,256 at the end of March 1990) has been set up with partners from OECD countries. Firms from Germany, Finland and the United States are leading investors in terms of numbers (with over 100 joint ventures each), while Italy and Germany have the top two ranks in terms of initial capital investment. In 1990, the Russian Republic had the largest share of foreign investment in the Soviet Union, accounting for some 70 percent of both registrations and foreign capital invested (International Monetary Fund et al., 1991). Cooper (1991) gives a sound assessment of the potential role of joint ventures in the CIS: "It is a flexible organizational form, well suited to the rapid exploitation of new technological possibilities and to insertion into niches in the administrative structures of the Soviet economy."

License purchases are increasingly used for the transfer of technology. The licensee gains rights through such purchases to produce and sell specific commodities using inventions and processes developed by the licensor. Soviet license

Table 6.1. Joint Ventures and Initial Capital in the Soviet Union, 1987–1990

End Period	Total Number of Joint Ventures	Increment	Total Initial Capital (in millions of rubles)	Increment
December 1987	23	23	159.37	159.37
December 1988	193	170	826.85	667.48
June 1989	689	496	2,122.73	1,295.88
December 1989	1,269	580	3,521.99	1,399.26
June 1990	1,754	485	4,000.00	478.00

Source: International Monetary Fund et al., 1991.

purchases have several goals: conserve convertible currency through the domestic production of goods; free Soviet R&D resources for other purposes of a higher priority; shorten the time to start new production through the use of imported technology; and decrease the cost of introducing innovations. Statistical information on Soviet license purchases from the West is at best sketchy. According to one estimate, the Soviet Union spent US$64 million in 1980 on license purchases, a modest amount compared to US$300 million spent by the Federal Republic of Germany, US$1.2 billion by Japan, and US$700 million each by the United States and the Netherlands in 1979 (Smeliakov, July 24, 1981, cited in Bornstein, 1985).

The Soviet Union has imported intermediate products as well as machinery and equipment from the West on a regular basis (Table 6.2). The OECD countries supplied a significant share of such imports, with machinery and equipment, metallurgy and chemicals as the top three categories. The imports of parts and accessories of machinery and equipment have also registered sharp increases in recent years. Western exports of machinery, equipment, metal products and chemicals have played a major role in the transfer of technology to the Soviet Union, especially in its energy, chemicals, engineering and mining industries (Bornstein, 1985).

Turnkey projects are another important conduit of technology transfer with wide-ranging possibilities. "In turnkey projects, foreign firms undertake to supply whole production systems and thus may provide feasibility studies, design of facilities, supervision of construction, delivery and installation of plant and equipment, licenses, training for Soviet engineers and technicians in the Soviet Union and abroad, and assistance in commissioning of facilities and the start-up of production" (Bornstein, 1985).

Turnkey projects are usually undertaken on a massive scale involving millions of dollars. Several such projects have been set up in the Soviet chemicals, ferrous metallurgy, motor vehicle, machine-building, and light and food industries. Given their versatility, they hold out much promise in the new economic milieu of the CIS.

Table 6.2. OECD Industrial Goods Exports to the Soviet Union, Selected Categories, 1970, 1976 and 1982 (values in US$ millions)

	1970		1976		1982	
	Value	% of Total	Value	% of Total	Value	% of Total
Machinery and equipment	889	36	4,002	36	5,239	28
Metallurgy	330	14	2,867	26	4,346	23
Chemicals	292	12	1,004	9	2,506	14
Paper, pulp and processed wood	193	8	503	5	1,112	6

Source: Adapted from Bornstein, 1985.

Industrial cooperation agreements (ICAs) contribute to technology transfer in many ways. The cooperation may take many forms and encompass production, investment, marketing, financing and/or research and development. The parties to the agreement can make payments in cash or in kind.

A quantitative evaluation of Soviet-Western ICAs is difficult because the available published information has been compiled chiefly from press reports and sample surveys of private Western firms. Table 6.3 presents a quantitative profile. The table reveals that the chemical industry, mechanical engineering and metallurgy together account for slightly over 60 percent of such agreements.

Technology is also transferred through "noncommercial" channels such as publications, trade and industrial exhibitions, exchange programs and illegal means like industrial espionage. These channels do not, however, constitute of a major source of technology transfer.

U.S.-SOVIET UNION TECHNOLOGY TRANSFER: CRITICAL ISSUES

The issues that have figured prominently in technology transfer between the United States and the Soviet Union are briefly examined in this section. The collapse of the Soviet Union and the formation of the CIS could be expected to impact on several of these issues in a positive fashion (discussed in the next section). The key issues concern the infrastructure, entrepreneurship and innovation, financing, technology choice, code of conduct, and ownership of property and privatization.

A country's infrastructure has a major impact on its ability to assimilate foreign or imported technology. The assimilation of Western technology in the Soviet economy has encountered many difficulties. These include problems with organization, supply, pricing, performance indicators and incentives, and delays in plant construction. Compared to Soviet technology, Western technology

Table 6.3. Soviet Industrial Cooperation Agreements with Western Firms, 1983

Branch of Industry	Percent of Total (%)
Chemical industry	36.7
Metallurgy	10.1
Transport equipment	6.9
Machine tools	5.5
Mechanical engineering	14.7
Electronics	6.0
Electrical equipment	2.7
Food and agriculture	4.1
Light industry	6.4
Other branches	6.9
TOTAL	100.0

Source: Bornstein, 1985

imposes more rigorous standards for complementary inputs of Soviet origin, including buildings, machinery and equipment, skilled labor, high quality materials and R&D services. These problems of complementarity have influenced the nature, amount and rate of technology imports from the West.

Entrepreneurship in the Schumpeterian sense (risk taking and decision making under conditions of uncertainty) was virtually absent in the Soviet Union. Innovation, both on the demand side (product innovation) and supply side (process innovation), gives central planning its greatest challenge. In the Soviet Union, technological change was predicted and embodied in economic plans. About 45 percent of the Soviet R&D budget was being spent centrally (Academy of Sciences or the ministries), in contrast to the United States, where most basic research was carried out by universities, with 65 percent of applied research conducted by corporations. Soviet R&D performance, apart from some patchy successes, has been very weak, as is evident in Table 6.4.

The infrastructure that fosters innovation did not exist in the Soviet Union. As Hanson (1981) put it, "The lure of profits, the threat of business failure, the pressures of competition (generally imperfect and oligopolistic competition but still competition of some sort), together with the far greater decision-making autonomy of the individual business unit made the whole setting of innovation quite different" (also see Kornai, 1982).

The Soviet Union's ability to finance imports of Western technology depended mainly on three interrelated factors: the Soviet convertible currency balance of payments—the excess of convertible currency earnings from exports over outlays for other high-priority imports like grain; Western credits—the availability and terms of Western credits to cover credits in the current account; and countertrade—the extent of Soviet success in making countertrade arrangements

Table 6.4. Estimates of Aggregate Soviet Factor Productivity (average annual percent growth)

	1961-65	1966-70	1971-75	1976-80	1981	1982
Increase in factor inputs	4.5	4.1	4.2	3.6	3.2	3.1
Increase in factor productivity	0.5	1.1	-0.5	-0.8	-1.0	-1.1
GNP Growth	5.0	5.3	3.7	2.7	2.2	2.0

Source: Buck and Cole, 1987.

for future exports to service the convertible currency debt (Bornstein, 1985). In all these respects, the Soviet record was far from adequate in stimulating technology transfer to any appreciable degree.

The Soviet Union's acquisition of certain kinds of Western technology has been significantly constrained by unilateral and bilateral export restrictions of Western governments. Export restrictions were imposed by the Coordinating Committee for Multilateral Export Controls (COCOM) for the NATO countries and Japan. As Bornstein (1985) points out, "It is clear that these restrictions have curtailed Soviet acquisition of Western technology for important branches of Soviet industry such as electronics, electrical machinery, oil and gas equipment and chemical equipment." Western restrictions were designed primarily to maintain the Soviet technology gap in certain key areas for strategic reasons. Sometimes restrictions were designed to curb the export of production know-how rather than sales of products.

The existence of a code of conduct is an important issue in the context of West-East technology transfer. It is worth noting that international trade in commercial technology has taken place in a market increasingly characterized by oligopoly, monopoly and monopsony. This has led to concentration in a small number of multinational corporations (MNCs) of a large percentage of operative and commercial technology that is urgently needed for the industrialization of less developed countries (LDCs). Thus, the primary justifications for a mechanism to regulate the international flow of operative and proprietary technology lie in the imperfections of the market itself.

The lack of private ownership of property constituted perhaps one of the most crucial constraints to West-East technology transfer. The economic monopoly of the Soviet state was all-embracing. All urban enterprises and land belonged to the state. Creation of collective farms and the abolition of private enterprise gained momentum under Stalin, continued under Khrushchev, and remained intact under Brezhnev, Andropov and Chernenko. The movement for some measure of economic decentralization began only in 1985 when Gorbachev assumed power. The pace of economic reforms under Gorbachev, however, was too slow and their scope was too limited to make any appreciable impact on the flow of Western technology to the Soviet Union.

OPPORTUNITIES FOR TECHNOLOGY TRANSFER IN THE MINING INDUSTRY

A survey of the state of technology in the mining industry of the former Soviet Union points to several glaring examples of technological backwardness in comparison with Western countries. The situation is especially serious in the metal mining industry. Therefore, this sector takes on special importance in terms of the opportunities and challenges for West-East technology transfer in the 1990s and beyond.

The former Soviet Union has enormous deposits of metallic and nonmetallic resources. Notwithstanding the use of inefficient and obsolete technology in many instances, the mineral production in the country has been significant. Table 6.5 presents a composite picture of the production of metallic minerals in the Soviet Union in 1990.

Despite the prominence of the metal mining industry in the former Soviet Union, the technology employed was inefficient and outmoded in most instances. Specific examples of technological deficiencies in the metal mining sector are cited below.[1] A case in point is the Soviet aluminum industry. The Sodenburg technology employed by most Soviet aluminum smelters is outmoded; the prebaked anode technology is universally preferred. Most of the Soviet aluminum plants are old and obsolete. The industry's capacity utilization rate is only 64 percent, a very low figure by Western standards. Western smelters normally operate at 95 percent of nominal capacity. The Soviet average for power consumption is between 15,000 and 16,000 kilowatt hours (kwh) per tonne of aluminum for smelters using 255 KA pots. This compares poorly with an average power consumption of 14,662 kwh per tonne in Western smelters. Furthermore, no Soviet smelter has a computerized control system, which is becoming increasingly common in Western smelters (International Monetary Fund et al., 1991).

Another example of backward technology relates to copper production. The former Soviet Union was known to have a large number of copper deposits. These included vast deposits in Kazakhstan, the Urals, the Norilsk region in Siberia, and Uzbekistan. The five mines in the Norilsk area all contain copper with grades ranging from 0.6 to 3.9 percent copper content—very low copper grades. The Norilsk complex as a whole, including its processing operations on the Kola Peninsula, produced one-third of the country's copper output (International Monetary Fund et al., 1991).

The heavy equipment (such as gushers and ball mills) in use at many concentrators is badly in need of modernization, according to Mekhanobr, which designed most of these plants. Five years is the natural life of most equipment of this type, whereas about 50 percent has been in service for 10–15 years (International Monetary Fund et al., 1991).

Although the need for imported technology was paramount in the Soviet mining industry, opportunities for technology transfer and other forms of technical assistance were limited largely because of its organization and structure. A brief description of the regulatory environment follows. All mining and smelter oper-

Table 6.5. Metallic Mineral Production in the Soviet Union, 1990

Metal		Production
Aluminum	('000 tonnes)	2,380
Gold	(tonnes)	285
Iron ore	(million tonnes)	550*
Lead	('000 tonnes of contained metal)	750*
Zinc	('000 tonnes of contained metal)	1,020†
Nickel	(tonnes)	251†
Copper	('000 tonnes)	107†

* Crude ore
† Smelter production
Source: International Monetary Fund et al., 1991.

ations were state-owned enterprises monitored by three types of national organizations with differing degrees of control and responsibility: ministries or quasi-ministerial bodies; research institutes; and All-Union industry associations. The ministries and quasi-ministerial bodies in the metal mining and smelting sector consisted of the Ministry of Metallurgy (responsible for all nonferrous mining and smelting operations with the exception of gold, diamonds and most nickel production); Glavalmazzaloto (this body, formed in 1988, had the responsibility for the mining and processing of gold and diamonds); Norilsk Nickel (which oversaw the activities of the huge Norilsk mining, smelting and refining complex in northern Siberia); Gosplan (which had a Department of Metallurgy with the sole function of planning production investment in the mining and metallurgical sectors); foreign trading organizations; and the Ministry of Geology (responsible for geological mapping and exploration).

The technical ore research institutes served individual metal industries and provided expertise in areas such as mine planning or mineral processing. The major institutes were VAMI (the All-Union Aluminum and Magnesium Institute), Giproruda (designed iron ore mines and pelletizing and concentration plants), and Mekhanobr (in charge of research and design of nonferrous mineral processing and concentration). Industry associations were responsible for representing the enterprises that produced a single metal or a group of metals. Examples are Sovaluminum, serving the aluminum industry since 1989, and Vtormet, in charge of the collection and treatment of ferrous and nonferrous scrap. An organizational structure, such as the above, with a diversity of agencies and multiple levels of decision making, rendered the process of technology transfer for all intents and purposes extremely difficult.

Creating the economic conditions for better use of achievements in science and technology will promote the successful development and exploitation of the mineral and raw material resources of each independent state, including Russia, particularly in the Far East. Here we identify five main areas in the mineral resources industry where there are great needs for foreign technology transfer:

(1) machinery and technology for prospecting; (2) machinery and technology for mining; (3) machinery and technology for mineral enrichment; (4) technology for processing; and (5) technology for transportation of mineral resources. In this context, it is crucial to recognize the close relationship between prospecting and mining and processing activities. Thus, the detailed study of the technology of the geological objects should be followed by a similar study of the technology of their further exploitation. Under the old Soviet system, this was seldom followed through.

At the beginning of the 1980s, objective prerequisites began to be developed: first, for improving the technologies for prospecting; and, second, for the development of new technologies and the assimilation of imported technologies for mining, processing and transportation of raw materials. However, the economic difficulties that arose by that time and their subsequent escalation into an economic crisis did not contribute to such activities, especially where imports of technology were concerned.

There is very little information about the volume, the dynamics or the effects of imported technology utilization in the mining industry of the Soviet Union. This does not mean that there were no imports of technology. The purchase of technologies, especially from the United States, certainly occurred on several occasions during the last 10 years, but probably in small quantities. The main imported items were machinery, materials and spare parts.

For example, the ministry in charge of the oil and gas industry of the Soviet Union bought machinery and materials worth an average of 1.5 million rubles each year during the 10-year period. The bulk of the purchases consisted of pipes that were imported at a rate of approximately 200,000 tonnes annually (with a considerable amount from Japan), at a cost of about US$200 million. Recently, joint ventures have been set up to develop new technical methods and technologies, especially drilling technologies for inclined and horizontal wells (with companies from Great Britain and Germany).

During the last 10 years, the Ministry of Ferrous and Nonferrous Metallurgy bought for the mining industry loading and transportation devices of three, five and eight tonnes capacity; equipment for drilling underground holes; hydraulic hammers to destroy nonstandard material; ball mills (300 cubic meters); chambers for flotation equipment; analyzers of chemical composition of ore and enrichment products in effluent (from Finland); cone-type crushers (from Bulgaria); hydraulic tool carriers for additional equipment; drilling devices (from Poland); and mining equipment (from the United States, Japan and Sweden). However, the needs for foreign equipment were far from being met. This circumstance now makes it necessary to increase the pace of internal development of new technologies for open-pit and underground mining.

Another example of the purchase of foreign technology relates to the mining branch of the coal industry of the Soviet Union. The items imported included mechanized purification complexes, enrichment equipment (centrifuges, sizing screens, and water removing sieves), conveyors, wagon-cars (from Poland), and cutting and sinking equipment (Bulgaria). Between 1980 and 1990, 700 million rubles were spent per Five-Year Plan period. Sometimes the equipment purchased was the result of joint designing and development.

The environmental standards of the Soviet smelting industry call for significant improvement. The prime source of atmospheric pollution from the aluminum smelting process is the gaseous emissions from the electrolytic pots. Another form of air pollution within the smelters results from inadequate pollution controls. The standards at the older smelters are very poor.

All copper smelters in the former Soviet Union have sulfuric acid plants (with the exception of the Norilsk smelters). However, the efficiency with which carbon dioxide is captured is low, so airborne pollution levels around the plants are still high. Considerable investment will be required to bring the environmental standards of the copper smelters up to an acceptable level (the Alaverdi smelter closed in 1989 because its sulfur emissions were 200 times the permitted limit).

We envision significant growth in West-East technology transfer in the mining sector in the years ahead, as each independent state strives to develop its own profitable mining industry. Thus, there will be diverse opportunities to meet the demands of individual states for equipment, materials and development of new technologies. This is especially true of Russia, which accounted for 46 percent of the former Soviet Union's output of ferrous ores, 58 percent of the copper, 64 percent of the bauxite and 90 percent of the tungsten. Another important reason for the development of new technologies in mining and prospecting is the growing need to better process the wastes of mining industries in order to protect the environment and to provide more rational utilization of mineral resources.

IMPACT OF NEW POLITICAL DEVELOPMENTS ON TECHNOLOGY TRANSFER

The collapse of the Soviet Union and the emergence of the CIS have resulted in an economic and political order far more conducive to effective and efficient technology transfer than that which existed under the previous system. The structural transformation of the political order has created an environment with abundant opportunities for West-East technology transfer. Perhaps the most important development is the elimination of an overarching central system that in the past was often at odds with the republics in the Soviet Union. The decentralization process that is the hallmark of the dramatic new developments has brought in its wake political autonomy and economic independence to the former Soviet republics to an extent that was unthinkable even a few short months ago. With the emergence of autonomy and independence, some of the formidable impediments to West-East technology transfer have vanished or significantly weakened.

As we have seen from a survey of past experience, one of the principal stumbling blocks in the implementation of technology transfer in the Soviet Union had to do with the relationship between the center and the republics. Often, the two had vastly different priorities and it was difficult to forge consensus. Many worthwhile projects have been the casualty of this dichotomy. The situation is very different today following the dissolution of the two-tier system and the loss of supremacy of the center in decision making.

One of the essential prerequisites for successful technology transfer is to have in place a legal, financial and physical infrastructure that promotes the development of a modern, efficient and resilient economy. This would include a modern banking and accounting system, a new commercial code and a stable and convertible currency. Free at last from the cumbersome regulations and stymieing restrictions of the central Soviet system, the states are now in a better position to carry out these reforms. Increased stability and accountability could lead to a significant increase in foreign trade, foreign investment and the inevitable inflow of foreign technology.

Another important issue in the context of technology transfer relates to the resource endowments of the different states in the new Commonwealth. The autonomy of the states under the new economic order gives each state the power to fashion its economic development programs taking into account its special needs and unique resource capabilities. The technology transferred would thus closely correspond to the specific needs and desires of the individual states and would add to the efficiency and effectiveness of the transfer process. The comparative advantages of the different states with respect to resource endowments could be more accurately factored into the agreements between the recipient and donor countries under the new system.

CONCLUSIONS

Very few countries offer as many potentially appealing opportunities for large-scale technology transfer as the CIS. These opportunities include a number of autonomous internal markets; a common pressing need for efficient and expanded consumer goods industries; vast amounts of natural resources whose efficient exploitation would benefit both the Commonwealth states and foreign investors; significant new investment opportunities created by the revamping of the former Soviet defense industries; a woefully inadequate infrastructure that would vastly benefit from the infusion of foreign know-how; and a critical need to import technology, managerial and entrepreneurial skills, and experience with a market economy.

Viewed in this perspective, it becomes clear that a rapid expansion in the infusion of foreign investment and technology is critical to the development of the Commonwealth and its integration into the global economy. By providing appropriate and timely incentives, the CIS could quicken the pace of economic reform and achieve success in attracting substantial flows of foreign investment so vital to the transition to a market economy. It is also clear that the obstacles to be overcome in achieving this are formidable.

NOTES

1. This discussion of the specific examples is based on information presented in "A Study of the Soviet Economy," International Monetary Fund et al. (Vol. 3), Chapter V.7.

REFERENCES

Bornstein, M., 1985, *The Transfer of Western Technology to the USSR,* OECD, Paris.

Buck, T. and J. Cole, 1987, *Modern Soviet Economic Performance,* Basil Blackwell, Oxford.

Cooper, J., 1985, "Western Technology and Soviet Economic Power," in *Technology Transfer and East-West Relations,* edited by E. Schaffer, St. Martin's Press, New York.

Cooper, J., 1991, "Soviet Technology and the Potential of Joint Ventures," in *International Joint Ventures: Soviet and Western Perspectives,* edited by A.B. Sherr et al., Quorum Books, New York.

Gopalakrishnan, C., 1989, "Transnational Corporations and Ocean Technology Transfer," *American Journal of Economics and Society,* Vol. 48, No. 3, pp. 373–384.

Hanson, P., 1981, *Trade and Technology in Soviet Western Relations,* Macmillan, London.

International Monetary Fund, IBRD, OECD, and EBRD, 1991, *A Study of the Soviet Economy* (Volumes 2 and 3), Paris.

Kornai, J., 1982, *Growth, Shortage and Efficiency,* Basil Blackwell, Oxford.

Ledin, H., 1978, "Methods of Transfer," in *Technology Transfer Practice of International Firms,* edited by F. R. Bradbury, Sijthoff & Noordhoff, Netherlands.

Mansfield, E., 1974, "Technology and Technological Change," in *Economic Analysis and the Multinational Enterprise,* edited by J.H. Dunnin, George Allen and Unwin, London.

Ruoho, S., 1990, "Soviet Foreign Trade and Joint Ventures," in *Gorbachev and Europe,* edited by V. Harle and J. Livonnen, St. Martin's Press, New York.

Schaffer, E., 1985, "Introduction," in *Technology Transfer and East-West Relations,* edited by E. Schaffer, St. Martin's Press, New York.

Smeliakov, N.N., July 24, 1981, "I Spros Zavisit Predlozheniia" (And Demand Depends on Supply), *Trud,* Moscow, p. 2.

PART TWO: REGIONAL ISSUES AND AFFAIRS

Chapter 7

DEVELOPMENT OF THE ARMENIAN MINING INDUSTRY: RATIONAL EXPLOITATION AND USE OF JOINT VENTURES

Yury A. Agabalyan, Stepan K. Mekhakian, and Levon M. Bagdasaryan

BACKGROUND

Armenia, despite its small land area, is rich in various mineral deposits, which have permitted the establishment of projects to mine and process copper, copper-molybdenum, gold-bearing ores, cement raw materials, bentonites, perlites, volcanic and felsite tuffs, basalts, granites, marbleized limestones, rock salt, gemstone materials, mineral waters, etc. Furthermore, Armenia has a number of deposits with proved reserves of iron ores, zeolites, diatomites, ceramic clays, and gypsum on the basis of which existing industries can develop and new ones might be set up.

This chapter focuses on a number of problems involving the present state of the mineral raw materials base of the Republic of Armenia, the republic's mining and processing industries, proposed methods of rational exploitation of deposits, and the foundation of joint ventures with foreign firms.

NONFERROUS METALS ENDOWMENT

The Kajaran deposit is one of the world's largest copper-molybdenum deposits whose explored profitable reserves can produce at the present level of output for more than 100 years, while its indicated reserves will be able to provide output for more than 200 years. Another similar project is the Agarak mining and concentrating integrated works with limited reserves of approximately 15 years. Because of departmental interests—that is, of overstated plans for molybdenum

J.P. Dorian et al. (eds.), CIS Energy and Minerals Development, 99–111.

concentrates output—only rich ores of the Kajaran deposit are mined, while the "poor" or lower-quality ores are dumped and lost, despite their being economically accessible. It should be noted that the recovery value of such "poor" ores exceeds the mean recovery value of the Agarak ore deposit that is located 30 km from the Kajaran field. Thus, two critical problems that need to be solved are the following:

- to discover ways to utilize the Agarak concentrating mill, taking into account its limited run period as per reserves; and
- to mine fully the Kajaran ores.

Investigations in recent years show that full rational mining of the Kajaran ores is closely connected with utilization of the Agarak concentrating mill capacities. It has been proved that it is advisable to put into prolonged storage an open pit of the Agarak integrated works and to begin processing the Kajaran deposit ores at the Agarak concentrating mill.

Besides conditioned molybdenum concentrate, the Kajaran and Agarak concentrating mills produce copper concentrate. The latter is also produced at the Kapan concentrating mill, where the copper ores of the Kapan deposit are processed.

The Alaverdi integrated mining and metallurgical plant had been the sole metallurgical project in Armenia and the only copper-smelting plant in the Caucasus at the time of its recent closure. It met the copper requirements not only for our republic but for a number of other regions as well.

The hasty and unreasoned decision to dismantle the Alaverdi integrated mining and metallurgical plant put Armenia into a difficult position. First, the aforementioned industries lost a reliable source of raw materials; second, copper concentrates produced in the republic are difficult to process in Commonwealth metallurgical works because of a lack of spare capacity in operating copper smelters; and third, it is expensive to process the concentrates in foreign smelters because of their low copper content (15–18%). In order to better utilize economic resources, it should be noted that the world market value of copper products that could be manufactured from locally procurable concentrates is estimated to exceed US$100 million dollars a year.

The need to restore copper smelting production in Armenia is thus obvious. To solve this important problem two alternatives are proposed:

- restoration of copper production at the Alaverdi integrated mining and metallurgical plant; and
- construction of a new integrated works for copper production.

The Alaverdi integrated plant was dismantled primarily because it polluted the air with sulfide anhydrides. The amount of harmful emissions depends on the bulk processing of concentrates, which in turn depends on copper content in the concentrates, provided all other conditions, such as steady-state processes and environmental control measures, are equal.

Hence, the main problem of the present studies is addressing improvements in processing copper-molybdenum and copper ores in order to increase sharply the copper content in a concentrate and providing a proportional decrease in bulk concentrates. To settle this problem, an agreement should be made between our republic's integrated mining and concentrating plants and the English firm, Mark-Rich, on buying new concentrating facilities in exchange for copper concentrates. Over recent decades, various organizations have raised the possibility of setting up in Armenia a plant for metallic molybdenum production. Construction of such a plant also will permit the extraction of rhenium and other valuable components.

In recent decades Armenia emerged as one of the world's noteworthy gold-bearing provinces. About ten gold ore and gold complex fields have been revealed. Extractable reserves of six gold fields have been already proved; two of them are under exploration now. The large Zod deposit in eastern Armenia is being worked by a combination of open pit and underground methods. Underground mining is carried out by the cut-and-fill stoping method. The Yerevan Polytechnic Institute and the Institute of Geological Sciences, Academy of Sciences, Armenia, have developed a more advanced method of underground mining which may be applicable to other mineral deposits throughout the republic. Compared with the existing system, the new system has a number of advantages: it increases labor productivity and excavation rates, and at the same time decreases expensive solidifying mixture consumption. A patent has been granted for the new method.

The Megharadzor deposit is represented by steeply dipping thin ore bodies containing loose unstable ore and rather unstable enclosing rocks. The ore is characterized by high gold content and recovery value. The most important issue in increasing mining efficiency of this deposit is the selection of an optimal mining method.

Studies carried out by the Yerevan Polytechnic Institute resulted in different alternatives for half-consolidating filling methods, whose key advantage was a decrease in consumption of expensive consolidating materials, and even more important, a drastic decrease in ore dilution, the economic benefit of which is commensurate with the ore mining cost. Ores of the two deposits are processed at the Ararat gold recovery plant where gold concentrates are obtained.

One of the major new developments in Armenia is the construction of a plant to produce gold and silver, which are to be fully utilized by the expanding Yerevan jewelry works. For an industry based on the ancient national traditions of Armenian skilled jewelers, it is desirable to establish foreign economic relations and offer competitive products for the world market.

Further development of the gold mining industry is associated with the exploitation of a number of gold ore and gold-complex deposits, where lead, zinc and copper are to be recovered along with gold and silver.

Armenia also possesses a small lead-zinc field. Putting into operation polymetallic and gold-complex deposits will permit the production of lead and zinc metal in amounts sufficient to meet the republic's requirements.

Given increases in production efficiency, Armenia intends to expand the processing of many complex metal ores, as well as wastes of mining, concentrating, and metallurgical plants such as overburden rocks, concentrate tailings and metallurgical slags. Usable material for the manufacture of facing blocks, slabs and heavy aggregates is, for example, contained in the gabbroic overburden rocks of the Zod deposit. After sand-washing, tailings of the Kajaran concentrating mill may be utilized in building, and alumina raw materials can be used for aluminum production. Tailings of the Ararat gold recovery plant are also noted because noble metals can be re-extracted from them; moreover, they may be used as raw materials for cement-making and ceramic industries.

FERROUS METALS

In contrast to the nonferrous mining and metallurgy industry of Armenia, there are no projects of ferrous metallurgy in the republic. Armenia does, however, have sufficient mineral raw materials to eventually develop an iron ore mining industry and ferrous metallurgical plants. Two explored deposits with proved economic reserves of iron ores in Armenia are Razdan and Abovian.

The Razdan iron ore deposit is located in a favorable geographical and economic region near the town of Razdan. Ore is found to be naturally alloyed and contain rare-earth elements as well as germanium, gallium, thallium, niobium, tantalum and zirconium. Geological conditions at Razdan are favorable for mining as the deposit is a bedded ore body, dipping at a low angle with an opening to the surface which contains massive and disseminated ores. The massive ores are considered economic reserves, while the disseminated ores are, at present, unprofitable to mine. Disseminated ores occur in the hanging and lying walls of the massive ores. Preliminary technical and economic studies show that simultaneous open-pit mining of the massive and impregnated ores could be profitable. This would enable an increase in profitable ore reserves by 40 percent. The overburden rocks are represented by granite, granite-epidotite skarns, hornfels, and granitoids that have not been studied extensively, but nevertheless, are considered suitable for building material manufacture. Furthermore, the granites may be used as abrasive materials and in the manufacturing of gem stones.

Numerous processing studies have yielded top-grade iron concentrates with a soluble iron content of 68 percent, permitting ecologically sound metallurgical production by means of direct sponge iron reduction followed by processing and refining of technically pure iron. Analyses have shown that iron produced in this manner is much more pure than that produced via commercial methods. On the basis of this technically pure iron, a number of experimental smeltings of special steel grades and precision alloys were carried out. The quality was found to be higher than that produced by full-scale operations. The presence of rare-earth elements in the metals increases electric resistance and alloy survivability.

The Abovian iron ore deposit is located five km from the town of Abovian. In addition to iron, the ores of this field contain disseminated rare-earth elements of the cerium and yttrium groups, as well as apatite. The iron ores occur under a layer of basaltic and andesitic-basaltic lavas with thicknesses ranging from 50 to 180 m. Abovian iron ores are explored and reserves are confirmed as mineable.

Processing studies of Abovian ores suggest the iron ore concentrates contain a soluble iron content of 65 percent. The concentrates were found suitable for obtaining sponge iron by means of the direct reduction process. Rare-earth concentrates were produced as well, significantly increasing the recovery value of the ore. With dressing, apatite is removed from tailings with one percent P_2O_5. Processing of the tailings yields an apatite semiproduct which requires further technological analysis. Powder metallurgy may develop in the republic on the basis of its available sponge iron resources.

Large-scale ferrous metallurgy development in Armenia may be dependent on the Svarants iron ore field. Prospected iron ore reserves at Svarants are approximately 400 million tonnes with an average iron content of 20 percent. With predicted reserves, the total amount of iron ore is estimated at one billion tonnes. In spite of its low iron content, a significant increase in the recovery value of each tonne of ore may be expected if complex mining techniques are used. The ores contain 22–28 percent magnesium oxide, 1.2–1.8 percent titanium oxide, and 0.13 percent vanadium pentoxide. In addition to these oxides, the ores also contain nickel, cobalt and other elements.

Preliminary technological analyses at Svarants yielded an iron ore concentrate with an iron content of 55 percent and a recovery rate of 67 percent. Unroasted and roasted magnesite-ferrite refractory items were obtained from dressed tailings. Slag magnesium-silicate cement was recovered as well. The gross value of useful by-products at Svarants is significant per each tonne of ore. It is therefore important to continue processing studies in order to identify and obtain marketable products, including concentrates, metals, and alloys. Advances in processing will predetermine the expediency of exploration and provide an estimate of profitable reserves and the viability of commercial exploration of the Svarants deposit.

NONMETALLIC MINERALS

Armenia is highly rich in prospected reserves of top-grade bentonites, perlites, diatomites, and zeolites. All prospected deposits are characterized by favorable geological and mining conditions suitable to exploitation, permitting open-pit mining with a low stripping factor. High-grade alkaline bentonites occur at the large Sarigyugh deposit, which contains profitable reserves exceeding 50 million tonnes. Prospected reserves of alkaline-earth bentonites are also found in Armenia's Noemberyan deposit.

Half of the Commonwealth requirements for bentonites are met by domestic supplies, while for bentonite products, only 20–25 percent. Armenia's annual economic gains from supplying bentonite and bentonite products to the Commonwealth are presently 250–300 million rubles. Some construction activities in the Commonwealth are hampered because of limited supplies of bentonites and natural bentonite powders obtained by grinding and milling. The share of bentonite powders in total output is around 26–27 percent. Such an imbalance in supplies is due to continuing delays in building new bentonite processing facilities.

In recent years, the Ijevan bentonite integrated plant, built to utilize the Sarigyugh deposit bentonites, has failed to achieve designed output and produce the 700,000 tonnes of bentonite powders required to manufacture 10,000 tonnes of activated bentonite powders used in the Soviet oil industry. In 1987, the mining of ball clay in the Sarigyugh open pit was 504,000 tonnes, and bentonite powder production reached 395,000 tonnes (or 56.4 percent of designed output).

Advances in processing technology worldwide suggest that with little additional expenditures, Armenia could improve significantly its grade of modified bentonite powders. Certain technical conditions for clayish drilling muds could increase the yield of clayish mortar 1.5–2 times for every tonne of bentonite powder. Economic benefits from increases in grade would be realized by way of improvements in drilling.

As a result of changes in production and sales methods, the technical and economic indices of the Ijevan bentonite integrated works were improved markedly in 1990 owing to an increase in product output stimulated by a 17 to 20 percent rise in wholesale prices for one tonne of modified bentonite powders and only a slight (2–3%) cost increase.

Assuming that full and complex utilization of the Sarigyugh deposit bentoclays is possible, and taking into account the present industrial capacity of the Ijevan bentonite integrated works, it is recommended by Armenian mining officials that the bentonite industry turn entirely to modified bentonite powder production for use in drilling, the production of iron ore pellets and foundry activities, and the food industry (activated bentonite powders).

Significantly, relatively little consumption of bentonite powders is required when pelletizing iron ore concentrates. Bentonite powders increase pellet strength while decreasing iron content in the concentrate. Bentonite powders manufactured from the Sarigyugh deposit permit low consumption levels of less than 0.7 percent while providing high pellet strength. Firms from Italy, the United States and other countries have expressed interest in further developing Armenia's bentonite industry.

As for perlite reserves, Armenia plays a leading role in the new Commonwealth. Three deposits in the republic contain total economic reserves of more than 150 million cubic meters. Two of them are in operation now. In 1990, the Aragats perlite mine produced more than 500,000 cubic meters of raw perlite, yielding the following products: broken stone and sand for construction (as concrete aggregates), expanded perlite, and filter-powder (1,900 tonnes). In 1990, more than 1.5 million cubic meters of perlite air brick were mined at Armenia's Jraber open pit and subsequently processed into broken stone and sand at a crushing plant. A major portion of the mined raw materials is used as light-weight concrete aggregate. Numerous analyses have proved that Armenian perlites are of high technological quality and they may be used successfully as raw materials for production of light-weight expanded materials.

The range of the swelling factor of Aragats perlites is 10–16 compared to 5–7 for other similar perlite deposits in the CIS; its swelling temperature is 850–1000 °C compared to 1050–1100 °C for similar deposits.

Armenia's expanded perlite varieties are characterized by high filtration rates which predetermine their use in certain production processes requiring high rates of filtration, such as beer brewing. In the United States, 15–20 percent expanded perlite is used for filtration materials production

The Institute of Organic Chemistry of the Armenian Academy of Sciences has developed a new method for filter powder production using Aragats perlite for which patents have been granted. Since 1966, a pilot production plant affiliated with the Institute of Organic Chemistry has been producing 200–300 tonnes of Aragats filter powder annually which is used in Armenia's chemical, petrochemical, medical and food industries. Pilot-plant tests have showed these powders to be competitive with similar products made internationally, such as Dekolit-4200, and Metazil-A. Many foreign companies recognize Aragats filter powder to be of high quality. Establishing a production facility for Aragats filter powder at the Aragats perlite mine site would provide substantial benefits to Armenia's national economy.

Owing to its highly reactive properties, high purity and homogeneity, Aragats perlite is found to be a valuable raw material for glass manufacturing. With hydrothermal processing, perlite yields complex semiproducts, Kanazit-1, used for digesting clear glass and crystal, and Kanazit-2, used for production of dark-green glass, water glass, sodium metasilicate, alkaline aluminum silicate and other products.

Water glass is used for the production of silicogel, silicates paints, synthetic cleansing agents, fertilizers, and catalysts. Sodium metasilicate is used as a bleaching material in the textile industry, in the production of cleansing powders, and in petroleum cracking.

Studies conducted by the research company Stone & Silicates confirmed the technological and economic feasibility of using expanded perlites from Armenia for production of effective heat-insulating building materials, such as perlite fibrous slabs (Armix). An increase in production efficiency is possible with assistance from foreign firms. Various companies from the United States, Italy, Japan and other nations are now considering taking part in shipments of technological facilities designed for expanded perlite production. Firms from the United States, Germany, and Belgium are also considering taking part in shipments of technological facilities designed for perlite fibrous slab production from expanded perlite.

Five diatomite deposits in Armenia contain profitable reserves of 16 million cubic meters. Two deposits currently being worked—Dradzor and Vorotan—possess seven million cubic meters of profitable reserves. The predicted diatomite reserves in Armenia are tremendous.

A majority of the sorbents produced in the CIS from diatomites are mineral raw materials which underwent primary treatment involving only breaking and milling. A smaller share of sorbents is derived from processing and activation (by means of acids and salts) which are capable of significantly improving the filtration and adsorption properties of diatomites. This downgrades the quality of mineral sorbent products produced in the CIS and does not promote increases in production. The use of diatomites is clearly inefficient. While more than half of world's mined diatomites are used as filter materials, Commonwealth diatomites are utilized only as raw or primary processed materials in the construction industry. In the CIS, only 10–15 percent of the mined raw materials are used for filter powder and other sorbent production. In contrast, in the United States the annual output of diatomites is 650,000–700,000 cubic meters, 60 percent of which is used for filter material production. Of the total quantity of manufactured diatomites exported by the United States, filter materials make up 25–30 percent. The United States exports processed materials to more than 50 countries at an average price of US$100–120 per tonne.

In the CIS, filter materials, manufactured only in Nor-Kharberd (Armenia) and Inzensk and Kirovograd (Russia), undergo sophisticated thermal processing. Since 1976, Armenia has been operating the Nor-Kharberd pilot production plant at the Dradzor deposit, and manufacturing sorbents designed for chromatic separation of gas and liquid mixtures, filtration of wine products, and so on. Diatomite powders of the Nor-Kharberd pilot production plant are considered among the best in the Commonwealth of all filtering substances.

In 1965, an Italian firm expressed interest in Dradzor diatomites and requested from republic officials diatomite samples (0.5–1.0 tonnes) from the deposit in order to carry out commercial tests. The firm had considered drawing up a contract for the receipt of diatomites in the amount of 5,000–20,000 tonnes per year, while at the same time guaranteeing the sale of these products to many European countries. Unfortunately, economic cooperation between the firm and Armenia failed to materialize at that time.

The Nor-Kharberd pilot production plant is expected to continue making advances in the processing and activation of Dradzor diatomites in order to increase output of high quality marketable products, such as filter materials. The scale of production of high-grade marketable products, such as filter materials, adsorbents, chromatography carriers, etc., will be boosted to meet the needs of foreign markets.

Natural zeolites, a class of hydrated silicates of aluminum and either sodium or calcium or both, are found in abundance in Armenia. The wide use of natural zeolites is made possible by their high adsorption, cation-exchange, and catalytic properties. Natural zeolites are not inferior in character to synthetic zeolites and, in some cases, are actually superior. The mining and processing costs of natural zeolites are 20–100 times less than the cost for manufacturing synthetic zeolites.

Armenia's zeolite reserves include the Noemberyan clinoptilolite deposit (Nor-Kharberd Territory) with prospected profitable reserves of 12 million tonnes. Recent exploration in the western and eastern parts of Nor-Kharberd Territory increased reserves to 25–30 million tonnes. Predicted reserves of the Noemberyan zeolite basin are 150–170 million tonnes.

Zeolites are generally used as gas adsorbents and drying agents as well as water softeners. Years of analyses of Armenia's Noemberyan zeolites have revealed many specific uses, including:

1. fodder additives used in raising cattle, hogs, sheep and poultry;
2. filling agents in synthetic detergents for drying and cleaning bases, for liquids and for oil purification;
3. cationite for artificial soil formation and for sewage purification; and
4. land reclamation agent in plant growing.

When zeolites are added to hog fodder, the average increase in weight is 6–18 percent. Utilization of Armenian zeolites in cow rations raises milk production by 6.6–8.7 percent. Tests on younger cattle of different ages have showed weight increases of 8.2 percent, with the total weight gain per head averaging 11.2 kg. Similarly, feeding zeolites to sheep and chickens has resulted in live-mass increases by 4.8 and 17.6 percent, respectively.

Noemberyan zeolites have also been used satisfactorily as filling agents in rubber mix and paste synthetic cleansing agents.

When used as soil additives in agriculture and horticulture, Armenian zeolites proved effective in raising crop capacity. The positive results include:

- improvement of soil structure and an increase in its permeability (especially important for clay and loamy soils);
- natural zeolites absorb main plant nutrients (nitrogen, potassium, etc.) in the form of exchange cations and progressively give them back during the plant growth;
- zeolites may decrease soil acidity by way of ion exchange;
- natural zeolites in soil increase its water-holding properties;
- clinoptilolite tuff selectively binds heavy toxic metals found in soil and prevents plants from intaking them.

Applying clinoptilolite tuffs of the Noemberyan deposit to soil results in a neutralization of dern-podsolic soils and increases barley crop yields by 20–50 percent. Noemberyan zeolites were used in cucumber hotbeds and crop yields increased by 33–48 percent. Analyses carried out on fodder lands showed an increase in hay crop yields up to 6.6 12.2 kg/ha on meadow-steppe lands by applying zeolite in amounts of 60–500 kg/ha against a background of applied nitrogen and phosphorus.

Noemberyan zeolite additives applied to soil decrease sharply (2.1–3.2 times) the nitrogen losses in drainage. This is important not only for efficient fertilizer application but also in order to significantly decrease the seepage of nitrogens from soil into water basins that results in pollution. This was confirmed by tests carried out on Armenia's Lake Sevan basin.

Despite low mining and processing costs, zeolites are not commonly used in the CIS today. A large number of operations extensively use artificial zeolites or foreign substitutes. Reasons for the inadequate use of zeolites in the CIS include the following: incomplete technological studies on the application of natural zeolites in different industries; a lack of quality control for zeolite raw materials and processed products as the present specifications available are not well grounded or approved; a lack of accurate prices for zeolites of different qualities and applications; the absence of a customer-consumer market; and a general lack of experience in utilizing zeolites.

Economic studies have confirmed that commercial mining of the Noemberyan deposit is possible. Capital outlays would be paid back in two years, providing an annual output of zeolite tuff of at least 100,000 tonnes. The long-term CIS requirements for Noemberyan zeolites are estimated at around 500,000–600,000 tonnes per year. The main areas of use would be agriculture, drying and gas cleaning, etc.

As a means of comparison, in the United States annual zeolite mining yields one million tonnes. Zeolites are used extensively in agriculture, air and sewage purification. Union Carbide conducts both mining and technological studies of zeolites to discover new areas of application for zeolite products. Firms like Union Carbide and Zeolite of Japan are welcome to participate in zeolite mining and processing activities in Armenia.

Economic development around Armenia's capital, Yerevan, has been facilitated, in part, by the massive Yerevan rock salt basin, which contains predicted reserves estimated at 200–250 million tonnes. Some parts of the Yerevan deposit are prospected, while some are under exploitation today by underground mining and well leaching.

The Avan salt integrated works located in Yerevan has been operating since 1967. Salt mining at Avan is carried out by both underground mining and leaching.

Avan contains two separate levels of rock salt. An upper salt-bearing level (210–270 m deep) is mined by underground techniques, while a lower level (910–1200 m) is worked by leaching. The salt mined from the upper level is used in the cattle raising industry, and salt brine extracted from the lower level yields high-quality common salt.

Remarkably, the Avan underground salt mine also serves as a medical facility in addition to being a republic-wide distributor of salt. Approximately 270 meters below the mine's surface, the upper level of Avan serves as a medical retreat for patients with severe respiratory illnesses. Up to 60 patients with pneumonia and other diseases visit Avan annually, spending several hours a day for about three months below the surface. The Avan "medical wing" comes complete with beds, ping-pong tables, drawers, and other recreational facilities, with a recovery rate reported to be near 100 percent.

The Elar brine works, under the supervision of the Research-Industrial Incorporation Nairit, has been operating since 1971 and supplying the integrated chemical works with salt brine to manufacture caustic soda, rubber of different trademarks, and a number of other types of products. Actual profitable salt reserves are sufficient for the chemical works for a period of more than 40 years at full capacity.

Prospecting for rock salt has been carried out in the Yeghvard Territory of Armenia where predicted reserves are estimated to be 27 million tonnes.

A feasibility study completed in 1991 revealed a new trend to form the underground chambers for natural gas storage in salt-bearing width. The main finding when evaluating the Yeghvard Territory was that the total volume of underground chambers formed for natural gas storage were closely dependent on annual brine consumption, i.e., common salt "extra" production, and on demands of the chemical industry and other industries.

Clearly, Armenia has all the preconditions to set up a large-scale rock salt industry and increase production of "extra" salt obtained from brines: a tremendous rock salt base and many years of experience in underground salt mining and evaporated salt "extra" production from salt brines. Initially, however it is necessary to estimate the demands for salt in the CIS and the countries of Europe and the Middle East.

Armenia has great prospected reserves of natural finishing and building stones, light-weight scoriae, mineral waters, gem stones, ceramic and porcelain-faience clays and glass raw materials.

The raw materials base of finishing stones is notable for its diverse genotypes and species of natural stones (more than 20). The share of profitable reserves is 43.7 percent of that of the former Soviet Union. Currently, 25 deposits of 42 are being mined. Natural finishing stones of Armenia that are in great demand worldwide include highly decorative felsite tuffs, travertines, colored breccia-conglomerates, colored marbles, and granitoids varying in color and pattern. Finishing and building stones used in many civil and industrial buildings in Armenia are mined by stone-cutting machines, drilling-blasting and drilling-wedging methods, as well as rope saws.

Despite its already significant contribution to Armenia's national economy, the republic's stone industry is plagued by some problems. Mining currently lags behind demand, and out-of-date technology is used in stone mining and processing, leading to inefficiency, low waste utilization, and a limited range of manufactured finishing products.

MOST CRITICAL PROBLEM: ENERGY SHORTAGES

Without exaggeration, the most critical problem facing Armenia is supplying the republic with adequate energy resources. Preliminary exploration in the republic has failed to reveal either petroleum or gas deposits. The possible existence of oil in Armenia surfaced as a topic of debate only recently as new research has confirmed that Armenia's geostructure lends itself to the accumulation of oil formations which has encouraged geophysicists in the republic to drill several wells. Oil reserves in Armenia's neighbors—Azerbaijan, Georgia, Iran, and

Turkey—provide some reason to speculate on the existence of oil in the Armenian plateau. Following the analysis of preliminary geophysical data in 1983, the drilling of Armenia's first oil well near the town of Voghtchabard in the same year led to oil gushing out at a depth of 4,571 meters (14,200 ft.). Drilling of a second well near Hoktemberian produced 30,000 cubic meters (900,000 cu. ft.) of oil daily, at a depth of 4,333 meters (13,550 ft.). The crude produced was a mixture of methane and nitrogen. With the introduction of advanced drilling technology and equipment, Armenia will likely expand its search for oil reserves.

As for other energy resources, exploration has identified a series of prospective deposits of solid fuels, including coking coals and oil shales. At present, a large-scale geological exploration program has been initiated in Armenia to identify energy resources, if possible, in the various territories of the republic. Considering Armenia's current acute energy crisis, viable deposits, if found, would be evaluated and exploited rapidly. Armenia is a country of young volcanism, indicating it is advisable to carry out large-scale geological-geophysical prospecting to find geothermal energy sources. A number of hydrothermal and petrothermal heat sources have already been identified in the republic. The former are of interest because of their location near populated areas, the latter are of interest for their potential to generate electric energy. Special attention is being paid to a recently discovered petrothermal anomaly, located at a depth of 1,500 m with a temperature above 300 °C. Additional exploration for geothermal and other energy resources will require close cooperation with foreign firms of Japan, the United States and other countries.

SUMMARY REMARKS

Armenia is the smallest of the former 15 Soviet republics, with a population approaching 3.7 million persons. The Armenian territory covers just 29.8 thousand kilometers. Armenia borders on two other former Soviet republics—Georgia and Azerbaijan—and the countries of Turkey and Iran.

Despite its small size, Armenia is well-endowed in mineral resources, which form the basis for industrial development in the republic. Armenia's mining industry accounts for about 18 percent of the economic output of the republic, or 360 million rubles (US$600 million) annually. It has the fourth largest mining industry among the former Soviet republics. Armenia is a significant producer of gold, copper-molybdenum, and salt in the Commonwealth, and the largest supplier of construction stone in the group of nations. Armenia produced 40 percent of the former Soviet Union's molybdenum resources, an ingredient required by industry to manufacture high-strength steel. Some iron ore and zinc is also mined in the republic.

Armenia's minerals endowment—like all of its natural resources—is considered vitally important for economic security reasons by government officials. As such, the recently issued proclamation of independence purposely emphasized that all future use of Armenia's resources would be determined by local authorities only. Government and mining officials plan to work together in the years ahead to ensure that minerals development plays an increasingly larger role in overall economic growth. With this as an objective, foreign investment is being sought to boost the efficiency of mining operations and increase output in the republic.

Chapter 8

KAZAKHSTAN'S MINERAL RAW MATERIALS INDUSTRY AND ITS POTENTIAL

Shakarim F. Zhansetov and Madenyat A. Asanov

INTRODUCTION AND BACKGROUND

Kazakhstan's mineral raw materials base is remarkable for its variety and size, and is capable of satisfying the diverse needs of a population numbering more than 16 million persons. Mineral deposits in Kazakhstan are known for their uniqueness, which allows for the establishment of huge industrial facilities. Kazakhstan occupies a leading position in the former Soviet Union with regard to major economic minerals, both in terms of explored reserves and the scope of extraction and processing. In particular, Kazakhstan accounts for significant shares of the former Soviet Union's mining and processing of the following strategic minerals: chromites, 97 percent; bismuth, 95 percent; tungsten, 52 percent; lead, 47 percent; zinc, 40 percent; copper, 36 percent; bauxites, 25 percent; phosphorites, 31 percent; ferrous ores, 14 percent; and coal and petroleum, 10 percent.

This chapter presents a brief overview of the characteristics of the main mineral raw materials resources in Kazakhstan, including petroleum, gas, coal, and metallic and nonmetallic minerals. Special attention is paid to such large-scale unique deposits as Tenghiz (petroleum and gas), Karaganchak (petroleum-gas condensates), the coal and ferrous metal deposits of central and northern Kazakhstan, the rich reserves of multicomponent nonferrous ores (lead, zinc, copper, tungsten, bismuth, and precious and rare metals), and the Karatau phosphorites.

In the restructuring of the republic's economy, the role of the mineral raw materials sector is considered along with projected trends for the development of the raw materials base. The success of economic restructuring will depend to a great extent on the creation of a long-term strategy for the rational utilization of the republic's mineral resources. One of the principal requirements in promoting rational exploitation of mineral resources is increasing the comprehensiveness of their utilization. A greater comprehensiveness in mineral raw

J.P. Dorian et al. (eds.), CIS Energy and Minerals Development, 113–125.

materials utilization can lead to lower environmental pollution and a reduction in the production costs per unit of output. In order to increase productivity and better protect the environment, Kazakhstan's mining industry needs to be reorganized and employ newer technologies, including state-of-the-art low waste and waste-free extraction and processing systems.

At the same time, it will be necessary to introduce a suitable regulatory regime that would correspond to the emerging market structures and ensure that taxes and royalties are aimed at stimulating a rational use of mineral resources. The mining industries should themselves be internally restructured in order to ensure an optimal yield of final products in the entire industrial sector.

A very complicated issue will be the establishment of a new pricing system for mining industry products which accurately reflects supply and demand. This development will require a radical transition of the industry to a system based on international rules of cost calculation and rational accounting for capital and other expenses. However, implementation of these steps should be gradual and well thought out.

In the transition to a market economy, it will be necessary to anticipate the influence of different socioeconomic factors on mineral resource development policies. The transition should lead to an environment where mineral producers can make rational decisions based on their own as well as national interests.

The Republic of Kazakhstan's unique endowment in various mineral raw materials coupled with fairly favorable geologic conditions for their development have allowed it to become one of the world's most important mining and mineral-processing regions. In a relatively short period of time, Kazakhstan has experienced an emergence of large-scale mining and processing industrial complexes that were established to exploit and utilize the raw materials found in hundreds of explored deposits of fossil fuels (oil, gas, coking and power-generating coals), nonferrous, rare and ferrous metals, and various minerals. Reserves of the major industrial raw materials will allow the currently operating mines and enterprises to produce well into the future. Moreover, expert forecasts of new discoveries of mineral deposits are quite optimistic. The effective exploitation of Kazakhstan's mineral raw materials resources is a major goal in the implementation of current plans for expansion and consolidation of the republic's export potential.

Overall, more than 3,000 occurrences of various economic minerals (excluding underground waters) have been explored in the republic, with about 1,000 of these being currently exploited. All elements in Mendeleyev's periodic table can be found in the subsoil of the republic, and of the 80 elements that are economic minerals, 53 are available in volumes sufficient for cost-effective extraction, dressing and metallurgical or chemical processing. The value of explored mineral reserves in the republic is estimated at more than two trillion U.S. dollars, with fossil fuels (oil, gas, and coal) accounting for half of this total and nonferrous, rare and precious metals for over 30 percent.

Mineral raw materials are mined and processed by 1,880 enterprises. Kazakhstan's subsoil annually yields material values to the amount of US$20 billion.

Important questions exist, however, about the efficiency of operations in Kazakhstan and how much value is lost by inadequate mining practices.

The rock spoil heaps, tailings and slag dumps of various mining enterprises in the republic have accumulated over 16 billion tonnes of wastes containing many valuable components, including nonferrous, rare and precious metals, and raw materials for the chemical and construction industries. Remarkably, the mineral contents in waste accumulations of certain types of geologic deposits sometimes exceed those in mined deposits or in deposits intended for future exploitation.

CHARACTERISTICS OF KAZAKHSTAN'S BASIC MINERAL RAW MATERIALS RESOURCES

Petroleum and natural gas

Kazakhstan's hydrocarbon reserves are contained in 153 occurrences, including 80 petroleum, 24 gas-petroleum, 21 petroleum-gas condensate, five gas condensate and 19 gas fields. The main hydrocarbon reserve base is concentrated in western Kazakhstan's Guriev, Mangistau, Uralsk and Aktyubinsk regions. At present, the hydrocarbon resources of these regions are almost equivalent to those of Western Siberia in terms of explored and extrapolated petroleum and gas reserves.

The proved petroleum and natural gas deposits in western Kazakhstan range stratigraphically from the Devonian to the Paleogene age. Sulfur-rich petroleum fields prevail among the known deposits. As far as oil density is concerned, light petroleums account for almost 80 percent of the reserves. The geologic structure of the majority of the hydrocarbon deposits is complex, and most of the deposits are very large. The ten largest deposits account for 85–90 percent of the explored petroleum resources in Kazakhstan, with almost one-third of these reserves concentrated in the giant Tenghiz deposit.

The Tenghiz deposit, discovered in 1981, is located in the northwestern Pre-Caspian area (Guriev District). Tenghiz is a unique petroleum and gas deposit, unrivaled in the size of its reserves by any other known deposit in the world. Even though the deposit has been explored to a depth of only 5,500 meters, its potential reserves, estimated at 4.0–4.6 billion tonnes of petroleum, ensure cost-efficient exploitation. Extractable oil reserves are estimated at 1.5 billion tonnes; however, some experts claim this figure should be raised to two billion tonnes which implies that these reserves are valued at US$270–330 billion in current world prices.

Another unique feature of that area is the Karaganchak petroleum-gas condensate deposit, which contains up to three-fourths of all nonassociated gas reserves in the republic. The deposit is located in the Uralsk region in the northern part of the Pre-Caspian depression. Discovered in 1979, test production began in 1984. Petroleum-gas condensates are produced in wells which yield nonassociated gas, condensates, and solution gas, but no petroleum. The output is then separated and transported by two pipelines (one for gas and the other for condensates) to the Orenburg gas-processing plant in the Russian Federation.

The lower horizons of older deposits of valuable petroleum reserves (Makat and Dossor in the Guriev region) may eventually prove to be rich in hydrocarbons once drilling reaches deeper strata. These deposits have not yet been drilled to depths exceeding 40 meters. The Tenghiz petroleum deposit, the Karaganchak petroleum-gas condensate deposit and, actually, all hydrocarbon deposits in western Kazakhstan are unique in the peculiar geological characteristics as well as the quality and composition of their respective raw materials. Consequently, the deposits require specialized technologies for extraction and processing. They are characterized by high sulfur content; a complex composition which yields gaseous sulfur, mercaptans and other products, in addition to fuel and lubricants; by anomalously high pressures, which reach hundreds of atmospheres; extreme temperatures; and high reactivity of the concentrated components. These conditions indicate that the development of new hydrocarbon deposits in western Kazakhstan and an expansion of the raw materials extraction and processing industry will require comprehensive analysis and carefully considered decisions on technological, economic, environmental and social issues that also take into account local interests and the needs of the republic.

Coal

More than 4,300 coal fields and deposits, of which about 50 have industrially significant reserves, are known to exist in Kazakhstan. Of the various types of coal, 60–65 percent of all reserves are bituminous (Karaganda and Pavlodar regions) and 35–40 percent are lignite (Kustanai, Tourgai and Pavlodar regions). Coking coals constitute up to 20 percent of the reserves and are found mainly in the Karaganda coal basin.

The Karaganda coal basin is situated in central Kazakhstan, and its coal-bearing capacity is found in structures from the Carboniferous and Jurassic ages. The formation of Carboniferous age comprises about 80 layers, 30 of which are of sufficient thickness to be mined. There are 26 mines in the basin with a total annual production capacity of 45.7 million tonnes and a yearly output of around 32.6 million tonnes of coal. Twenty-three mines produce coking coals. Mining occurs at depths of between 349 and 835 meters.

The bulk of the coking coal is supplied to the Karaganda metallurgical works in Temirtau, where it is processed for consumption. Coal and coke concentrate supplies from Karaganda are of great importance to the operation of the metallurgical plants in the southern Urals (Russia) as well as those in the more remote areas of the Russian Federation, namely Magnitogorsk, Orsko-Khalilovski, Chelyabinsk, Novolipetski and other regions. The Karaganda basin contains explored reserves of high-grade coking coals which makes development of new mines feasible; a number of promising new locations have been singled out for prospecting.

Domestic steaming coals in Kazakhstan are excavated partially by open-cast mining. The largest occurrence of steaming coal is the huge Ekibastuz basin in the Pavlodar region of Kazakhstan. The basin's economically mineable reserves total more than 10 billion tonnes. The coal is excavated in open-cast collieries (strippings) at Severny Bogatyr and Vostochny at maximum excavation depths of 185 meters (Severny), thus production costs are among the lowest in the industry. The main consumers of the high ash content (41.4%) Ekibastuz coals are the thermal power stations of Kazakhstan (40.9 million tonnes), and the Urals (29.7 million tonnes) and Western Siberia (3.9 million tonnes) in Russia.

The Shubarkol coal field is of great importance to utilities in central and northern Kazakhstan. Discovered in 1983, it has about 1.7 billion tonnes of economically mineable reserves and relatively favorable geological conditions for excavation. One of the field's open-cast collieries, Tsentralny, was designed to have an annual coal output capacity of 14 million tonnes.

In the immediate future, there are plans to begin open-cast mining at various lignite coal fields, including Tourgaiskoe (Tourgai region), Maikyubenskoe (Pavlodar region), and the Yubileynoe deposit (Semipalatinsk region).

METALLIC MINERAL RESOURCES

Kazakhstan's metal deposits can be grouped as follows: ferrous (iron, manganese, chromium, and titanium); nonferrous (copper, lead, zinc, nickel, and cobalt); rare or strategic (tin, tungsten, molybdenum, bismuth, antimony, mercury, vanadium, tantalum, niobium, cadmium, beryllium, and zirconium); trace rare (indium, gallium, and germanium); light (aluminum and magnesium); and precious (gold, silver, and platinum).

Ferrous metals

Kazakhstan's ferrous metal ore reserves are sufficient to support the rapidly growing needs of not only the republic's metallurgical enterprises but also of those in the southern Urals and Western Siberia. Total reserves of Kazakhstan's iron, chromium, manganese and titanium metals ores exceed those of Western Europe and are almost twice as large as those of the United States. Deposits with industrial reserve capacities, which consist basically of easily dressable magnetite ores, are concentrated in northern Kazakhstan (Kustanai region) and in the Atasui and Karsakpai fields (Dzhezkazgan region).

Out of the 37 known ferrous metal ore fields in the republic, seven are currently being exploited: Sokoloyskoe (average iron content of 39.6%), Sarbaiskoe (43.2%), Korzhunkulskoe (42.7%), Kacharskoe (40.8%), Lisakovskoe 35.2%),

Zapadny Karazhal (52.1%), and Kentobe (where the average iron content in profitably mineable reserves is 49.6%). The main consumers of the ore, besides the Karaganda metallurgical works, are five plants in Russia, including the Magnitogorski, Orsko-Khalilovski, Zapadno-Sibirski, Satkinski and Zlatoustovski metallurgical plants.

The manganese ores of central Kazakhstan supplied the primary raw material for ferromanganese production in the former Soviet Union. There are 11 explored manganese deposits in the republic. They are located in the Dzhezkazgan region and form two groups: Atasui and Dzhezdin-Ulutan. All of these manganese ores have a low phosphorous content and are primarily represented by the carbonate-silicate-oxide type. Currently, three deposits are being developed: Bolshoi Ktai, by open-cast mining, and Kzhezdinskoye and Zapadny Karatai, by underground mining. Zapadny Karatai is a complex iron-ore and manganese deposit where manganese ores are extracted in association with iron ores. The ores with a manganese content exceeding 19 percent are processed at the Dzhezdin dressing mill. The extracted high-grade manganese concentrates are used in the chemical industry and for production of standard brands of ferromanganese.

Chromium-iron ores are concentrated in the southern Kempirsai group of deposits located in the Aktyubinsk region. These ores have a 20–60 percent chromium-trioxide content. Kazakhstan leads the world both in explored deposits and output of chromium-iron ores, which in the past accounted for 98 percent of Soviet production.

Out of the 17 deposits of the southern Kempirsai group, four are currently being processed by the Donski ore-dressing plant. Until quite recently, only rich ores, which do not require dressing, had been used. Such ores have a high chromium-oxide content (over 45%) in a favorable proportion to iron oxide. The ores are suitable for the production of ferrochromium, metallic chromium, and chromic salts, while ores with a lower chromium-oxide content are used for the production of refractory materials. Now that a dressing process has been established at the Donski plant, poor quality ores with a 30–45 percent chromium-oxide content are beginning to be utilized in the technological production process.

Lower quality ores with a 10–30 percent chromium-oxide content can also be used if sufficient processing is undertaken. These ores can be easily dressed by the gravitational-flotation technique, resulting in concentrates with a 50–52 percent chromium-oxide content.

Primary consumers of Kazakhstan's chromite production are the Aktyubinsk and Pavlodar (Yermakovski) ferroalloy plants, the Chelyabinsk ferrochromium plant, various chromium-compounds plants, and others. Some of the output is also exported.

Titanium is found in Kazakhstan in the form of complex zirconium-rutile-ilmenite placers. The metal and its alloys are produced at the Ust-Kamenogorsk titanium-magnesium works. Raw titanium (especially when it contains zirconium minerals) can be used for the production of refractory materials, for example, refractory concrete for glass-making factories.

Nonferrous, rare and precious metal ores

It is a well-known fact that Kazakhstan has a highly developed nonferrous metallurgical industry with access to reliable raw materials supplies. The republic possesses considerable bauxite reserves of varying quality with an alumina content of 40–60 percent. Reserves are mainly concentrated in northern Kazakhstan in the bauxite-rich Amangeldynski and Zapadny Tourgaiski districts. Mined bauxites are processed at the Pavlodar alumina plant, with the recovered alumina transported to metallurgical plants outside the republic. Deposits of other types of aluminum raw materials, such as nephelites and alunites, are also found in the republic.

Occurrences of copper ores of various types are also numerous in Kazakhstan. Major deposits, including Dzhezkazgan, Zhezkent, and Kounrad, have been exploited for many years. The deposits served as the basis for the construction of several high-capacity mining-metallurgical works with complete copper-smelting production cycles. Development of a new copper-bearing area in the eastern Balkhash geographic region will allow for a narrowing of the gap between mining and processing capacities which has emerged in recent years in the republic's copper industry.

Kazakhstan occupies a leading position in the former Soviet Union in the mining and processing of lead-zinc ores. Mines and processing facilities are concentrated mainly in the three ore-bearing regions: Rudny Altai, central Kazakhstan, and southern Kazakhstan. At present, there are 52 industrially exploited deposits; among them are large and giant deposits such as Ridder-Sokolnoe, Zyrianovskoe, Tishinskoe (Rudny Altai), Zhairem, Ushkatyn and Karagaily (central Kazakhstan). The ores of these deposits are mainly polymetallic with various ratios of lead, zinc, and copper. Apart from the primary industrially significant components (lead, zinc, copper, barium, and sulfur), the ores contain recoverable quantities of gold, silver, cadmium, bismuth, selenium, tellurium, thallium, gallium, indium, germanium, cobalt, antimony, and molybdenum. Intensive exploration for polymetallic deposits in prospective areas, especially in those adjoining the operating industrial facilities, has resulted in the discovery of a number of new large deposits, particularly in the Rudny Altai area. It is estimated that the exploitation of these deposits can be undertaken in a cost-effective manner. The development and exploitation of these deposits will contribute to the strengthening of the mineral raw materials resource base of Kazakhstan's mining and metallurgy facilities.

The republic has significant potential to develop a rare-metals industry, since it possesses considerable resources of tungsten, molybdenum, bismuth, tantalum, zirconium, niobium, vanadium, and beryllium. Unfortunately, the output of tungsten and molybdenum ores in Kazakhstan has not reached full capacity as the development of the largest deposits, Verkhnee Kairakty and Koktenkol, has come to a standstill due to an acute shortage of capital resources.

Gold mining is carried out in the metallogenic regions of northern, eastern and southern Kazakhstan. The explored gold reserves, including both gold-ore deposits and considerable amounts of gold by-products from complex copper, copper-polymetallic and lead-zinc deposits (also the only source of silver in the republic), will be able to provide a valuable boost to the consolidation of Kazakhstan's economic and political sovereignty.

Nonmetallic mineral resources

The republic is developing a wide variety of explored nonmetallic mineral resources, including phosphorites, borates, barites, carbonates, asbestos, china clay and various other clays (refractory, high-melting, potter's), common salt, dolomites, fluxing limestones, construction raw materials, and others. The first diamond deposit to be discovered and explored in Kazakhstan is Kumdy in the Kokchetav region.

The republic's explored phosphate rock reserves are concentrated in two unique areas: the Karatau deposit of fine-grained phosphorites and the Aktyubinsk concretion-ore field. The ore consists of the leading geological-industrial types of phosphorites; however, the particular quality of the ore requires extensive processing.

Kazakhstan has substantial potential to develop its fluorspar industry on the basis of the Solnechnoe (Dzhezkazgan region) and Taskainar (Dzhambul region) fluorspar deposits. Kazakhstan's deposits of semiprecious and ornamental stones are especially promising from the viewpoint of export potential. The republic has significant deposits of semiprecious stones (jadeite, chalcedony, opal, jasper, agate, and chrysoprase), facing and ornamental stones (marbles and granites) and shell rock.

WASTES OF THE MINERAL RAW MATERIALS INDUSTRY

At present, over 16 billion tonnes of various wastes from mining, dressing, metallurgical and other production processes are accumulated in the republic. Total wastes increase annually by about one billion tonnes.

This considerable volume of waste accumulation is due to a large amount of processed rock bulk, the considerable share of open-cast mining in the industry, an inadequate level of comprehensive utilization of mineral raw materials, and nonexistent or extremely slow introduction of effective technologies for processing wastes, slag, and dressing tailings. A crucial factor in waste accumulation has been the bureaucratic practice of the sector in writing off as wastes "usable mineral raw materials if they were not the responsibility of a particular department." Another factor is the absence of an effective economic mechanism for stimulating rational and comprehensive utilization of mineral raw material resources.

Table 8.1. Material Wastes in Kazakhstan's Mining Industry

Production Stage/Category	Quantity (million tonnes)
Overburden	
Nonferrous metallurgy	4,276.0
Ferrous metallurgy	4,130.2
Chemical industry	613.1
Dressing Tailings	
Nonferrous metallurgy	1,534.7
Ferrous metallurgy	301.5
Slags	
Nonferrous metallurgy	51.3
Blast furnace	19.3
Steel-making	21.4

Table 8.1 lists the amounts of wastes accumulated in the basic branches of Kazakhstan's mineral raw materials sector in millions of tonnes. Dressing tailings and slag from metallurgical plants contain precious and rare metals and rare-earth elements in industrially significant quantities. Every year, millions of tonnes of pyrite-containing raw materials are dumped in dressing tailings without any further processing, when in fact almost all of Kazakhstan's ore-dressing plants could produce pyrite concentrates. Thus, this segment of the industry also has economically viable reserves that could be exported and serve as a basis for the creation of joint ventures for the commercial production of mineral concentrates from industrial wastes.

PROBLEMS OF EFFECTIVE UTILIZATION OF MINERAL RAW MATERIALS RESOURCES

Kazakhstan's resource wealth was a key factor in predetermining the orientation of its raw materials development in the past. At present, with the formation of a sovereign state and the restructuring of the national economy under way, decreasing the mining industry's share in the economy in proportion to other industries has become a priority. Nevertheless, Kazakhstan's foreseeable economic future will depend largely on the scale and efficiency of the utilization of its minerals wealth.

With the effective development of its resource potential, especially petroleum and natural gas, Kazakhstan will be able to expand and consolidate its exporting activities, increase hard currency receipts, and introduce state-of-the-art equipment and technologies. However, it would be a mistake to overexploit the republic's minerals wealth to realize these plans. An economy based largely on a single industry, such as mining, is not likely to succeed in the long run as it remains dependent on markets in other countries. Profits from Kazakhstan's raw materials exports have been declining in recent years, and raw material exports are not capturing the value added from processing.

In addition to the above, the following trends are likely in the development of the republic's mineral raw materials sector:

- a growth in demand for raw materials, on the one hand, and a trend towards preservation of resources, on the other;
- general exhaustion of the best, most convenient and accessible deposits, as well as increased exploitation of those deposits with relatively inferior economic, geographical and geological conditions;
- a decline in the quality of the mined raw materials in terms of major economically valuable components;
- increasing environmental damage resulting from the operations of mining and processing (dressing), metallurgical and chemical industries and from the wastes they produce;
- a growing awareness that technological progress alone cannot reverse the deterioration of mining conditions and mineral raw materials quality; and
- rising production costs for prospecting, mining, and processing of mineral raw materials and for environmental protection, resulting in higher prices for a majority of mining products.

It is quite evident that in restructuring the Kazakhstan economy, the current and future potential of the mineral raw materials sector must be considered and, consequently, a rational strategy must be worked out to develop the republic's mineral raw materials over both the short term and long term. Such a strategy must be aimed at finding solutions for today's as well as tomorrow's problems, which are closely related and interdependent in their technological, environmental, social and economic aspects.

One of the key strategies in the rational exploitation of the mineral endowment is increasing the comprehensiveness of its utilization which can be attained through measures that include:

- comprehensive utilization of each deposit's reserves (including thorough extraction and utilization of all mineral resources of a particular deposit, and practical use of overburden rocks, mining waters and openings);
- comprehensive development and utilization of a particular type of deposit in a given territory;
- comprehensive utilization of raw materials (notably, the efficient recovery of industrial components and industrial utilization of dressing, metallurgical and chemical wastes); and
- comprehensive development and utilization of all mineral raw materials resources of a particular region carried out in accordance with predetermined optimum periods and scales of exploitation of particular deposits.

More thorough mineral raw materials utilization will lead to an acceleration of productivity in Kazakhstan's mining industries, a reduction in production costs per unit of output, and, of particular importance, a lower rate of environmental pollution. Numerous enterprises in the minerals sector are significant polluters of the republic's air and water. The ecological capacity of some industrial mining regions (for example, Dzhambul-Karatau, Ust-Kamenogorsk, Temirtay-Karaganda, Pavlodar-Ekibastuz and other territories) has reached its limits, and environmental damage in these regions may well exceed direct economic benefits from the exploitation of the mineral resources. Such a situation calls for the technical re-equipment of the mining industry, the introduction of state-of-the-art, low-waste and waste-free technologies, and the cleaning of wastes and harmful components produced in the extraction process. Currently, with the exception of some nonferrous metallurgy technologies, the mining industry operates with outdated technologies, machinery and materials. Much of the relevant machinery, equipment and materials are not produced in Kazakhstan at all, and depreciation of equipment often exceeds 50 percent.

In the near future, it will be necessary to introduce suitable regulatory mechanisms that will coordinate with emerging market structures in formulating taxes, payments, and fines aimed at stimulating the rational use of mineral resources. However, this process should be gradual, since in the current situation, which has emerged as the result of past practices including budget subsidies for the majority of mining industries and centralized price setting, very few enterprises be able to afford to pay fees and taxes in full.

It is well known that, as in many other sectors of the economy, the manufacture of finished products in the mining industry (metal, rolled stock, mineral fertilizers, petroleum products, gasoline, residual fuel oil, and lubricants) is much more profitable than production of raw materials or primary products (concentrates and semiproducts). Until very recently, however, large quantities of metal ores, including alumina, concentrates of various nonferrous and rare metals, and gold, were sent out of Kazakhstan in semifinished form for further processing. Natural gas also is traditionally piped out of the republic for further processing. As is obvious, Kazakhstan has not been using its significant economic potential for the production of high-grade transportable finished goods for domestic use as well as for export. Therefore, it is essential for the mining industries to be restructured through the establishment of new facilities and the expansion of existing processing capacities in order to ensure a yield of finished products for the entire mineral raw materials industrial complex.

CONCLUSIONS

The mining industry, certainly one of the key industries in the republic, contributes significantly to Kazakhstan's economy both in terms of output volume and production revenues. Most of the industry is, however, plagued with low efficiency that reduces profitability. The majority of goods producers in the republic now have been given a considerable degree of freedom in price-setting, and while prices and tariffs for a number of widely used mineral raw materials remain fixed, the enterprises of the mineral raw materials complex are not likely to remain economically profitable in the future unless prices are adjusted.

Perhaps the most difficult and sensitive problem facing the mining industry is price formulation. On the one hand, the republic has considerably underestimated its national wealth for a long time. Republican statistics, in contrast to those of other countries, did not take into account the value of natural resources in general and mineral resources in particular. This is the result of not only the former ideological approach to natural resources, but also the absence of a scientific method for their economic evaluation. On the other hand, capital investments were not duly considered in price formulation, since they were almost completely financed by the state budget. Cost/price calculations of production did not consider expenses such as social security, health care, education and personnel training. In addition, deductions for depreciation were considerably underrated.

Thus, due to various factors, both artificially created and conditioned by historical circumstances, prices were not only distorted but also below cost. Introduction of actual production costs of natural resources into the workings of economic mechanisms, imposition of varying payment schedules, and elimination of the abovementioned factors that in the past led to the underestimation of other expenses will steeply increase the cost and prices of the mining industry's output. This, in turn, is likely to bring about a price increase of the products in the international economy as a whole.

The process of Kazakhstan's integration into the world economic system and creation of joint ventures will require a transition to international rules of cost accounting. Therefore, in order to reconcile the current accounting and price-setting systems with principles of a market economy, the steps taken must be carefully analyzed and executed, since errors in price setting may result in heavy economic and sociopolitical losses. Transition to a market economy will bring into play various incentives and possibly contradictory policies whose combination will affect the overall strategy of mineral raw materials utilization. This stems from different motivations in mineral resources utilization. Consequently, the actions of mining enterprises and companies striving to maximize profits may diverge from the position of the government and central official bodies that aim to maximize benefits for a region or the state as a whole. Economists may promote policies that call for the efficient use of a region's resources in the interests of future generations.

Economically developed countries can offer Kazakhstan enterprises export opportunities on which to base decisions, that comply with company interests, and are also in the national interest. Countries such as Kazakhstan, with an underutilized economic potential and an acute need for foreign investment, may need to assess the social need for profits against the benefits of strict economic and environmental regulations.

Because of the importance of minerals to Kazakhstan's economy, Kazakhstan will have to closely monitor the evolution of newly independent mining enterprises in the republic and the first examples of foreign investment, specifically those aimed at the prospecting and development of the republic's petroleum and gas resources.

If Kazakhstan fails to structurally and technologically reorganize its mining industry in particular and the economy in general and allow economic mechanisms to stimulate rational resource utilization, the current reforms will not succeed, and blame for the ensuing misfortunes will be laid upon the market economy and its present proponents.

Chapter 9

GEOLOGY, MINERAL RESOURCES AND THE MINERAL DEVELOPMENT POTENTIAL OF THE RUSSIAN FAR EAST

Allen L. Clark and Genady V. Sekisov

INTRODUCTION

The present and future development of the Russian Far East will take place within the broader scope of the newly constituted Commonwealth of Independent States (CIS) in general and within the Russian Federation in specific. More importantly, the development of the Russian Far East will occur within a much more extensive and competitive international marketplace in which individual commodities, areas and specific mineral deposits of the Far East must compete for exploration and development monies worldwide, in many cases with the newly formed states of the former Soviet republics and the previous CMEA nations.

In particular, it must be noted that the mineral resource sector of the Soviet Union was the world's largest, with an annual commodity production value for mineral raw materials of US$180 billion in 1991. Not only did the Soviet Union have the world's largest minerals industry, it also ranked number one in the world in terms of proved reserves of barite, cobalt, fluorspar, iron ore, nickel, tin, titanium, and tungsten, and in the energy sector, in brown coal, natural gas and oil.

This vast mineral potential comprised over 18,000 individual deposits as recorded in the State Balance of National Wealth records. Approximately 5,000 of the recorded deposits are under industrial exploitation, 4,700 are set aside for future mining and nearly 9,000 are planned for mining in the nearest future. According to Borisovich (1991), the proportion of proved reserves set aside for future development by mining enterprises is relatively large, and for certain minerals it may comprise as much as 25–40 percent of total proved reserves. For the majority of these deposits, their geological and economic characteristics, although the same as those of deposits currently being mined, require large-scale capital

J.P. Dorian et al. (eds.), CIS Energy and Minerals Development, 127–156.

investment for their future development. The State Balance additionally included a large number of deposits where prospecting work was carried out 20–30 years ago as well as deposits not scheduled for mining until after 2015. These latter deposits include 30 percent of the proved reserves of iron ore and copper, 14 percent of lead-zinc, 39 percent of bauxite, 24 percent of titanium, 32 percent of phosphate, 16 percent of fluorspar, and 45 percent of brown coal reserves.

Equally important as background for assessing the mineral development potential of the Russian Far East is an understanding of the uneven distribution of mineral reserves among the former Soviet republics. Kazakhstan, as an example, has a strong mineral resource base representing significant shares of the CIS's mineral reserves, including asbestos (20%), barite (82%), bauxite (22%), chrome ore (99%), copper (28%), iron ore (10%), lead (38%), molybdenum (29%), phosphate (65%), tungsten (53%) and coal (12%). More importantly, Kazakhstan produces over 50 percent of the CIS's total barite, chrome ore, lead, phosphate, and zinc and approximately 20 percent of the asbestos, bauxite, coal, and copper output. However, the Russian Federation, of which the Far East is a major portion of the total area, contains the vast majority of proved reserves in the CIS.

The Russian Far East enjoys a comparative advantage over other resource-rich areas in that over 120 different types of metallic, nonmetallic and energy resources are known, and the region is endowed with large reserves of antimony, boron, brown coal, diamonds, fluorite, gold, platinum, silver, tin, and tungsten—mineral commodities which are not as abundant in other areas of the CIS in general and in the Russian Federation in specific. It is particularly noteworthy that the area contains the largest reserves of gold and diamonds which represent an obvious comparative advantage for the region in terms of attracting foreign investment.

Overall mineral development in the Russian Far East would appear to have a promising long-term potential, a modest intermediate-term development potential and a limited short-term potential for significant development in all but very few commodities. It is the purpose of the remainder of this chapter to discuss the general geology, known mineral potential and deposit types of the region, to assess some of the areas and commodities with high development potential, and to conclude with a discussion of the issues affecting development potential of the region.

GENERAL GEOLOGY

The geology, tectonics, and mineral resources of the Russian Far East are, as would be expected for a region roughly the size of the United States, complex, incompletely studied and, in many areas, only partially evaluated in terms of mineral potential. As an example, the region contains the Siberian platform to the west, complex fold belts to the north and south, volcanogenic zones to the east as well as the shelf areas of the Pacific and Arctic oceans. Similarly, the rock units of the region range in age from the earliest Precambrian to the Quarternary and in type and composition from kimberlite pipes to massive limestones and recent volcanics. As a result, any discussion of the detailed geology of the Russian Far East and its mineral deposits is beyond the scope of the present chapter. Therefore, the following discussion will concentrate on the basic characteristics of the major geologic zones of the Russian Far East and only focus broadly on associated mineral deposits which will be covered in greater detail later in this chapter. Throughout the following discussion, the reader is referred to Figure 9.1 for the general location and distribution of the geologic zones examined.

Siberian platform

The Siberian platform is one of the world's most extensive geologic platforms, formed upon a basement of gneisses and crystalline schists of early Precambrian (Archean) age which, within and marginal to the platform, occur as elevated massif areas separating large depressions. The large depressions serve as the main basins for rivers on the platform. The main structural elements defined by the massifs and depressions are the elevated Aldan shield and Lena Yenisei platform area, the eastern portion of which forms the Viljui depression (syncline) and the southern portion of the Angara-Lena depression. The Angara-Lena depression is in turn separated from the Yenisei depression by the Pelelui uplift on the western margin of the platform. These sequences of massifs and depressions and their associated river basins are hosts for the occurrence of several of the richest and largest gold placer deposit areas in the Far East.

Seven complexes comprise the platform mantle rocks: (1) the Riphean complex of terrigenous-carbonaceous rocks with a thickness of 4,000–5,000 m; (2) the Vendsko-Cambrian complex also composed of terrigenous-carbonaceous sediments, which in the Angara-Lena depression is represented by the slate-bearing strata with a thickness of approximately 3,000 m; (3) the Ordovician-Silurian complex composed of terrigenous sediments, limestones and dolomites with a thickness of 5,000–6,000 m; (4) the Devonian-Lower Carboniferous complex represented in the south by predominantly continental strata with "traps" (lava flows and sills often referred to by the general name of "Siberian traps") and, in the north, by a thick series of trap basalt and salt-bearing formations with a thickness of 5,000–6,000 m; (5) the Middle Carboniferous (Middle Triassic) complex which developed in the Tungus syncline and is represented by a thick sequence (up to 1,000 m) of coal-bearing strata of middle Carboniferous-Permian age overlaid by a Triassic volcanoclastic (tuff) and lava sequence 3,000–4,000 m thick; (6) the Upper Triassic-Cretaceous complex composed of continental and local marine sandy-argillaceous and carbonaceous sediments which occur exclusively at the margins of the platform; and (7) the Cenozoic complex which developed locally and is represented by continental sedimentation, weathering products and glacial materials and is of highly variable thickness.

It is particularly noteworthy that in Devonian, Triassic and Cretaceous times,

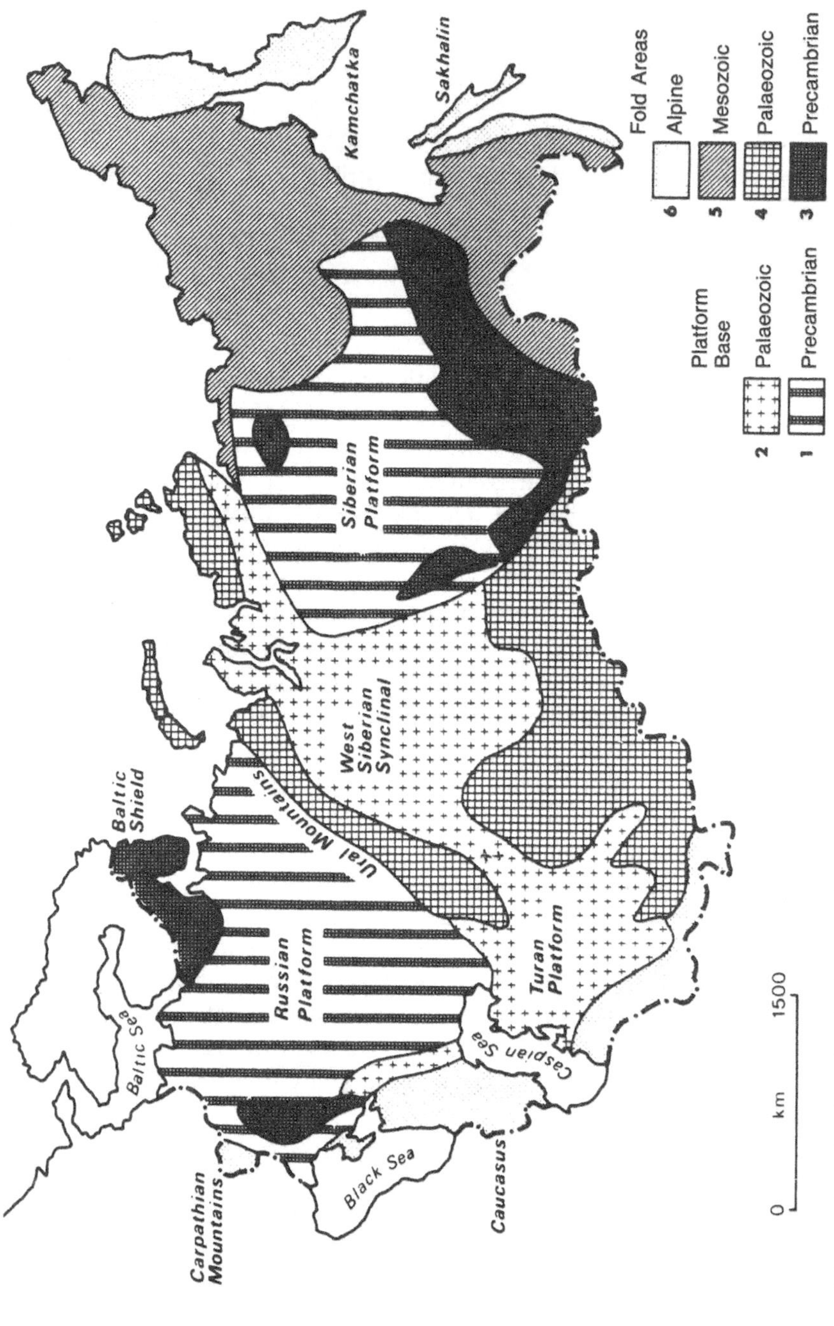

Figure 9.1. Major Geologic Provinces of the Far East and Adjacent Areas

kimberlite pipes, which are the source of the extensive diamond bedrock and placer deposits of the region, were formed in the northeast portion of the platform.

The mineral deposits associated with the Siberian platform include the world-famous Norilsk Cu-Ni-PGE deposits associated with the "Siberian trap" basalts and feeders; extensive siderite ores associated with the Precambrian massif areas; copper-bearing red bed sandstone deposits of Upper Cambrian and Ordovician ages; extensive gold and platinum placer deposits associated with the massif and "trap" areas, respectively; diamond placer deposits associated with kimberlite pipes; and bauxite deposits occurring both as lateritic deposits and as metamorphosed diasporic deposits intercalated with Lower Cambrian algal reef limestones. Additional mineral occurrences include deposits of mica, phlogophite, graphite and rock salt.

Mongolian-Okhotsk fold area

Adjacent to and south-southeast of the Siberian platform lies the Mongolian-Okhotsk fold area, which is inferred to be separated from the main Siberian platform area by a major geosuture zone. The Mongolian-Okhotsk fold area is in fact composed of three individual fold zones, i.e., the Eastern-Zabaikal, Upper Amur and Priokhotsk zones. The Mongolian-Okhotsk fold area in the Far East is of particular importance in that it is the northernmost extension of the Angara geosyncline area which extends into southern Kazakhstan, where it is the host province for the world-famous Dzhezkazgan porphyry copper deposit.

Within the Russian Far East, the Mongolian-Okhotsk fold area is known for the occurrence of tin, tungsten, molybdenum, arsenic, antimony and diverse polymetallic (Cu-Pb-Zn) deposits. Additionally, a broad zone of placer and smaller lode gold deposits are concentrated just south of the area in association with the Archean and Lower Proterozoic gneisses in the Bureja massif. Large jasperlite deposits of iron also occur in association with the Riphean strata of the area but to date are not of economic importance.

Pacific fold belt

The Pacific fold belt lies east of the Siberian platform and, although it contains rock units of all the geologic eras, the most characteristic feature of the region is the predominance of Mesozoic and Cenozoic rock units of diverse origin, including marine sediments, terrestrial coal-bearing units and complex sequences of volcanic rocks. In particular, marine deposits of limestone, shale and sandstone characteristic of the Triassic age and intercalated marine, continental (sometimes coal-bearing) and extensive extrusive rocks (largely quartz porphyries) characteristic of the Jurassic and Cretaceous ages are both widespread and voluminous.

Mesozoic sedimentation and volcanism took place within a well-defined geosynclinal area adjacent to and parallel to the pre-existing Hercynian Pacific fold belt. The most characteristic feature of the Pacific geosynclines development in the northern segment was the formation of a later Mesozoic (Middle Jurassic to Late Cretaceous) fold structure almost exclusively adjacent to the pre-existing structures of the Pacific fold belt. Conversely, in the southeast, where the Pacific geosyncline developed adjacent to Mesozoic and older fold structures, the Mongolian-Okhotsk volcanic belt was formed.

To a great extent, the periods of significant sedimentation and eruptive activity were largely initiated by Hercynian orogenesis at the end of the Paleozoic age, and continued with large-scale magmatism throughout the Mesozoic and into the Cenozoic age. Particularly noteworthy with respect to the Mesozoic magmatism was the formation during the Cretaceous age of many large porphyritic granitic bodies which have associated gold, tin, tungsten, copper and molybdenum deposits.

The major mineral deposits associated with the Pacific fold belt are lode and placer gold deposits associated with both Cretaceous and Neogene granitic porphyries, tin-bearing granites and placers, polymetallic lead, zinc, copper and silver vein and disseminated deposits, molybdenum-tungsten greisen and disseminated deposits, as well as minor deposits of antimony, mercury, arsenic, jasperlite iron and sedimentary manganese. Graphite, fluorite, and native sulfur (associated with extinct volcanoes in Kamchatka) also occur.

MINERAL DEPOSITS OF THE RUSSIAN FAR EAST

The mineral deposits of the Russia Far East are numerous and diverse and, given that the area has been prospected primarily by reconnaissance methods, both the number and types of deposits are expected to increase in the future. In this section, the authors wish to first address the distribution of the major metallogenic areas and their association with specific geological and structural units; second, to discuss the types of mineral deposits known to occur in the area and their attributes; and to conclude with a subjective assessment of the development potential of various deposit types.

Geological-structural-metallogenic zones

Although the Far East is rather easy to classify in terms of broad geologic areas, the details of the region's geology, structure and mineral deposits are both complex and, in many instances, less understood than is desirable. Nevertheless, evaluations of the geologic and structural divisions (Figure 9.2) and the metallogenic zones (Figure 9.3) of the Far East, based on the works of Kunaev (1984), demonstrate a clear association of individual metallogenic zones and the individual geological-structural zones of the region.

Table 9.1 represents a compilation of data, based on Figures 9.2 and 9.3, which equates the individual metallogenic regions, provinces and subprovinces with their corresponding geological-structural terranes as defined in Figure 9.2. An analysis of Table 9.1, Figure 9.2 and Figure 9.3 shows a number of interesting aspects of Far East mineralization.

- Specific mineral associations are strongly bounded by individual geological-structural terranes.
- Although mineralization occurs throughout the geologic record of the Far East, there are strong tendencies for gold occurrences to be associated with older Precambrian and Paleozoic units, chrome with Cretaceous units, molybdenum with rocks of Jura-Cretaceous age, tungsten and tin with Paleozoic units, and complex gold, mercury, and antimony deposits with younger Cenozoic and Recent rock units.
- Based on an analysis of rock units in Table 9.1, Figure 9.2 indicates that massif units of Precambrian and Lower Paleozoic ages are major source areas for a majority of the Far East's lode and placer deposits.
- Jura-Cretaceous geosynclinal fold systems and zones containing ophiolite sequences are primary host rocks not only for chromite (as would be expected), but also for mercury, antimony and tungsten.
- Mercury occurrences are common virtually throughout the Far East region, closely paralleling the widespread mercury distribution in adjacent Alaska.

Although many additional observations can be made regarding the close association of Far East mineralization with specific geological and structural terranes, the relationships discussed are sufficient to allow the formulation of two major conclusions which will affect long-term exploration and development within the region: first, that regional tectono-stratigraphic analysis would be an invaluable tool in selecting exploration areas; and second, that areas of extensive Precambrian and Paleozoic massifs are primary exploration targets.

Deposit types

Table 9.2 is a partial compilation of the known mineral deposit types which occur in the Russian Far East. The compilation of deposit types occurring in the region is based on the authors' personal knowledge of the region, the numerous descriptions of mineral deposits in the literature and technical reports and, in some instances, a geologic interpretation of deposits which may occur because of favorable geologic environments known to exist in the region. Deposit criteria with respect to favorable host rock depositional environments and tectonic settings are given for all deposit types in Table 9.2. Additionally, tonnage and grade estimates are given, when available, for similar deposits worldwide using data taken from Cox and Singer (1986).

Data given for tonnages and grades have been taken from distribution curves given in Cox and Singer (1986). Values for small, medium and large tonnages

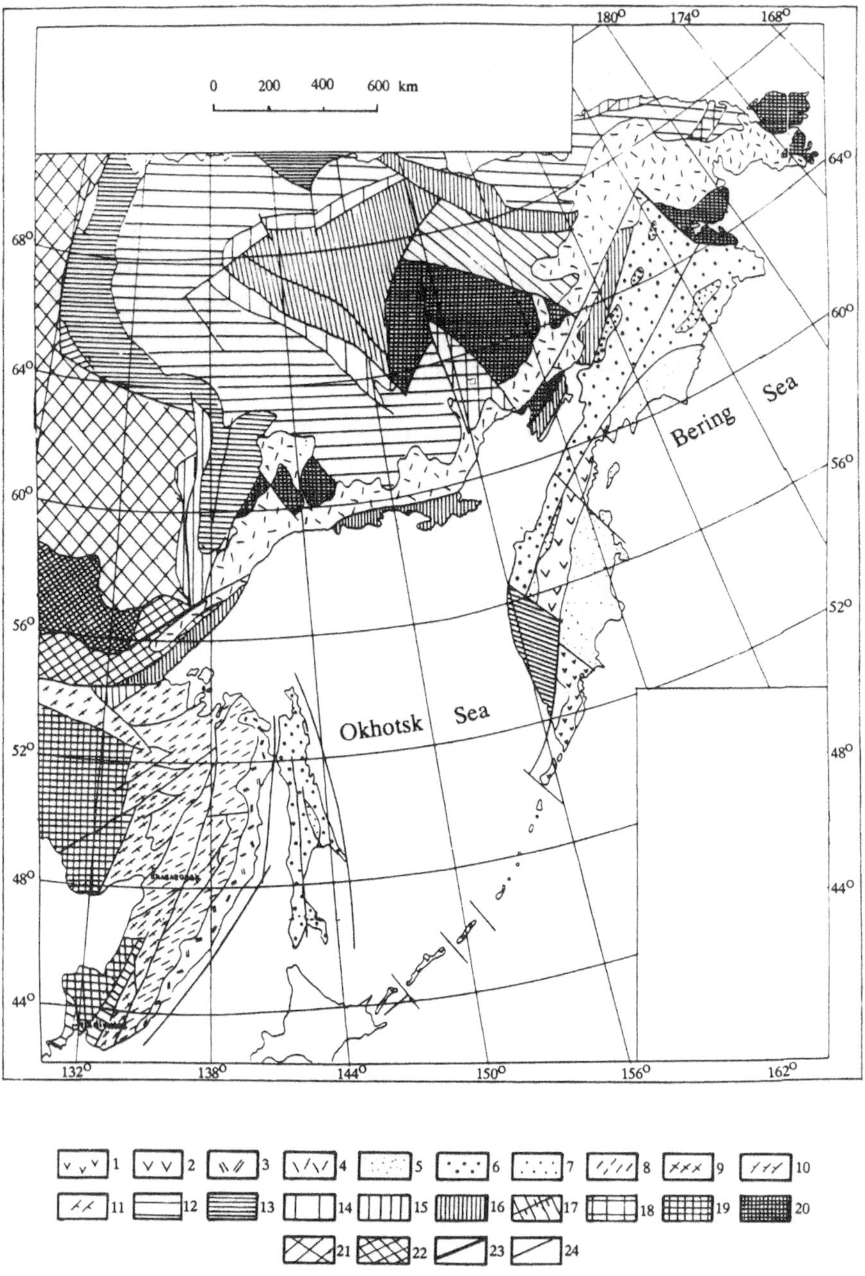

Figure 9.2. Geological and Structural Zones of the Russian Far East (after Kunaev, I.V. 1984) (See following for legend)

Legend for Figure 9.2.

No 1–4	Volcanogenic belts representing marginal oceanic "Island Arc Andesites" (1) and marginal continental "rhyolite rich" sequences (2), (3), (4). Ages K_1-N
No. 5–7	Geosynclinal folded systems and zones consisting of volcanogenic and silicious-terrigenous complexes with zones of ophiolite complexes. Ages J_3-N
No. 8–10	Geosynclinal volcanogenic and silicious-terrigenous complexes without ophiolites. Ages Pz_2-K
No. 11–12	Geosynclinal terrigenous complexes. Ages Pz-J_3
No. 13–14	Pericratonal aulocogenic terrigenous-silicious complexes. Ages Pr_3-Pz
No. 15	Undivided geosynclinal complexes, volcanogenic-silicious-terrigenous and volcanogenic-terrigenous complexes. Ages Pz Mz
No. 16	Massif complexes near geosynclinal basins and foredeeps. Ages Pz-Mz
No. 17–19	Basement and cover complexes. Ages Pz, Pz_2 and Pr_2, respectively.
No. 20	Platform of terrigenous-silicious complexes. Ages Pr_3-Pz_2
No. 21–22	Basement complexes. Age Ar_2
No. 23–24	Faults.

and for low, medium and high grades represent for each grouping the 90th, 50th and 10th percentiles of deposits, respectively.

For example, for tonnage, small indicates that 90 percent of all deposits will have a greater tonnage than the figure given; medium means that 50 percent will have a higher tonnage and 50 percent will have a lower tonnage; and large refers to deposits where only 10 percent will have a larger tonnage than the deposit in question. For grade, the same rule applies. For low-grade deposits, 90 percent of all deposits will be of a higher grade; medium indicates that half will be of a higher grade and the other half of a lower grade; and high means that only 10 percent of all similar deposits will be of a higher grade.

As stated previously, mineral deposits in the Far East will have to compete in terms of tonnage and grade with similar deposits worldwide that are considered attractive for exploration and development funding. As a result, the purpose of the presented tonnage and grade figures is to provide a very rough basis for assessing the development potential of mineral occurrences in the Russian Far East against the international norms the figures represent.

As can be readily seen, the Far East has a broad range of mineral deposit types upon which future exploration and development can be based. As significant amounts of data were or will soon be available on the known deposits of the region, Table 9.2 provides a ready means to both evaluate the types of deposits being developed and assess their general economic potential given the tonnages and grades of deposits developed elsewhere in the world. The simple presence of individual deposit types, however, provides no guarantee of development, even if deposits match or exceed world averages. Therefore, to further refine the

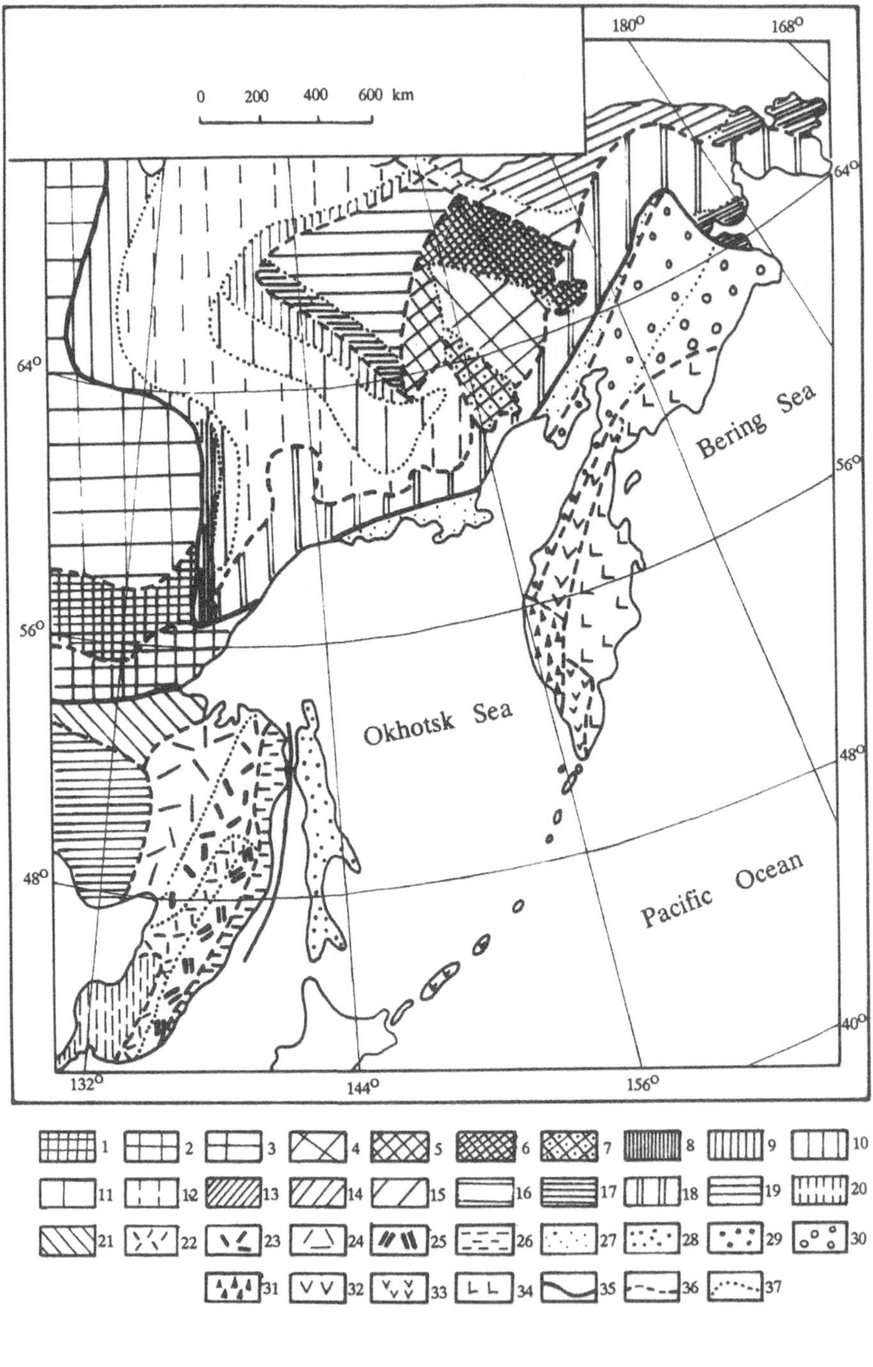

Figure 9.3. Metallogenic Regions, Provinces and Subprovinces of the Russian Far East (after Kunaev, I.V., 1984) (See following for legend)

Legend for Figure 9.3.

Eastern Siberian Region:

1. Aldano-Amginskaya Province: Fe-Ar; Pk-Pr_3; Au (Mo, Pb, Zn)-J_3-K_1;
2. Stanovaya Province: Au, Ti, Fe-Ar_2; Pk-Pr_3; Mo, Pb, Zn-J_3-K_1; Au, Ag, Cu (Pb, Zn, Sb)-K_{1-2};
3. Leno-Viljuiskaya Province.

Verkhoyano-Chukotskaya Region:

4–7. Prikolymo-Omolonsk Province:
4. Omolonskaya Subprovince: Fe-Ar, Au, Ag, (Cu)-D_{2-3};
5. Prikolymskaya Subprovince: Cu, Pb, Zn, Fe-Pr_3-Pz_2;
6. Omolonskaya Subprovince: Cu, Mo, Au, Ag (Pb, Zn)-K;
7. Sugoyskaya Subprovince: Mo, Pb, Zn-K_2.

8–12. Yano-Kolymsk Province:
8. Sette-Dabanskaya Subprovince: Pk-Pr_3 Pb, Zn, Cu, Fe-Pr_3-Pz_1;
9. Omulevo-Polousnenskaya Subprovince: Cu, Pb, Zn, Sb, Hg-Pz_2; Mo, Au-J_3-K_1;
10. Verkhoyanskaya Subprovince: Au-J_2-K_1;
11. Indigiro-Kolymskaya Subprovince: Au, Sb, (W, Mo)-J_3-K_1;
12. Yano-Balygychanskaya Subprovince: Sn, W, Pb, Zn, (Sb)-K_{1-2}; Hg(Sb)-K_2-P_1.

13–15. Chukotsk Province:
13. Chukotsko-Anadyrskaya Subprovince: Au-J_3-K_1;
14. Anjuisko-Chukotskaya Subprovince: Au(Sb)-K_1; Sn, W-K_2; Hg(Sb)-K_2-P_1;
15. Yuzhno-Anjuiskaya Subprovince.

16–17. Alazeysk Province:
16. Alazeyskaya Subprovince:
17. Ilyn-Tasskaya Subprovince: Pb, Zn, Ag;
18. Okhotsko-Chukotsk Province: Au, Ag, Cu, Hg, (Pb, Zn, Sb)-K_2-P_1.

Amurskaya Region:

19–20. Hanka-Burejinsk Province:
19. Burejinskaya Subprovince: Fe, Bp-Pz_2; Mo-Pz_3; Sn, Sb-K_2; Hg-P;
20. Hankajskaya Subprovince: Sn, Sb-Pz;
21. Amursko-Okhotsk Province:
21. Amursko-Okhotskaya Province: Mn, Fe-Pz_2; Au(W)-Pz_3 -Mz_1; Hg-P;

22–25. Sikhote-Alynsk Province:
22. Zentralno-Sikhote-Alynskaya Subprovince: W(Au)-K_2;
23. Nizhne-Amurskaya Subprovince: Au(W, Mo, Sb)-K_2;
24. Primorskaya Subprovince: Sn(W, Cu)-K_2; Hg-P;
25. Vostochno-Sikhote-Alynskaya Subprovince: Sn, Pb, Zn-K_2P;
26. Pribrezhnaya Province: Au, Ag, Hg, Sb, Au-P.

Near the Pacific Region:

27. Taigonossko-Murgalsk Province:
27. Taigonossko-Murgalskaya Province: Mo-K_1, Au, Ag, Cu-K_2.

Legend for Figure 9.3. (continued)

28.	Sakhalinsk Province:
28.	Sakhalinskaya Province: Cr-K_1; Hg(Sb, W)-N.
29–30.	Zapadno-Kamchatsko-Koryaksk Province:
29.	Zapadno-Koryakskaya Subprovince: Cr-K_1; Sn(Ag)-P_3-N_1; Hg-N_1;
30.	Khatyrskaya Subprovince: Cr-K_1; Hg, W-N.
31.	Zentralno-Okhotsk Province:
31.	Zentralno-Okhotskaya Province: Cu, Ni-K_2; Cu(Mo)-N.
32–33.	Kurilo-Kamchatsk Province:
32.	Zentralno-Kamchatskaya Subprovince: Au, Ag, Pb, Zn, Hg, C-N;
33.	Kurilo-Yuzhno-Kamchatskaya Subprovince: Au, Ag, Cu, Pb, Zn, Hg, C-Q.
34.	Vostochno-Kamchatsk Province:
34.	Vostochno-Kamchatskaya Province: Cr-K_2; Hg-N;
35–37.	Borders of:
35.	Minerogenic Regions;
36.	Minerogenic Provinces;
37.	Minerogenic Subprovinces.

assessment of mineral development potential in the region, the authors undertook the following subjective development potential evaluation.

Subjective development potential evaluation

Clearly, no mineral development program will take place within the Russian Far East without thorough and complete feasibility studies of the economic potential of individual deposits. There is, however, considerable value in initially assessing the development potential of the major deposit types which are known to occur in the Russian Far East in order to establish some basic guidelines and perspectives with regard to the likelihood of specific types of development.

It is self-evident that certain types of deposits have an intrinsic advantage with respect to attracting development capital, e.g., diamonds, platinum, gold and gem stones, whereas others such as iron ore, manganese, bauxite, lateritic nickel and many nonmetals often require very unique and large deposits to attract investment, and still others, perhaps the majority, such as copper, lead, zinc, titanium, cobalt, and molybdenum, may require an intermediate deposit quality to attract investment. Given this wide range of variables, the authors developed the following checklist of variables against which to evaluate the relative development potential of mineral deposit types in the Russian Far East.

High — High geologic favorability (type, area, access); deposit known or estimated to occur with sufficient size and grade (10th percentile or better) to be internationally competitive; infrastructure available or cost acceptable for development; external markets available or anticipated; mining costs in lower 25 percent internationally; environmental costs minimal/acceptable; deposit meets development objectives.

Medium Medium geologic favorability (constrained type, area, access); deposit inferred to occur with sufficient size and grade (50th percentile or better) to be internationally competitive; infrastructure available or cost acceptable for development; external markets constrained or probable; mining costs in lower 25 percent internationally; environmental costs minimal/acceptable; deposit meets development objectives.

Low Low geologic favorability (uncertain type, limited area, access); deposit inferred to occur with marginal tonnage and grade (in 50th percentile or more); uncertain international competitiveness; infrastructure cost sensitive; constrained external or predominantly internal market; mining costs above 25 percent internationally; environmentally cost-sensitive; present and future (2000) deposit not development priority.

Based on the above checklist, the authors evaluated the development potential of individual deposit types in the Russian Far East.

It should be reemphasized that the evaluation of development potential given is subjective and was made using a limited set of criteria. Indeed, basic issues of political stability, appropriate legislation, and ownership—all key determinants in investment—were not considered. Nevertheless, the analysis does again focus attention on the fact that only a small subset of deposit types can be viewed as desirable for foreign investment and development. Although the resource potential of the Far East is diverse and large, the near-term prognosis for mineral development is optimistic only for a limited number of deposits.

MINERAL DEVELOPMENT TARGET AREAS/DEPOSITS

In the preceding section, a subjective overview of the development potential for specific deposit types has been given and the general evaluation criteria is discussed. Such an analysis provides a useful decision-making tool with respect to evaluating which types of deposits may occur that will have a competitive advantage either economically, within market niches or in support of overall development goals. The analysis does not address the fundamental tenet of mineral exploration and development which is "if the deposit is big enough and rich enough, it will be developed." One only needs to look at mineral developments in some of the world's most remote and primitive areas to see the truth of the statement.

Table 9.1. The Scheme of Minerogenic Division of the Russian Far East

Metallogenic Regions/Provinces/Subprovinces[a]	Geologic-Structural Unit[b] Unit Number	Age
Eastern Siberian Region		
1. Aldano-Amginsk Province: Fe-Ar; Pk-Pr_3; Au (Mo, Pb, Zn)-J_3-K_1;	20	Pr_3-Pz_2
2. Stanovaya Province: Au, Ti, Fe-Ar_2; Pk-Pr_3; Mo, Pb, Zn-J_3-K_1; Au, Ag, Cu (Pb, Zn, Sb)-K_{1-2};	22	Ar_1
3. Leno-Viljuisk Province.	22	Ar_2
Verkhoyano-Chukotskaya Region		
4. Omolonsk Subprovince: Fe-Ar, Au, Ag, (Cu)-D_{2-3};	20	Pr_3-Pz_2
5. Prikolymsk Subprovince: Cu, Pb, Zn, Fe-Pr_3-Pz_2;	20	Pr_3-Pz_2
6. Omolonsk Subprovince: Cu, Mo, Au, Ag (Pb, Zn)-K;	17	up to Kz
7. Sugoysk Subprovince: Mo, Pb, Zn-K_2.	18	up to Pz_2
Yano-Kolymsk Province:		
8. Sette-Dabansk Subprovince: Pk-Pr_3 Pb, Zn, Cu, Fe-Pr_3-Pz_1;	15	Pz-Mz
9. Omulevo-Polousnensk Subprovince: Cu, Pb, Zn, Sb, Hg-Pz_2; Mo, Au-J_3-K_1;	14	Pr_3-Pz
10. Verkhoyansk Subprovince: Au-J_2-K_1;	13	Pz_2
11. Indigiro-Kolyma Subprovince: Au, Sb, (W, Mo)-J_3-K_1;	12	Pz_3
12. Yano-Balygychansk Subprovince: Sn, W, Pb, Zn, (Sb)-K_{1-2}; Hg(Sb)-K_2-P_1.	12	Pz_3
Chukotsk Province:		
13. Chukotsko-Anadyrsk Subprovince: Au-J_3-K_1;	20	Pr_3-Pz_2
14. Anjuisko-Chukotsk Subprovince: Au(Sb)-K_1; Sn, W-K_2; Hg(Sb)-K_2-P_1;	12	Pz_3
15. Yuzhno-Anjuisk Subprovince.	16	Pz_3-Mz
Alazeysk Province:		
16. Alazeysk Subprovince:	16	Pz_3-Mz
17. Ilyn-Tassk Subprovince: Pb, Zn, Ag;	16	Pz_3-Mz
18. Okhotsko-Chukotsk Province: Au, Ag, Cu, Hg, (Pb, Zn, Sb)-K_2-P_1.	4	K_1-R_1

Table 9.1. *(continued)*

Metallogenic Regions/Provinces/Subprovinces[a]	Geologic-Structural Unit[b] Unit Number	Age
Amursk Region:		
Hanka-Burejinsk Province:		
19. Burejinsk Subprovince: Fe, Bp-Pz_2; Mo-Pz_3; Sn, Sb-K_2; Hg-P;	19	up to Pr_2
20. Hankajsk Subprovince: Sn, Sb-Pz;	19	up to Pr_2
21. Amursko-Okhotskaya Province: Mn, Fe-Pz_2; Au(W)-Pz_3,Mz_1; Hg-P.	15, 9, 10	Pz-Mz
Sikhote-Alynsk Province:		
22. Zentralno-Sikhote-Alynskaya Subprovince: W(Au)-K_2;	8	K
23. Nizhne-Amursk Subprovince: Au(W, Mo, Sb)-K_2;	8	K
24. Priamur'e Subprovince: Sn(W, Cu)-K_2; Hg-P;	9	Mz_1-Z
25. Vostochno-Sikhote-Alynsk Subprovince: Sn, Pb, Zn-K_2P;	8	K
26. Pribrezhnaya Province: Au, Ag, Hg, Sb, Au-P.	3	K_2P
Near the Pacific Region:		
27. Taigonossko-Murgalsk Province: Mo-K_1, Au, Ag, Cu-K_2;	16	Pz_3-Mz
28. Sakhalinsk Province: Cr-K_1; Hg(Sb, W)-N.	6	J_3-K_3
Zapadno-Kamchatsko-Koryaksk Province:		
29. Zapadno-Koryaksk Subprovince: Cr-K_1; Sn(Ag)-P_3-N_1; Hg-N_1;	6	J_3-K_2
30. Khatyrsk Subprovince: Cr-K_1; Hg, W-N;	6	J_3-K_2
31. Zentralno-Okhotsk Province: Cu, Ni-K_2; Cu(Mo)-N.	2	N
Kurilo-Kamchatsk Province:		
32. Zentralno-Kamchatka Subprovince: Au, Ag, Pb, Zn, Hg, C-N;	2	N
33. Kurilo-Yuzhno-Kamchatka Subprovince: Au, Ag, Cu, Pb, Zn, Hg, C-Q;	1	N-Q
34. Vostochno-Kamchatka Province: Cr-K_2; Hg-N.	5	K_2-P

a. Metallogenic Regions/Provinces/Subprovinces of the Far East (according to Kunaev, I.V., 1984; Conventional signs to V. Ya Shesternev) index numbers key to map patterns and zones shown in Figure 9.3.

b. Geological-Structural Units of the Far East (according to Kunaev, I.V., 1984; conventional signs to V. Ya Shesternev) unit numbers and associated age designations key to map patterns and zones shown in Figure 9.2.

Table 9.2. Mineral Deposit Types of the Russian Far East

Deposit	Rock Type	Depositional Environment	Tectonic Setting(s)	Tonnage (million tonnes)			Grade (percent)		
				small	med	large	low	med	high
Noril'sk Cu-Ni-PGE	Flood basalts, picritic intrusive rocks, picritic gabbro, norite, olivine gabbro, dolerite, intrusive and volcanic breccias. Associated with evaporites or some external source of sulfur.	Magma has intruded through evaporites or pyritic shale and formed sills in flood basalts during active faulting.	Rift environment.						
Alaskan PGE	Dunite, wehrlite, harzburgite, pyroxenite, magnetite-hornblende pyroxenite, two-pyroxene gabbros, hornblende gabbro, hornblende clinopyroxenite, hornblende-magnetite clinopyroxenite, olivine gabbro, norite. Post-orogenic tonalite and diorite are commonly spatially related.	Deposits occur in layered ultramafic and mafic rocks that intrude into granodiorite, island arc or ophiolite terranes. Evidence indicates shallow levels of emplacement.	Unstable tectonic areas.						
Placer PGE-Au	Alluvial gravel and conglomerate and heavy minerals indicative of ultramafic sources and low-grade metamorphic terrane. Sand and sandstone of secondary importance.	Marine (near shore), rivers and streams (medium to low gradient), desert (eolian) sand dunes, in-situ weathering.	Paleozoic to Mesozoic accreted terranes, Tertiary conglomerates along major fault zones; low terrace deposits; high-level terrace gravels.	.011	.11	1.8	330	2500 .048[b]	6500[a]

Diamond pipes	Kimberlite diatremes. Olivine lamproite (K-rich Mg-lamprophyre) and leucite lamproite.	Pipes intruded from mantle source under high pressure but with rapid quenching.	Most pipes intrude cratonal areas, stable since Early Proterozoic age. Some intrude folded cover rocks that overlie deformed cratonal margins. Pipes are not correlated with orogenic events but occur in areas of epeirogenic warping or doming and along major basement fracture zones. Some pipes occur at intersections of regional zones of weakness visible in LANDSAT or SLAR.						
Low-sulfide Au-quartz veins	Greenstone belts; oceanic metasediments: regionally metamorphosed volcanic rocks, graywacke, chert, shale, and quartzite. Alpine gabbro and serpentine. Late granitic batholiths.	Continental margin mobile belts, accreted margins. Veins are generally post-metamorphic and locally cut granitic rocks.	Fault and joint systems produced by regional compression.	.001	.03	.91	6	16	43
Porphyry Cu	Tonalite to monzogranite or syenitic porphyry intruding granitic, volcanic, calcareous sedimentary, and other rocks.	High-level intrusive rocks contemporaneous with abundant dikes, breccia pipes, faults. Also cupolas of batholiths.	Rift zones contemporaneous with Andean or island-arc volcanism along convergent plate boundaries. Uplift and erosion to expose subvolcanic rocks.	19	140	1100	.31	.54	.94

Table 9.2. *(continued)*

Deposit	Rock Type	Depositional Environment	Tectonic Setting(s)	Tonnage (million tonnes) small	med	large	Grade (percent) low	med	high
Cu skarn deposits	Tonalite to Monzogranite intruding carbonate rocks or calcerous clastic rocks.	Miogeosynclinal sequences intruded by felsic plutons.	Continental margin late orogenic magmatism.	.034	.56	9.2	.7	1.7	4.0
Porphyry Cu-Mo	Tonalite to monzogranite stocks and breccia pipes intrusive into batholithic, volcanic, or sedimentary rocks.	High-level intrusive porphyry contemporaneous with abundant dikes, faults, and breccia pipes. Cupolas of batholiths.	Numerous faults in subduction-related volcanic plutonic arcs. Mainly along continental margins but also in oceanic convergent plate boundaries.	120	500	2100	.26 .0072	.42 .016	.69[c] .035[d]
Sediment-hosted Cu	Red-bed sequence containing green or gray shale, siltstone, and sandstone. Thinly laminated carbonate and evaporite beds. Local channel conglomerate. Some deposits in thinly laminated silty dolomite.	Epicontinental shallow-marine basin near paleoequator. Sabkhas. High evaporation rate. Sediments highly permeable.	Intracontinental rift or aulacogen—failed arm of triple junction of plate spreading. Passive continental margin. Major growth faults.	1.5	22	330	1.0	2.1	4.5
Volcanic-hosted Cu-As-Sb	Andesite, dacite, flows, breccias, and tuffs.	Volcanic terrane, uppermost levels of intrusive systems.	Continental margins and island arcs.						
Algoma Fe	Mafic to felsic submarine volcanic rocks and deep-water clastic and volcaniclastic sediments.	Volcano-sedimentary basins (greenstone belts of Precambrian shields) generally with rapid turbidite sedimentation and thick volcanic accumulation.	Tectonically active submarine volcanic belts, most commonly preserved in Precambrian shields.						

Superior Fe	Commonly interlayered with quartzite, shale, dolomite.	Stable, shallow-water marine environment, commonly on stable continental shelf or intracratonic basin.	Now commonly preserved in forelands of Proterozoic orogenic belts.	11	170	2400	30	53	66
Fe skarn deposits	Gabbro, diorite, diabase, syenite, tonalite, granodiorite, granite, and coeval volcanic rocks. Limestone and calcareous sedimentary rocks.	Contacts of intrusion and carbonate rocks or calcareous clastic rocks.	Miogeosynclinal sequences intruded by felsic to mafic plutons. Oceanic island arc, Andean volcanic arc, and rifted continental margin.	.33	7.2	160	36	50	63
Sn veins	Close spatial relation to multiphase granitoids; specialized biotite and (or) muscovite leucogranite common; pelitic sediments generally present.	Mesozonal to hypabyssal plutons; extrusive rocks generally absent; dikes and dike swarms common.	Foldbelts and accreted margins with late orogenic to postorogenic granitoids which may, in part, be anatectic; regional fractures common.	.012	.24	4.5	.7	1.3	2.3
Sn skarn deposits	Leucocratic biotite and (or) muscovite granite, specialized phase or end members common, felsic dikes. carbonate rocks.	Epizonal intrusive complexes in carbonate terrane.	Granite emplacement generally late (post orogenic).	1.6	9.4	58	.13	.31	.76
Tungsten skarn deposits	Tonalite, granodiorite, quartz monzonite; limestone.	Contacts and roof pendants of batholith and thermal aureoles of apical zones of stocks that intrude carbonate rocks.	Orogenic belts. Syn-late orgenic.	.05	1.1	22	.34	.67	1.4

Table 9.2. *(continued)*

Deposit	Rock Type	Depositional Environment	Tectonic Setting(s)	Tonnage (million tonnes)			Grade (percent)		
				small	med	large	low	med	high
Polymetallic veins	Calcalkaline to alkaline, diorite to granodiorite, monzonite to monzogranite in small intrusions and dike swarms in sedimentary and metamorphic ricks. Subvolcanic intrusions, necks, dikes, plugs of andesite to rhyolite composition.	Near-surface fractures and breccias within thermal aureol of clusters of small intrusions. In some cases peripheral to porphyry systems.	Continental margin and island arc volcanic-plutonic belts. Especially zones of local domal uplift.	.0003	.0076	.2	140	820	4700[e]
							2.4	9	33[f]
								2.1	7.6[g]
								.89[c]	
Sedimentary exhalative Zn-Pb	Euxinic marine sedimentary rocks including: black (dark) shale, siltstone, sandstone, chert, dolostone, micritic limestone, and turbidites. Volcanic rocks, commonly of bimodal composition, are present locally in the sedimentary basin.	Marine epicratonic embayments and intracratonic basins, with smaller local restricted basins (second- and third-order basins).	Epicratonic embayments and intracratonic basins are associated with hinge zones controlled by synsedimentary faults, typically forming half-grabens.	1.7	15	130	2.4	5.6	13[g]
							1.0	2.8	7.7[f]
Southeast Missouri Pb-Zn	Dolomite; locally ore bodies also occur in sandstone, conglomerate, and calcareous shales.	Host rocks are shallow-water marine carbonates, with prominent facies control by reefs growing on flanks of paleotopographic basement highs. Deposits commonly occur at margins of clastic basins.	Stable cratonic platform.						

Sandstone-hosted Pb-Zn	Continental, terrigenous, and marine quartzitic and arkosic sandstone, conglomerate, grit, and siltstone. Local evaporates.	Host rocks deposited in combined continental and marine environments including piedmont, fluvial, lagoonal-lacustrine, lagoonal-deltaic, lagoonal-beach, and tidal channel-sand bar environments. Commonly succeeded by marine transgressions.	Deep weathering and regional peneplanation during stable tectonic conditions, accompanied by marine platform or piedmont sedimentation associated with at least some orogenic uplift. Sialic basement, mainly "granites" or granitic gneisses.	.47	5.4	62	.89	2.2	5.2[f]
							.23		3.0[g]
Polymetallic replacement deposits	Sedimentary rocks, chiefly limestone, dolomite, and shale, commonly overlaid by volcanic rocks and intruded by porphyritic, calcalkaline plutons.	Carbonate host rocks that commonly occur in broad sedimentary basins, such as epicratonic miogeosynclines. Replacement by solutions emanating from volcanic centers and epizonal plutons. Calderas may be favorable.	Most deposits occur in mobile belts that have undergone moderate deformation and have been intruded by small plutons.	.24	1.8	14	1.2	5.2	21[f]
							.82	3.9	19[g]
								150	690[e]
Zn-Pb skarn deposits	Granodiorite to granite, diorite to syenite. Carbonate rocks, calcareous clastic rocks.	Miogeosynclinal sequences intruded by generally small bodies of igneous rock.	Continental margin, late-orogenic magmatism.	.16	1.4	12	2.7	5.9	13[g]
							.87	2.8	7.6[f]
Simple Sb deposits	One or more of the following lithologies is found associated with over half of the deposits: limestone, shale (commonly calcareous), sandstone, and quartzite.	Faults and shear zones	Any orogenic area.	6.7×10^{-5}	1.8×10^{-4}	.0049	18	35	66

Table 9.2. *(continued)*

Deposit	Rock Type	Depositional Environment	Tectonic Setting(s)	Tonnage (million tonnes)			Grade (percent)		
				small	med	large	low	med	high
Serpentine-hosted asbestos	Serpentinites, dunite, harzburgite, pyroxenite.	Usually part of an ophiolite sequence. Later deformation and igneous intrusion may be important.	Unstable accreted oceanic terranes.	4.6	26	150	2.7	4.6	8.0
Volcanogenic Mn	Chert, shale, graywacke, tuff, basalt; chert, jasper, basalt (ophiolite); basalt, andesite, rhyolite (island-arc); basalt, limestone; conglomerate, sandstone, tuff, gypsum.	Sea-floor hot spring, generally deep water; some shallow water marine; some may be enclosed basin.	Oceanic ridge, marginal basin, island arc, young rifted basin; all can be considered eugeosynclinal.	.0028	.047	.91	24	42	49
Sedimentary Mn	Shallow marine sediments, most commonly carbonates, clay, and glauconitic sand, commonly with shellbeds, in transgressive sequences associated with anoxic basins.	Shallow (50–300 m) marine, commonly in sheltered sites around paleoislands. Most deposits overlie oxidized substrates, but basinward, carbonate deposits may be in chemically reduced settings.	Stable cratonic interior basin or margin.	.19	7.3	280	15	31	49
Upwelling type phosphate deposits	Phosphorite, marl, shale, chert, limestone, dolomite, and volcanic materials.	Marine sedimentary basins with good connection to the open sea and upwelling, areas highly productive of plankton. Deposition occurs mostly in warm latitudes, mostly between the 40th parallels.	Intra-plate shelf, platform, miogeosynclines, and eugeosynclines.	26	330	4200	15	25	32
Laterite type bauxite deposits	Weathered rock formed on aluminous silicate rocks.	Surficial weathering on well-drained plateaus in region with warm to hot and wet climates. Locally deposits in	Typically occurs on plateaus in tectonically stable areas.	.87	25	730	35	45	55

		poorly drained areas low in Fe due to its removal by organic complexing.							
Lateritic Ni	Ultramafic rocks, particularly peridotite, dunite, and serpentinized peridotite.	Relatively high rates of chemical weathering (warm-humid climates) and relatively low rates of physical erosion.	Convergent margins where ophiolites have been emplaced. Uplift is required to expose ultramafics to weathering.	7.8	44	250	1.0	1.4	1.9
Carbonatite deposits	Apatite-magnetite deposits tend to be in sovite (calcite carbonatite); RE types tend to occur in ankerite carbonatite; most deposits have both. In general pyroxenite, nepheline and feldspathic pyroxenite, carbonatite, fenite, ijolite, dunite, picritic-porphyrites, gneiss and alkalic fenitized gneiss, and locally alkaline volcanic rocks.	Multiple stages of igneous, deuteric and metasomatic crystallization in carbonatite magma.	Continental shields. Spatially related to fault lineaments such as East African rift system. Locally related to alkaline volcanism.	16	60	220	.18	.58	1.9
Shoreline placer Ti	Well-sorted medium- to fine-grained sand in dune, beach, and inlet deposits commonly overlying shallow marine deposits.	Stable coastal region receiving sediment from deeply weathered metamorphic terranes of sillimanite or higher grade.	Margin of craton. Crustal stability during deposition and preservation of deposits.	11	87	690	.23	1.3	6.9

a. Platinum grade in parts per billion
b. Gold grade in grams per tonne
c. Copper grade
d. Molybdenum grade
e. Silver grade in grams per tonne
f. Lead grade
g. Zinc grade

Source: Cox and Singer, 1986.

Irrespective of the previous analysis, therefore, there have been and will be in the future a number of specific areas and target deposits identified in the Far East which are believed to have significant economic potential, not all of which will have been previously selected by the analyses (Table 9.3). Therefore, although many deposits may fall within a high probability of development range, based on the preceding discussion, several will not, but are still worthy of consideration under the above fundamental tenet. It is the purpose of this section to identify specific areas and/or deposits of the Russian Far East in order to focus attention on them as potential development priorities. The coverage is by no means complete but does represent some of the most favorable development targets.

Tin

The Far East ranks first in known reserves of tin within the CIS, with more than 90 percent of total reserves concentrated in over 180 placer and bedrock occurrences. Within the Far East, Yakutiya has approximately 42 percent of total reserves, with the Primorye (24%) and Magadan (21%) areas in second and third place, respectively. Within the region, the Pyrkakajsky, Pravourmijsky and Tigrinsky occurrences have the highest potential for future development as bedrock tin resources. Nevertheless, the largest present and future tin development targets in the Far East are placer deposits. The tin placers of the Vanka Inlet, averaging 2.5 m in thickness and 600 gm/m^3 in concentration, with associated ilmenite, zircon, rutile and monazite, represent a primary development project. Additionally, exploration for large tin placers is being conducted in the basins of the Bureja, Niman, Selemja and Amur rivers, and in Chukotka, a new tin area in the Koryak upland is being actively explored.

Tungsten

The Far East contains approximately 10 percent of the CIS's tungsten reserves, which are concentrated in approximately 40 placer and bedrock deposits. As tungsten is widely associated with tin, it occurs primarily in the same territories but with a slightly different resource distribution, with 43 percent of reserves in Primorye, 34 percent in Yakutiya and 16 percent in Magadan. At present, large tin and tungsten deposits, mainly placers, are being outlined in Budjal, and particular interest is presently being shown in the molybdenum-tungsten stockwork deposit at Bugdainskoe which is both a large and unique deposit.

Copper

Although the Far East is neither a major copper producer nor the site of significant reserves, recent discoveries and work on the stratiform copper-sandstone

Table 9.3. Far East Mineral Deposit Types and Estimated Present and Future Development Potential

	Present Develoment Potential			Future Development Potential		
	High	Medium	Low	High	Medium	Low
Noril'sk Cu-Ni-PGE	X			X		
Alaskan PGE	X			X		
Placer PGE-Au	X			X		
Diamond pipes	X			X		
Low-sulfide Au-quartz veins	X			X		
Porphyry Cu		X		X		
Cu skarn deposits			X		X	
Porphyry Cu-Mo		X		X		
Sediment-hosted Cu	X			X		
Volcanic-hosted Cu-As-Sb			X			X
Algoma Fe			X		X	
Superior Fe			X			X
Fe skarn deposits			X		X	
Sn veins		X			X	
Sn skarn deposits		X				X
Tungsten skarn deposits		X				X
Polymetallic veins			X		X	
Sedimentary exhalative Zn-Pb		X		X		
Southeast Missouri Pb-Zn			X	X		
Sandstone-hosted Pb-Zn			X			X
Polymetallic replacement deposits			X		X	
Zn-Pb skarn deposits		X				X
Simple Sb deposits			X			X
Serpentine-hosted asbestos			X			X
Volcanogenic Mn			X		X	
Sedimentary Mn			X		X	
Upwelling type phosphate deposits			X			X
Laterite type bauxite deposits		X			X	
Lateritic Ni			X			X
Carbonatite deposits			X		X	
Shoreline placer Ti		X		X		

deposits of Yakutiya (the Udokan ore body with proven reserves of 700 million tonnes of ore grade 1.5% Cu) have turned this area into a primary development target. Similarly, recently discovered Cu porphyry and Cu-Mo porphyry systems in Khabarovsk Territory, Magadan Province and the Yakutiya Autonomous Region are additional target areas for future development.

Precious metals

The Far East is responsible for approximately 25 percent of the CIS's total gold output and has the largest gold reserves in the Commonwealth, with most of the deposits found in Yakutiya and Magadan. At present, most gold production is from placer deposits, particularly from the tributaries and main streams of the Amur, Indigirka and Kolyma rivers. To a great extent, the gold potential of the Far East, particularly with respect to lode gold deposits, is unknown and untested. However, the close association of gold mineralization with older massif structures and with acidic intrusives within the Pacific fold belt point to a very large potential.

Large gold placers have been discovered in several areas within the Khabarovsk Territory, in Chukotka and on the shelves of the Japan and the Okhotsk seas. In particular, more than 80 prospective areas have been discovered in the southern portion of the Primorye Territory, in the Shantar Islands, along the Okhotsk coast of the Lower Amur, on the western coast of Kamchatka, along the shore of eastern Chukotka, in the Chaun Inlet and the Vanka Inlet. The gold content in the placers is 250–600 mg/m^3, with some areas having up to 22 g/m^3. Placers are traced at depths to 400 m and far out into the sea. The thickness of mixed river and marine sands is up to six meters.

In assessing the gold potential of the Far East, particularly with respect to placer gold, it must be emphasized that many gold placers also produce substantial quantities of platinum group metals, cassiderite, scheilite, ilmenite, magnetite, zircon, monazite, rutile, and xenotime. From the latter grouping, a rare earth extraction industry could perhaps be developed in the future.

The Far East is a leader both within the CIS and worldwide with respect to explored silver reserves. In the CIS, Magadan Province, Khabarovsk Territory, and Yakutiya of the Russian Far East are ranked, respectively, first, second and third in terms of silver reserves. Some discoveries of large deposits of the Serra-Potozy type are anticipated owing to the discovery of silver concentrations in phosphorite deposits and in oxidation zones of polymetallic deposits.

The Far East is rapidly expanding and developing its platinum mining and development activities, particularly with respect to placer deposits associated with ultramafic massifs (Khonotor, Inogli and Chad in the Aldan shield; the Uzbelsky massif in Chukotka; and the Kuril massif in Kamchatka). In addition, the bedrock sources of platinum associated with the complex copper porphyry ores of Chukotka and Yakutiya and the massive copper-nickel ores of Shanuch in Kamchatka provide not only interesting development targets for the present, but indicate both areas and deposit types for future exploration and development.

Diamonds

The majority of diamond reserves (>75%) of the CIS occur in the Far East, principally within four main kimberlite provinces in the northwestern portion of Yakutiya, i.e., Malobotuobinsky (Mirninsky), Daldyno-Alakitsky (Ihalo-Udachnensky), Nizhnelensky, and Anabarsky. Over 97 percent of all diamond reserves are in hard rock deposits and only three percent are concentrated in placer occurrences.

Overall, CIS diamond mines hold reserves expected to last for 25–40 years, with reserves in open excavations lasting 15–17 years (Borisovich, 1991). The Udachnoye deposit in Yakutiya produces nearly 90 percent of all CIS diamonds and contains proved reserves expected to last for 10–12 years. Recently, an extremely rich diamond deposit was discovered in the Arkhangelsk region.

Ferroalloy metals

Currently, the Far East is not a major producer of either ferroalloy metals or steel products; nevertheless, the resource endowment of the region would appear to be favorable for future development of either a steel industry or for the exploration and development of ferroalloy ores for export.

Iron ore deposits of the Far East are concentrated in four major deposits/districts (Table 9.4) and, more recently, new deposits of high grade magnetite-hematite ore have been discovered in the region but are incompletely evaluated. As can be noted from Table 9.4, the average grade of iron ores in the major deposits/districts averages approximately 35 percent Fe_2O_3 and, therefore, would be considered low grade when compared against major deposits worldwide. However, the ores of the Malo Khingan occurrence are characterized by a low content of phosphorus and sulfur and are readily enriched (iron extraction is 68–87% and its content in the concentrate is up to 65.8%).

Manganese ore occurrences in the Far East Territory have been explored only in the Malo Khingan District. The ore balance reserves in the Southern Khingan occurrence total 8.6 million tonnes, with an average manganese content of 21 percent; tin, 8.6 percent; and phosphorus 0.05 percent. Braunite ores prevail (90% of reserves).

Chromite ore occurrences have not been discovered to date; however, chromite is contained in the Khonder platinum metals placer occurrence. Sand reserves of C_2 category are 36 million m^3, with a chromite content of up to 70 kilograms/m^3.

Titanium resources of the Far East are poorly quantified but exceed 1,890 million tonnes in the P_1 and P_2 categories. The majority of the resources are found in placer deposits; however, the titaniferous ore of the Dzhugzhur ridge area is mined by open pit. Similar deposits occur in the Amur Province.

Most of the future marine placers of titanium, associated with iron, rare earth-bearing minerals and locally gold, occur as coastal marine placers adjacent to Sakhalin, the Kuril Islands and the coast of the Primorye Territory. In the latter

Table 9.4. Main Ferruginous Deposits of the Far East and Their Indices

Occurrences	Reserves (million tonnes) Categories $A + B + C_1$	P_1	Average Content of Ferrum in Ore (%)
I. Garinskoye	389	NA	41.7
II. Occurrences group of Malo-Hingansky district (Sytarskoye, Kimkanskoye, Kabalinskoye)	591	NA	35.1
III. Milkanskoye		2,700	31.5
IV. Itmatinskoye	NA	3,700	42.6
Occurrences of the region	4,085	20,000	NA

NA Not available

case, marine placers of titaniferous magnetite have been traced for over 600 km. In the Kuril Islands, near Iturup Island, the Ruchar occurrence has reserves of 10.5 million tonnes of TiO_2 of which only 15 percent has been adequately explored. Resource estimates place the total resource potential at approximately three billion tonnes.

Fertilizer minerals

Development of fertilizer minerals in the Far East is quite low (approximately 15% of domestic demand); nevertheless, the region has substantial reserves and resources of fertilizer minerals. For example, the apatite-titanium-magnetite ore occurrences of Khalaro-Haninsky (Amur Province), Dzhugazhursky and Baladeksky (the Khabarovsk Territory) ore districts represent potentially large sources of both titanium and mineral fertilizer materials. The Moymakan, Gajuim, Dzhaninsk apatite occurrences are highly promising for future development, with geological reserves of 3.6 billion tonnes, with an average P_2O_5 content of six to nine percent.

Building stones

The Far East possesses tremendous reserves of various building materials and stones, including brick earth and loam, specialty clays, building and ornamental stones, perlite, and sand and gravel. More than 1,400 occurrences of building rocks have been discovered in the region. Within Khabarovsk Territory alone there are more than 560 occurrences.

MINERAL DEVELOPMENT OF THE FAR EAST

The mineral development of the Far East, as in the remainder of the CIS, must await the resolution of many issues before a clear vision of the future is possible. Nevertheless, several factors favor the development of a large and more vital minerals industry within the Far East. Among the most important are the following:

- The Far East has an established minerals industry based upon a large reserve base of several minerals.
- The Far East is richly endowed with precious metals (Au, Ag, PGE) and diamonds, which are high value commodities requiring less infrastructure than many other commodities.
- The Far East has a large and diverse energy base upon which to plan development of the region and to support industrialization.
- The Far East potentially has access to the rapidly growing markets for minerals in the Asia-Pacific region and the large economies of Japan and the United States.
- Preliminary evidence indicates that the Far East may contain a few giant deposits (gold, copper, diamonds) which could provide the catalyst for development of the region.
- The diverse minerals resource base of the region allows consideration of the development of integrated mineral industries based on steel, aluminum, and base metals.

Offsetting these very positive factors are the well known and often recited problems associated with a harsh climate, a lack of infrastructure, inefficient and outmoded technology, a lack of capital, a lack of political stability, and a regressive minerals policy and legislation. All of these factors present greater or smaller impediments to the current and near-term development of minerals within the Far East.

For the near term then, it would seem that the minerals industry of the Far East should build upon its considerable strengths. The following are examples of possible development paths for the minerals industry:

- Development of a mining-chemical industry such as the large Jaroslavsky Mining Enterprise, which produces over 75 percent of the CIS's fluorite from its Voznesensky deposit in the Primorye Territory.
- Expansion of the production of major mineral deposits such as the Dukatskoye gold-silver vein deposit, while simultaneously increasing production in surrounding and nearby areas.
- Improvement of the metallurgical processing capabilities of the existing facilities such as those of Pyrkakajsky, Pravourmijsky and Tegrin in the tin industry where ore extraction levels are 40 percent in Krasnorechenskaya,

50 percent in Solnechnaya and 70 percent in Hrustalnenskaya. Marginal improvements in recovery would substantially improve profitability. Similarly, within the Primorsky Mining Enterprise scheelite recovery is only 67 percent and that of associated copper is approximately 50 percent.

For the future, the Far East must, of course, look toward the further development and expansion of its mining activities. However, it must also look toward the larger goals of developing more integrated industries, such as steel and aluminum, and to the more complete utilization of its resources to produce value added through gold smelting and refining and jewelry or the development of gem-cutting capabilities, as but a few examples. Overall, the development of the mineral industry of the Far East must be both coherent and steady with the objectives of improving the well-being of the peoples of the Far East and demonstrating to the foreign investor that the Far East is a preferred place to invest. The region has enormous resources upon which to build, a strong will within its people and the support of many. Success is not a question of if, but only of when.

REFERENCES

Borisovich, Vitaly T., 1991, "Mining in the USSR and Conditions for Foreign Investment," paper presented at the Metal Mining Agency of Japan Forum, October, London, United Kingdom.

Cox, Dennis P., and Donald A. Singer (editors), 1986, *Mineral Deposit Models,* U.S. Geological Survey Bulletin 1693, Washington, D.C.

Kunaev, I.V. (compiler), 1984, Map on Geological and Structural Zones of the Russian Far East and Map on Metallogenic Regions, Provinces, and Subprovinces of the Russian Far East (in Russian), Khabarovsk, Russian Federation.

Chapter 10

ENERGY SUPPLIES OF THE RUSSIAN FAR EAST: PROSPECTS FOR USING NATURAL GAS AND NUCLEAR POWER

Nikolai I. Tsvetkov

INTRODUCTION

At present, the energy supply situation of the Russian Far East region is extremely strained. In 1990, the requirements for boiler-and-stove fuel exceeded the region's energy supplies by 4.5 million tonnes of oil equivalent (toe). Requirements will rise to 10 million toe, or 18 percent of demand in 1995.

Clearly, the energy needs of the region will increase further in the future. According to forecasts of various energy organizations for the period 1991–2005, requirements for electric power will grow by a factor of 1.8; for heating, 1.4; and for boiler-and-stove fuel, 1.35–1.4 (see Tables 10.1, 10.2 and 10.3). The projected demand of the region's economy for electric power, heating and boiler-and-stove fuels for the period up to 2005 will not be satisfied by further production increases in coal, which dominates in the region's current energy balance. Boiler-and-stove fuels and hydroelectric resources provide the rest of the current balance.

Under the new economic and political conditions prevailing in the Russian Far East, it is necessary, first of all, to dramatically restructure the energy industry. The region's traditional energy development strategy focused on the use of coal and hydroelectric power. Should that strategy be continued, the annual demand for boiler-and-stove fuels would exceed regional energy resources output by 20–25 million toe. In the near future, changes to the composition of the region's energy supply balance should aim at using more natural gas and, after 2000, more nuclear power. Increases in gas production in the Far East Economic Region (FEER) could be facilitated by the development of gas reserves on the Sakhalin shelf and oil and gas reserves in the Yakutiya and Chukotka provinces.

J.P. Dorian et al. (eds.), CIS Energy and Minerals Development, 157–162.

Table 10.1. Requirements of the Russian Far East for Electric Power (billion kwh)

	Years			
Directions of use by industries	1990	1990	2000	2005
All industry including	16.92	21.51	27.92	32.28
Fuel industry	2.08	2.56	3.60	4.12
Ferrous metallurgy	1.32	2.37	2.54	2.69
Nonferrous metallurgy	4.28	4.50	4.96	5.11
Chemical and petrochemical industry	0.49	0.85	2.34	4.13
Engineering industry	2.47	3.42	4.91	5.37
Timber and pulp and paper industry	2.20	2.32	2.95	3.39
Construction materials industry	1.57	2.00	2.49	2.67
Light industry	0.16	0.18	0.26	0.35
Food industry	0.85	1.05	1.23	1.27
Other industries	1.50	3.26	3.64	3.18
Construction work	2.62	3.56	4.10	4.60
Transport	4.90	5.30	5.70	6.10
Agriculture	3.60	4.20	4.60	5.00
Public services	11.47	14.97	17.46	21.03
Total useful consumption	39.50	49.54	57.78	69.01
Own needs of electric power stations	4.27	5.20	7.00	9.16
Losses in power network	4.63	5.86	7.42	9.20
Total demand	48.40	60.60	74.20	87.30

NATURAL GAS

Regionally produced natural gas supplies should become available in 1994–1995. Opportunities for using gas supplies for fuel and possibly export from Sakhalin and Yakutiya as well as Kamchatka and Chukotka are still being defined (Table 10.4). Whatever the circumstances may be, however, a gas pipeline network should be established to provide gas to such cities as Komsomolsk, Khabarovsk, Vladivostok and probably Glagoveshchensk. It will be necessary to draw on foreign capital and technologies to accelerate the use of natural gas resources in the region's economy.

Intensive development of the large gas reserves on the Sakhalin shelf will be of critical importance to ensure a reliable fuel supply to the FEER and improve air quality in the main industrial areas of the Sakhalinskaya oblast, and Khabarovsky and Primorsky krais. Should these reserves not be developed because of conflicting energy priorities, full exploitation of practically all of the proved coal deposits in Amurskaya oblast, Khabarovsky and Primorsky krais will be required. Moreover, it would be necessary to rely on coal supplies from other regions of the country, including large amounts of coal from the Kansko-Achinskoe deposit, the use of which would require the reconstruction of a num-

Table 10.2. Requirements of the Russian Far East for Heating (million hectorcal)

Directions of use by industries	Years			
	1990	1995	2000	2005
All industry including:	38.1	44.4	54.0	67.7
Fuel industry	3.6	4.0	4.8	4.9
Ferrous metallurgy	1.0	1.2	1.3	1.4
Nonferrous metallurgy	2.2	2.3	2.6	2.6
Chemical and petrochemical industry	3.1	3.3	4.8	6.1
Engineering industry	8.6	11.0	14.1	16.7
Timber and pulp and paper industry	7.3	8.8	10.8	17.1
Construction materials industry	6.5	7.7	9.2	11.1
Light industry	0.5	0.6	0.6	0.7
Food industry	4.8	5.0	5.5	6.5
Other industries	0.4	0.5	0.5	0.7
Construction work	3.5	4.0	4.3	4.8
Transport	2.9	3.1	3.2	3.4
Agriculture	4.5	4.9	5.3	5.8
Public services	46.5	49.6	52.7	54.5
Other kinds of use	9.3	9.5	9.8	10.0
Losses	2.5	2.9	3.3	3.6
Total demand	107.2	118.4	132.6	149.7

Table 10.3. Requirements of the Russian Far East for Boiler-and-Stove Fuel (according to variants I and II) (million tonnes of fuel equivalent)

	1990	1995		2000		2005	
	I	I	II	I	II	I	II
Russian Far East	42.79	50.73	49.41	58.18	66.27	63.27	63.01
Primorsky krai	11.83	14.36	13.85	16.66	15.73	18.24	17.34
Khabarovsky krai	12.06	14.78	14.47	17.41	16.60	20.29	19.17
Amurskaya oblast	4.70	5.53	5.35	5.91	5.75	7.06	6.74
Kamchatskaya oblast	1.39	1.84	1.69	2.06	1.97	2.39	2.17
Magadanskaya oblast	2.68	2.71	2.60	3.57	2.87	4.37	3.57
Sakhalinskaya oblast	4.59	5.35	5.35	5.91	5.91	6.78	6.71
Yakutskaya SSR	5.54	6.10	6.10	6.66	6.66	7.39	7.39

ber of operating power stations and boilers. The development of gas resources on the Sakhalin shelf, and notably the Chivo and Lunskoe deposits, would allow the region to eliminate the boiler-and-stove fuel deficit as early as 1991–1995, and reduce coal deliveries from other regions of the country. In addition, gas

Table 10.4. Possible Distribution of Sakhalin Gas (according to variants I and II) (billion cubic meters)

	1995		2000		2005		2010	
	I	II	I	II	I	II	I	II
Distributed resources of gas	5.8	8.0	17.0	17.0	25.0	25.0	25.0	30.0
Consumption by regions	5.8	5.4	8.2	10.5	10.6	13.4	13.0	15.0
Khabarovsky krai	4.5	4.1	4.8	6.0	6.2	7.0	7.7	7.0
Primorsky krai	—	—	2.0	3.1	2.7	4.7	2.7	5.4
Sakhalinskaya oblast	1.3	1.3	1.4	1.4	1.7	1.7	2.6	2.6
Export	—	2.6	8.8	6.5	14.4	11.6	12.0	15.0

development would allow the reduction of coal output in the region to sustainable levels and eliminate the need for construction of a number of costly coal mining operations.

COAL

As noted earlier, high priority development of oil and gas reserves in the FEER will allow for a substantial decrease in the exploitation of local coal resources and will eliminate the need for a number of expensive hydropower projects. However, it should not lead to neglect of the coal industry, as the region's shortage of coal will soon be aggravated by an expected closing of 14 coal mining operations where reserves are nearly depleted. These mines represent a total annual capacity of 19.4 million tonnes of coal which includes five open-pit mines containing a capacity of 16 million tonnes. A considerable drop in coal output will occur between 1992 and 1995 when the Raichikhinskoe deposit of brown coal is depleted. To compensate for the reduction in coal output from that deposit, the open-pit mine No. 1 at the Yerkovetsky deposit is currently being developed. It has a rated capacity of 4.5 million tonnes. Yerkovetsky open-pit mines No. 2 and No. 3 are also being considered for development; however, the construction of these open-pit mines would require use of large tracts of arable land.

Plans for development of the Svobodnenskoe deposit of brown coal in Amurskaya oblast which has with a rated capacity of 12.5 million tonnes need to be further analyzed as estimates indicate that it would be more efficient to focus efforts on developing a large open-pit mine at the Ogodzhinskoe brown coal deposit, which is similar to the coal of Urgal. The development of the Ogodzhinskoe deposit would require the construction of a branchline from the Baikal-Amur Mainline (BAM) railroad near the Fevralskaya station. A final decision on development of the deposit can be made only after an economic feasibility study is undertaken.

Coal output at the Urgalskoe deposit in Khabarovsky krai is 2.2 million tonnes annually. Serious attention is being paid to expanding operations at the Urgalskoe deposit. The region's coal demand for electric power stations, boiler stations and residential/commercial needs in 1992 will be 11 million tonnes, of which 10 million tonnes are delivered from Yakutiya, Amurskaya oblast, Primorsky krai, Chitinskaya oblast and eastern Siberia. The mining and geological conditions of the Urgalskoe deposit would permit an output of high-quality coal (4,800–7,000 kcal/kg) at a volume of 10–12 million tonnes per year. At this rate of output, reserves of the Urgalskoe deposit alone would be sufficient for 150–200 years of production. Total reserves at Urgalskoe are estimated at between 1.7 billion and 2.0 billion tonnes.

NUCLEAR POWER

An important element of the reorganization of the energy sector of the FEER will be the construction of nuclear power stations. By putting nuclear power plants into operation beginning in 2005, it will be possible to balance power generation capacity and electricity demand in Khabarovsk in particular and in the Far East as whole. The share of power produced at nuclear stations in Khabarovsky krai, for example, could constitute more than 40–50 percent of capacity requirements by the year 2010.

NEED FOR JOINT VENTURES

The improvement of the energy supply situation of the FEER will depend on a substantial restructuring of the region's economy and energy sector and efforts at improving export specialization. The promotion of international cooperation through joint ventures in the extraction, transportation and refining of hydrocarbons will increase the likelihood of oil and gas development in such areas as the Sakhalin shelf. The creation of a Far East marine oil and gas base would contribute significantly not only to solving the energy supply problems of the region, but also to restructuring and upgrading its economy. Joint ventures to develop oil and gas resources on the Sakhalin shelf could become a starting point for the economic reorganization of the region and lead to rapid integration with Asia-Pacific economies.

The current trend towards a reorganization of the FEER energy sector will help solve both old and new problems of energy supply in the region. All of the pending projects should be undertaken within a market economy framework—that is, private domestic and foreign financing should be sought. This is of particular importance to the FEER as the development and exploitation of new energy resources will require large investments because of its remote location and severe natural and climatic conditions.

CONCLUSIONS

In summary, the two most important tasks for the efficient development of the Far East's energy sector are a continuation of efforts to increase extraction of natural gas in the region and the preparation for construction of nuclear power stations in the area. The need to reduce pollution in the Far East, the transition to a market economy, and the growing economic and political independence of both the entire Far East region and its administrative territories (krais and oblasts) will require close coordination between energy policies and efforts to finance development and operations. Energy sector reorganization will not solve many problems related to inadequate supplies and excessive consumption in the region, but it will serve as a necessary first step in creating a rational and well-balanced industry.

Chapter 11

MINERAL RESOURCES OF THE RUSSIAN FAR EAST: PROSPECTS FOR EXPORT

Yuri I. Bakulin and Vitaly T. Shishmakov

INTRODUCTION

The Far East is Russia's outlying area. The eight thousand kilometer (km) railway that connects it with Moscow exerts noticeable influence on the economic development of only one-fourth of the Far East's territory. The remaining area is dependent on ship-borne cargo traffic on the two major rivers, the Lena and the Amur, the Northern Sea Route and a much longer southern route operating around Asia. Transportation links with the countries of the Asia-Pacific region can be easily maintained, but have been underutilized.

The following section analyzes the characteristics of the Far East that are decisive in planning a strategy for economic development in the region.

GEOGRAPHICAL AND ECONOMIC FEATURES OF THE FAR EAST

The Far East of the Russian Federation (referred to hereafter as Far East Economic Region, or FEER) includes the Republic of Sakha (Yakutiya), the Khabarovsk and Primorye territories, and the Amur, Kamchatka, Magadan and Sakhalin provinces.

The Republic of Sakha has two access routes to the south (the port of Osetrovo on the Lena River and the railway station at Berkakit where the Baikal-Amur Mainline connects with the Trans-Siberian railway), and one from the north (through the port of Tiksi on the northern sea route). Transport and communication links are well developed in the southern region of the Khabarovsk Territory, in the Primorye Territory and in the Amur and Sakhalin provinces (Figure 11.1), although the infrastructure is inferior to the transportation facilities in the

J.P. Dorian et al. (eds.), CIS Energy and Minerals Development, 163–176.

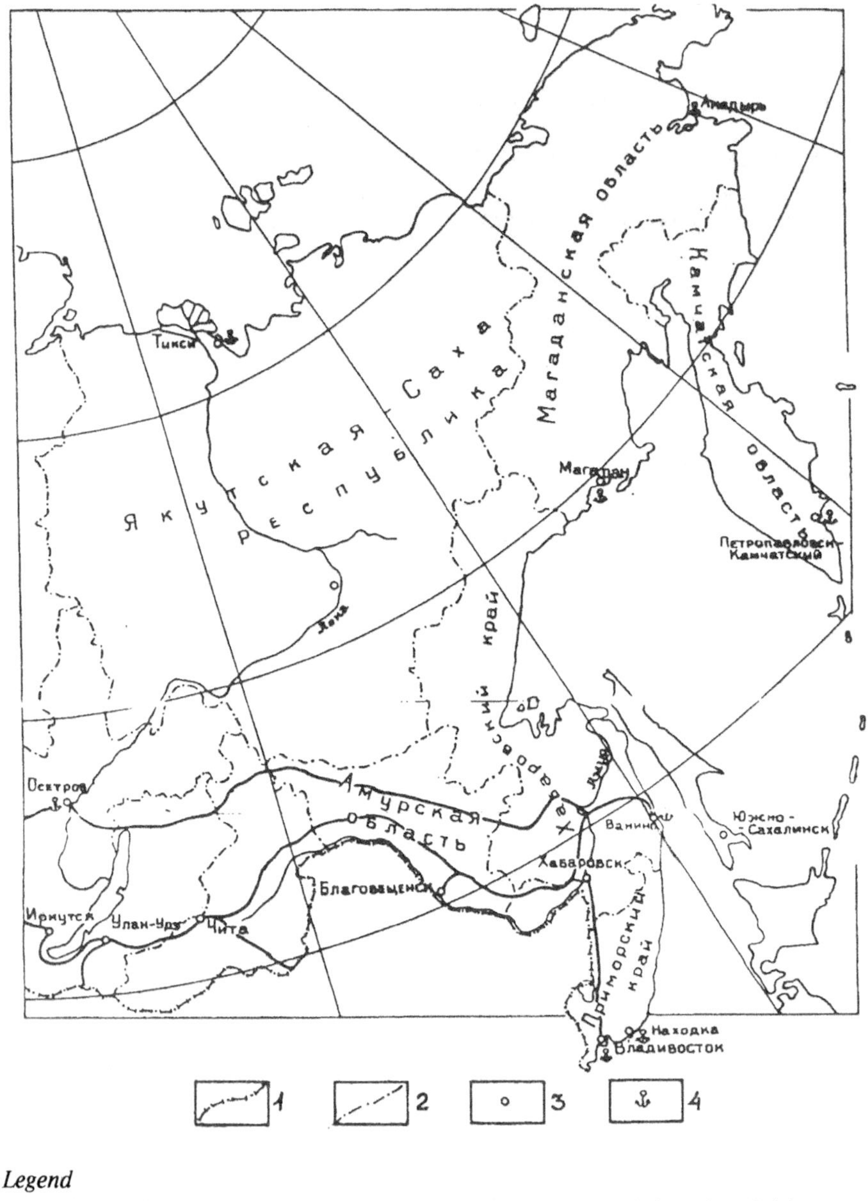

Legend

1. Border of the RSFSR.
2. Borders of administrative divisions.
3. Centers of administrative divisions.
4. Ports.

Figure 11.1. Russian Administrative Divisions of the Far East

European parts of Russia. The northern parts of the Khabarovsk Territory and the Magadan and Kamchatka provinces are accessible to shipping through the region's ports. The major type of transportation in the northern region is the automobile for which there are both winter roads and all-season highways. The most important motor route in the region connects Yakutsk with Magadan. For many towns, aircraft are the only link to the outside world.

Natural resources—timber, fish and minerals—are essential to the economy of all Far East territories. Lumber mills operate on a moderate scale, while fish catches are dwindling. Khabarovsk Territory's share of the national machine-building and metal-working gross output is 36 percent. However, these industries once served the military-industrial complex almost exclusively and now are undergoing conversion.

The mining industry of the FEER is one of the most important in Russia, as the area leads in the output of many natural resources. All of the brucite, nearly 100 percent of the diamonds, 98 percent of the tin, 90 percent of the boron, 80 percent of the fluorspar, over one-third of the antimony, nearly one-third of the tungsten, two-thirds of the polymetals and a considerable quantity of the precious metals produced in the Russian Federation are extracted in the region. Brown and bituminous coals, oil, natural gas and construction materials are also extracted in the Far East in large quantities. Yet the minerals and raw materials potential of the area suggests that a much higher level of productivity is possible. The reserves of many deposits remain untapped, although many of them are assessed as unique. There are numerous proved and explored deposits that remain idle.

New raw material complexes have been established for tin extraction in the Magadan Province of the Khabarovsk and the Primorye territories; for gold extraction in the Kamchatka Province; and for the development of commercial deposits of mercury in the Magadan Province, and iron and manganese ores in Khabarovsk Territory, Yakutiya and Amur Province. In addition, all types of materials necessary for metallurgy, including flux limestones, fire-resistant materials, fluorspar and others have been explored and evaluated. Raw material complexes are being developed for the phosphate-titanic (apatite-ilmenite) industry in Khabarovsk Territory and Amur Province and for unconventional types of raw materials (zeolites and perlites) in Primorye and Khabarovsk territories as well as in Amur and Sakhalin provinces. During the last few years, large quantities of germanium reserves in brown coal-bearing fields in Shkotovskoye and Rakovskoye and potentially commercial deposits of tantalum and niobium in the Voznesenski ore-bearing area have been identified and proved in the Primorye Territory. Geological and economic feasibility studies have been undertaken for raw material complexes for nickel in Kamchatka; for aluminum in Khabarovsk Territory; for copper in Yakutiya, Magadan Province, and Khabarovsk Territory; for molybdenum in Magadan Province and Khabarovsk Territory; as well as of new resource development complexes development for tin, tungsten, lead and zinc, oil and natural gas, coal, and other mining industries (Figure 11.2).

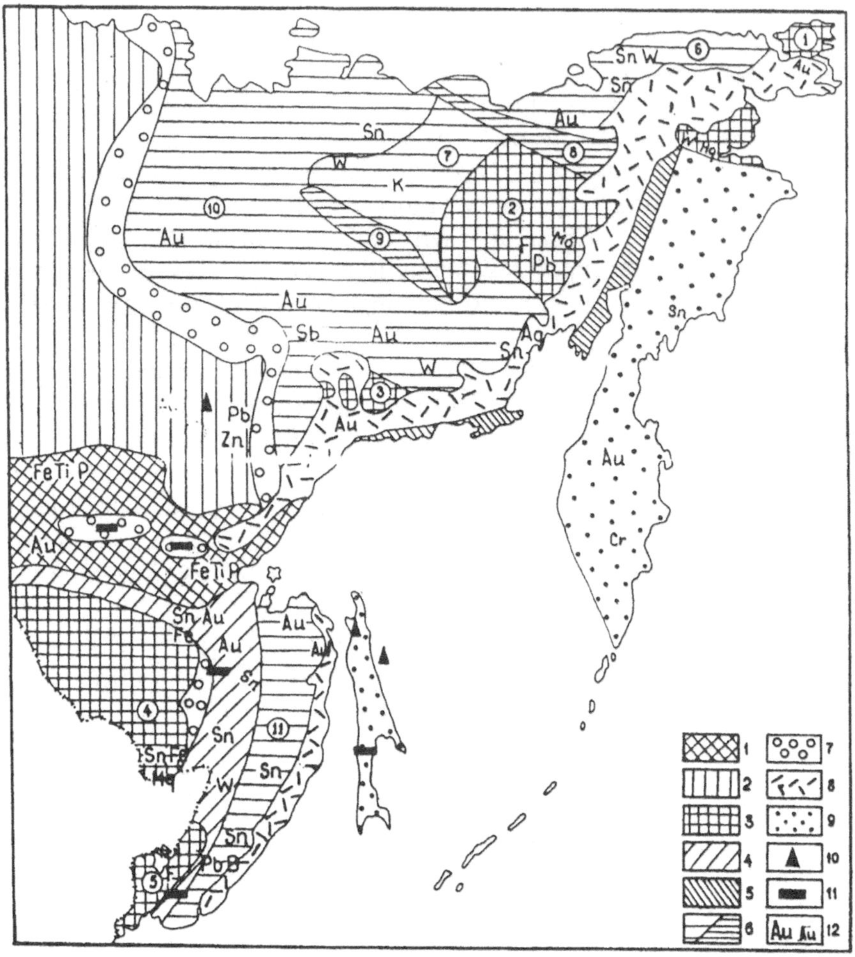

Figure 11.2. Mineral Resource Distribution Pattern of the Far East

Legend for Figure 11.2

1. Aldano-Stanovoi Shield.
2. Siberian platform.
3. Medium massifs: (1) Eskimosski.
   (2) Omolonski.
   (3) Okhotski.
   (4) Burejinski.
   (5) Khankàiski.
4. Amur-Okhotsk Paleozoic geosynclinal-folded system.
5. Early-Mesozoic geosynclinal-folded system.
6. Late-Mesozoic geosynclinal-folded system:
   (6) Chukotskaya.

(7) Alazeiskaya.
(8) Anyuiskaya.
(9) Ilin-Tasskaya.
(10) Verkhoyano-Kolymskaya.

7. Marginal and overlying troughs.
8. Marginal-mainland volcanic belt.
9. Cainozoic geosynclinal-folded system.
10–12. Major deposits of natural resources:
(10) Oil and gas.
(11) Bituminous coal.
(12) Symbols of major useful components in ore-bearing fields (ore-bearing regions are designated with large type symbols; other deposits are designated with small type symbols).

MINERAL AND RAW MATERIAL RESOURCES

Fuel and power-generating resources

Oil and natural gas. Oil reserves have been registered and explored at 40 deposits throughout the Far East; predicted reserves are estimated as being seven times greater than known reserves. In the future, the value of oil production may, at a minimum, triple. Growth in the region's oil industry can be stimulated by developing the proved fields in Yakutiya and Sakhalin and prospecting for new deposits.

Natural gas reserves have been registered in 53 fields, namely in Yakutiya and Sakhalin. The gas industry currently uses gas extracted from 20 fields, but its level of output does not meet the requirements of the area. The developed deposits contain enough gas for a doubling of output. Expected gas reserves are nearly twenty times the size of the developed reserves.

The development of oil and gas deposits in Yakutiya and Sakhalin requires large capital investments amounting to billions of rubles. In Khabarovsk Territory two oil refineries are basically dependent on crude supplies imported from other regions for feedstock. The economic future of the region will very much depend on the realization of the development potential of hydrocarbon deposits on the Far East mainland, including the Verkhnee-Burejinski basin, the Mid-Amurskaya and the Zeye-Burejinskaya depressions and the late-Mesozoic troughs of the Komsomolski region. The Tatar Straits offshore oil- and gas deposits may also contain commercially profitable resources; therefore, the oil and gas-bearing troughs in the region may become priority projects in new field prospecting.

Coal. About one hundred brown and bituminous coal deposits, which represent 55 and 45 percent of all reserves, respectively, have been explored in the Far East.

Nearly 60 coal-extracting enterprises (mines and stripping facilities) are in operation. The volume of extracted coals amounts to around 40 million tonnes per year, but the demand for coal in the area is much greater. Kamchatka and northern Khabarovsk Territory utilize coal delivered from other regions. There are plans to construct nine additional mines and sixteen stripping operations. Upon their completion the region's annual output of coal would double. Unfortunately, the implementation of these plans is being held back by a shortage of funds. Predicted reserves of coal within the FEER amount to 255 billion tonnes which is about twenty times the amount of the reserves in the developed deposits. The major deposits of bituminous coal (over 80% of the reserves) occur in Yakutiya, Khabarovsk Territory and Magadan Province.

In recent years, owing to a decline in coal output, coal shortages have caused acute problems for power generation in the southern areas of the FEER. A possible solution to the problems includes the immediate introduction of technological facilities that will expand the production capacities of the Erkovetskoye and Ogodzhinskoye coal-bearing fields in Amur Province and in the Burejinski basin in Khabarovsk Territory.

Ferrous metals

Russia's state mineral balance includes the iron-ore fields of Khabarovsk and Primorye Territories, and the Amur Province and Yakutiya; the manganese ore fields at Khingan in Khabarovsk Territory; and magnetite and brucite ore fields in Khabarovsk Territory. The predicted reserves of the FEER's ferrous metals are estimated at dozens of billions of tonnes, but they are concentrated mainly in regions that are difficult to access: Charo-Tokinski (Yakutiya) and Udsko-Selemdzhinski (Khabarovsk Territory). The iron-ore bearing fields are distributed more conveniently, such as in the Malo Khingan ridge near the Trans-Siberian railway; this deposit contains the Sutarskoye, Kimkanskoye and Kostenginskoye iron-ore fields. Total capacity of these iron-ore deposits is estimated at nearly one billion tonnes, or more precisely, 934 million tonnes together with the Garinski iron-ore field. The amount of ores in these fields is sufficient to meet regional metal demands for a number of years as well permit the production of concentrates for export.

The Kuldurski brucite-bearing field in the Malo Khingan ridge is a supplier of high-quality raw magnesium for ferrous metallurgy. The plans for the region envisage a significant increase in the output of brucite. Other fields have not yet been explored thoroughly because of an absence of users. The magnesium minerals of the Khingan Ridge also include magnesites in the Safonikhinskoye, Samarskoye and other deposits that are not yet being extracted.

A manganese-ore bearing field with a manganese-oxide content reaching 21 percent has been prospected in Yuzhno-Khinganskoye. The explored resources amount to 13.2 million tonnes.

Titanium. In northern Amur Province and in Khabarovsk Territory, the Kolaro-Dzhugdzhurskaya titanium-bearing zone has been identified. It is confined to an anorthosite massif belt that stretches in the sublattitudinal direction in the southern part of the Aldano-Stanovoy shield. Although it is estimated that the potential output of the zone may be enough to meet the world's demands for titanium, the deposits have not yet been thoroughly explored. The ores are complex and contain iron, vanadium, and apatite.

The complex apatite-titanium ore-bearing fields in Kolaro-Khaninski (Amur Province), Dzhugdzhurski and Baladekski regions together could support a raw material complex for titanium and a mineral fertilizer industry of world significance. Preliminary geological and economic feasibility studies of the mineral resources testify that the Bolshoi Seyim ore-bearing field (Kolaro-Khanski ore region) contains ilmenite and apatite concentrates, while the Maimakan and Gayum ore fields (in Dzhugdzhurski region) may be sufficient for obtaining ilmenite and apatite concentrates. With recent advances in processing technology for ilmenite-magnetite-vanadium-bearing sands in the placer deposits of Iturup Island (the Kurils), it may be possible to use the ilmenite and magnetite placers to obtain ore for steel, ferrovanadium and pigment titanium dioxide. Geological and economic feasibility studies of the ilmenite-magnetite-vanadium-bearing sands of Iturup Island have been promising.

Nonferrous and rare metals

Tin is the principal metal currently produced in the Far East. With few exceptions, tin mining enterprises in the region are operating at their full capacities. The Solnechny, Khinganski, Khrustalnenski, Pevekski, Iultinski, Komsomolski and Dukatski ore-dressing complexes and the Amalgamated Plant Dahlpolimetal will have available ores from prospected resources until the years 2000 and 2010, respectively, and perhaps another 10 or 20 years of predicted resources. Prospective tin reserves of the Pyrkakaiski stockworks and the Pravourmiyski and Tigrini ore-bearing fields are sufficient to support construction of a large and economically viable ore-dressing complex which could increase the output of tin concentrates in the Far East by 2.3 times.

A major problem of the tin extraction industry of the Far East is the use of primitive technology for ore processing which can only recover about 58 percent of the main components on average, while total utilization of capacity reaches 30 to 40 percent. The employment of more advanced technologies could provide for an increase of tin output by one-third without expanding the existing raw materials base.

Tungsten. The major enterprise producing tungsten concentrates in the Far East is the Primorye ore-dressing complex in Primorye Territory; the second important enterprise is the Iultinski ore-dressing complex in Magadan Province.

Tungsten concentrates are also extracted as by-products from complex ores at the Solnechni ore-dressing complex in Khabarovsk Territory and at a number of other enterprises in Yakutiya. Geological prospecting carried out in the zone of the operations of the ore-dressing complexes may ensure supplies of tungsten for another 25 to 30 years.

Lead, zinc and copper. The only enterprise producing these metals in the area is the Amalgamated Plant Dahlpolimetal that has been in operation since before the Russian Revolution. Explored reserves are sufficient to meet manufacturing requirements for 30 more years. There are also enterprises in the Komsomolski region and in Magadan Province (the Dukatskoye ore field) that produce these metals as by-products during the extraction and processing of ores.

Research and surveying work completed in the area during the past few years suggests that large ore-bearing deposits of commercial value may exist in a number of regions, including copper-porphyry-bearing fields in Khabarovsk Territory and the Amur and Magadan provinces; a pyrite-bearing massive sulfide field in Kamchatka; and a carbonate-hosted stratiform lead, zinc and copper-bearing field on the border between Yakutiya and Khabarovsk Territory.

Aluminum. Around 95 percent of the world stock of aluminum is produced from bauxite imported primarily from equatorial zones. This is also the case with the new Commonwealth of Independent States (CIS).

In Gyandzha, Azerbaijan, for example, both bauxite imported from Guinea and alunites extracted in a nearby field are being processed at the same integrated plant. The manufacturing cost of alumina produced from bauxite amounts to 160 rubles per tonne, while the cost of alumina produced from alunites amounts to 120 rubles per tonne, with a disbursing price of 165 rubles. During the processing of alunites, potassium sulfate (manufacturing cost, 150 rubles; disbursing price, 180 rubles per tonne) and sulfuric acid (38 and 48 rubles, respectively) are produced. The Far East's reserves of alunites are estimated at 3.3 billion tonnes. Among the largest deposits are the Shelekhovskoye (500 million tonnes), Iskinskoye (336 million tonnes) and Krugly-Kamen (208 million tonnes).

The development of alunite-bearing fields may also support the production of potassium fertilizers, sulfuric acid and coagulants for purifying sewage.

Precious metals. Nearly 50 percent of Russia's gold, 20 percent of its silver and a considerable share of its platinum group metals are extracted in the Far East. The level of extraction of these metals is expected to grow at a stable rate, provided that new primary deposits are explored on a regular basis as the operating placer deposits become depleted. Every administrative area of the FEER has large deposits of gold; however, their development is being held back by a lack of funds. Taking this into consideration, authorities are granting licenses to foreign companies for the development of a number of gold deposits in Khabarovsk Territory, and in Amur, Kamchatka and Magadan provinces.

Agrochemical raw materials

The local agrochemical raw materials resource base includes phosphate minerals (represented by apatites, phosphorites and ground phosphate-carbonate), potassium (represented by alunite), carbonaceous raw materials (limestone, dolomite and cockleshells), silicate raw materials (zeolites, pyrites, bentonites, vermiculites, diatomites and opokas), and organic raw materials (peat).

All agrochemical raw materials occur in the FEER in sufficient quantities, but only some of them are being extracted and not to the full capacities of the deposits. In Amur Province, small amounts of ground phosphate limestone are extracted. The perlites extracted in the southern regions are used only in the construction industry. Carbonates are used in agriculture for acidification of soils; demands are completely satisfied by the reserves existing in the area. Raw materials used for animal feed are brought in from other areas, although there is a prospective deposit at Vyazemskoye in the FEER.

Phosphate fertilizers are brought into the region from the western part of Russia, but these deliveries constitute only 30 to 40 percent of what is necessary. The case is similar for potassium fertilizers. The demands for phosphate and potassium fertilizers are expected to remain at around 2,000 and 710,000 tonnes, respectively, until the year 2000.

The Khabarovsk Territory alone requires 630,000 tonnes of zeolites each year. Annual demand for zeolites in the entire Far East is estimated at not less than 2.5 million tonnes.

In general, the Far East possesses sufficient mineral raw materials to nearly satisfy the needs of its agricultural industry. Shortages of phosphate and potassium fertilizers must however be addressed given the area's high demand for the materials. Further activities must be initiated to find additional sources. All other agrochemical raw materials are locally produced in abundance.

MINERAL RESOURCES UTILIZATION IN THE FAR EAST

At present, the Far East has an inefficient economic structure with undeveloped basic and social infrastructures. The mining industries are characterized by outdated processing technologies and a virtual absence of metallurgical processing of nonferrous metal ores. With the economy centered around prospecting and mining, the FEER forgoes significant potential revenues from the added value created by the processing of commodities. As a result, the level of profitability of the mining sector is minimized while costs are relatively high.

Prices of raw materials in the Far East are generally in the form of accounting prices that do not always cover production costs. While sales taxes are collected and paid at the site of the production of a commodity, they are imposed on an optional basis. As such, major mineral products of the FEER remain untaxed in most cases. The situation is aggravated by a state order which prohibits internal or external barter activities.

Owing to decades of inefficient central planning, resources are distributed in an irrational manner among state-run and locally-run enterprises in the FEER. Clearly, even under the most favorable scenario, the Far East cannot survive without foreign ties, as nearly 50 percent of its gross social product is based on import-export turnover.

The establishment of links with other areas in the Commonwealth and elsewhere will be based on the natural resources sector; therefore, after the introduction of a commodity exchange, the state's exclusive right to use natural resources will need radical revision. It is critical that the existing mineral and raw materials potential of the Far East be used not only for the economic benefit of the entire country, but also to the advantage of the area.

The past policy of the Soviet Union regarding the monopolization of mineral resource extraction and the creation of a monopolistic economy in the Far East, as well as the remittance to Moscow of the area's income earned from natural resources, resulted in an economic and social crisis and a reduction of mineral reserves within the FEER. There are numerous conditions endemic to the mining industry of the FEER that adversely affect the economic potential of the area and its environmental conditions. The main problems are as follows:

- The development of mineral deposits in the Far East lags behind the demand of the area due mainly to a low level of extraction output. As has already been discussed, expansion of the region's mineral and raw materials base would help alleviate this problem, particularly if output of tin, polymetals, tungsten, coal, oil and gas were increased.
- Although the raw materials base is reliable and generally sufficient to satisfy the needs of the area and to promote commercial and economic ties with the Asia-Pacific region, there is no systematic realization of its potential.
- Wastage of mineral materials is severe in the extraction and processing sectors of the Far East. Operational losses often amount to over 20 percent, while working losses average around 50 percent. The level of all-around utilization of nonferrous and ferrous metal ores hardly reaches 20 percent (while the value of important by-products often reaches the value of the basic component).
- Geological prospecting within the Far East, especially detailed surveying studies, also lags behind in present-day requirements. Only about 20 percent of the area has been geologically surveyed in detail.
- Although the convenient geographical location of the area could facilitate establishing international economic links with the Pacific Basin countries, it is often ignored and considered a problem rather than an opportunity.
- Raw minerals are not utilized efficiently in the Far East as mining and processing equipment are not manufactured in the area.

Solutions to these problems would aid progress in the entire mineral and raw materials production sector of the Far East; but this would certainly require much more capital than Russia has at its disposal. Important options would therefore involve searching for foreign partners who may be interested in the rarity and high quality of Far East minerals; establishing nominal delivery costs; developing stable and long-term economic ties; building up profitable enterprises; and many other factors. A great variety of minerals occur in the Far East in numerous geographic and geologic settings. Many minerals can be conveniently marketed within Russia. Under present conditions, the most suitable form of economic cooperation is through joint ventures, but a foreign partner can acquire 100 percent ownership of an enterprise, though production sharing is often preferred.

The proved oil and gas fields of Sakhalin and Yakutiya need capital investments of several millions of rubles. Therefore, despite the general shortage of energy supplies in the area, it is unlikely that these fields will be developed in the short term. In any case, their development should be studied more thoroughly. A more probable strategy, although it is not without certain risks, is the financing of oil and gas prospecting operations during more favorable economic conditions.

Exploration of the bituminous coal deposits in Yakutiya and Khabarovsk Territory also involves considerable capital investments to compensate for the remoteness of the resources from transportation systems (Yakutiya), for costly underground mining operations, and the preliminary dressing to diminish ash content (the Burejinski basin in Khabarovsk Territory).

Gold-ore and gold-silver deposits in the Far East require investments of 500 to 700 million dollars, depending on their sizes, while payback is expected in four to eight years. Capital investments of approximately the same magnitude are necessary for carrying out the exploration studies of large tin-ore deposits, including Pravourmiyskoye, Sobolinoye, Tigrinoye, and Pyrkakaiskoye.

The convenient location of iron-ore deposits in the Malo Khingan ridge suggests that open-cast mining as well as the establishment of a dredging factory for pellets production may be economically viable. Annual production of pellets can be increased from three to twelve tonnes. In the Malo Khingan ridge it is also possible to establish an integrated plant for producing magnesia materials of high purity and suitable for various uses, with brucite used as the technological raw material.

Despite the present unfavorable state of the world titanium market, it is recommended that an analysis be conducted of the economic feasibility of the development of complex deposits in Amur Province and Khabarovsk Territory that contain not only ilmenites, but also apatites and vanadium. The geologic variety of the apatite-titanium-magnetite ores, confined to layered intrusions of anorthosites, suggests the existence of considerable amounts of gold and platinum intrusions in the near-bed parts.

Of special interest to the Far East is the adoption of efficient mining technologies and the processing of tailings in those enterprises where complex ores are processed, notably the Solnechny ore-dressing complex (tin, tungsten, lead, zinc, copper and silver) and the Yaroslavski ore-dressing complex (fluorite, lithium, beryllium, cesium and rubidium). Technology used at the Solnechny complex—containing perhaps the world's largest tin deposit—is from the 1950s.

Economic reform to expand minerals output

To maintain production and promote expansion of the mining industries in the Far East, it is necessary to dramatically increase capital expenditures and introduce new technologies and operational procedures suitable to the characteristics of Far East mineral resources. It is also necessary to further consider the growing role and importance of infrastructure in the area and increasing costs related to environmental problems arising from mineral resources development.

Economic development of the Far East should be based on the natural resources base, particularly mineral and raw material resources, and on the economic and geographic position of the area. The development of the industries responsible for extraction and processing of minerals and raw materials should be coordinated with economic measures that take into account domestic and foreign customers, as well as the environmental and economic restrictions adopted in the region. Greater emphasis should be placed on the maximization of profits to investors developing mineral and raw material bases in the area and on establishing appropriate payment schedules for using the resources.

It is necessary to promote the development of the raw materials sector and modernize it using direct and indirect financial contributions from Russia's interregional and state funds. The support of this sector should be given a high priority in Russia's Far East policy. Its rational development will provide future stability for the area's economy and the implementation of joint ventures and economic cooperation projects, particularly with Asia-Pacific countries.

Economic relations of the Far East with the Russian Federation government should be based on taxation. Tax payments should be remitted from the local to the republican level and minerals production in the area should be allowed to realize as big a profit as free markets would allow at the time. The republic must provide funds for the most important priorities set by state policy. It is also necessary to use other financial resources, including untapped funds in the area, entrepreneurial capital, borrowed funds, and foreign credits.

It is necessary to reorient the needs of the Far East minerals industry in conjunction with the transition from a centrally planned to a market economy. Mineral utilization policy should address: preparation of the mineral and raw materials base 10–15 years prior to its industrial exploitation; scientifically based methods; environmentally sound utilization of natural resources; the introduction of new and unconventional sources of raw materials into the industrial operations; the relocation of mineral extracting and processing enterprises to outside major populated centers; and the use of world market prices. Moreover, the importance and the benefits of integration in the world minerals and raw materials market to the Far East should be considered. It is likewise necessary to reevaluate the quality of all the deposits in the area in light of newer methods of extracting and processing and concerns for environmental protection. The economic reassessment of the FEER's mineral base should be done taking into account world and local market prices, which will make it possible to appropriately recalculate the reserve sector of the base.

The exclusive rights of the state mining sector to prospect and extract minerals in the Far East should be abolished. All natural resources explored using state funds should be included in a single republican-territorial balance that considers licensing rights to explore using a differentiation of payments schedule according to the significance of the deposits to the territory or republic. Enterprises based in the FEER as well as foreign firms must have equal rights to prospect and extract natural resources, with the exception of certain strategic types of raw materials. Mining should be effected by granting the deposits on lease. The lease terms should include provisions for land reclamation. The process of leasing deposits should ensure that damaged lands will be restored. The rehabilitation of lands should last for two to five years.

A greater emphasis on the import of foreign advanced technologies and the introduction of foreign investments to the region's economy will make it possible to create a network of joint ventures and mining companies with sufficient capital for prospecting and extracting natural resources. The introduction of a payment scheme for using the resources and a differentiation of rates depending on the significance of the deposits to the republic or territory should help indicate the investment levels required for individual mineral commodities. The following distribution of investments into mining appears to be adequate until the year 2015: the republican budget, 45 percent; local budget, 30 percent; the funds of the consuming firms, 20 percent; and the internal budgets and credits of the geological and prospecting organizations and mining enterprises, five percent. Of course, the investment balance will depend on the pace of the transition of the Russian and Far East economies to the free market, on the extent of privatization, and on the amount of foreign capital invested.

The creation of large and small prospecting and mining enterprises with various forms of ownership will facilitate the formation of a prospecting and mineral mining market in 1993 and beyond. The contractor (enterprise) will draw up one- and five-year production programs independently, on the basis of contracts with customers or in terms of operation risk. It is necessary to introduce a system of economic benefits and incentives for developing and mining deposits in the area and for complete extraction of basic and secondary components of mineral raw materials at presently operating facilities.

The existing information base on the mineral and raw materials industry requires dramatic improvement. In a free-market environment, all information about the resources and the technical-economic characteristics of the geological and mining sectors must be open. A two-level republic-wide and territorial-based information system should be established which would sharply increase the effectiveness and efficiency of negotiations with foreign firms.

Specialized services are being created in the Far East to promote the rebuilding of the raw minerals base. These services are to provide for a coordinated policy in the mineral and raw materials industries; assist in the analysis of all mineral markets and exploration and mining activities; support in the evaluation of local ecological conditions; and promote the effective use of prices. They are to define and coordinate the priorities of technological policies set forth in the Far East area, including the computerization of manufacturing processes; the purchasing and introduction of updated technologies for carrying out more complete extraction of valuable components from mines; the processing of greater amounts of reserves in structurally complex deposits of gold, tin, and other minerals; and the utilization of unconventional types of raw materials.

Chapter 12

THE RUSSIAN FAR EAST: ECONOMIC CONDITIONS AND PROSPECTS FOR COOPERATION IN NORTHEAST ASIA

Pavel A. Minakir

INTRODUCTION AND BACKGROUND

The decline of the economy of the Russian Far East began in 1991. The slump in winter output was somewhat mitigated by subsequent price increases. Under the pressure of trade unions' and strikers' demands, the government finally attempted to match price increases with wages. By the beginning of 1991, the prices of commodities and services in the Far East had risen 190 to 210 percent, paralleling the growth in the population's monetary income. The level of credit granted in the region was even higher—270 to 300 percent. This, in turn, triggered a process of latent hyperinflation in the middle of 1991. The rapid rise of prices occurred at the same time as production fell steadily. The rising cost of products in virtually the entire industrial sector of the Far East (with the exception of fisheries and nonferrous metallurgy) was accompanied by a decrease in the availability of manufactured goods.

A disintegration of the consumer market followed, as there were deficiencies in essentially all consumer goods. The rise in personal income allowed consumers to purchase all goods offered for sale, but the decreased output and high inflationary expectations stimulated a steady growth in the buyers' demands.

Still more serious were the aftereffects of inflation in the capital goods market. Here the prices rose even higher than on the consumer market: prices rose by factors of 2.1 and 2.3 over the price increases in the consumer markets in Russia and the Far East, respectively. These factors reflected an alarming drop in the competitiveness of the Far East industries in the Russian market. Nonetheless, however, the region did manage to take advantage of inflation in 1991 by stimulating competition by year's-end.

J.P. Dorian et al. (eds.), CIS Energy and Minerals Development, 177–187.

Price liberalization

A sharp rise in prices on January 2, 1992 spurred further competition for the consumer. The competition was mainly centered around industrial and technical production, where importance is attached to a comparative price competitiveness. Competition has always been limited in the Far East, but in 1992 it had become an obstacle to sustaining current production output levels. The consistent rise in prices throughout the region was essentially the result of heightened production and transportation costs. At present, these higher costs are not being subsidized by an income redistribution policy of the central government. The products become noncompetitive as the consumer chooses cheaper products and not just those which are affordable with the funds received from the government. Besides, the financial strength of the enterprises in the Far East is considerably less than in other regions of the country.

After the aborted August 1991 coup, the disruption of economic ties due to the disintegration of the Soviet Union became a dominant factor in Russia's production crisis. As early as 1991, enterprises in the Far East experienced difficulties in concluding contracts involving deliveries of materials and construction items. At the beginning of the year the enterprises could conclude only 15 to 20 percent of all necessary contracts. Later, the situation became stabilized at around 80–85 percent.

In November 1991, the promulgated policy of price liberalization in Russia caused panic among the consumers, and they literally ignored the legalized market. As soon as the government announced its intentions, the black market became very active in raising prices, capitalizing on the fact that the population was prepared for the price hikes and ready to spend its savings. Meanwhile, manufacturers began stockpiling their ready-made products, expecting the price rise and hoping to earn excess profits. This led to a severe deterioration of the economy in November–December 1991. The official average cost accounts for the Far East economy in 1991 do not look extremely alarming; they show a six to seven percent decline in industrial output volume, although in per unit terms declines are much greater.

Enterprise profits

The profits of industrial enterprises increased during that period from 12 kopecks per ruble of sales to 31 kopecks. The rise in incomes, on the one hand, reflects a rise in prices, while on the other hand, it encourages enterprises to continue reducing output. The rise in prices in January 1991 once again confirmed that the strategy pursued by the enterprises was both efficient and popular.

Enterprise profits in 1991 grew at a faster rate than price increases. The former grew by 168 percent (108 percent in Russia), while the latter rose by 117 percent (105 percent in Russia). This is evidence of the fact that the enterprises take advantage of their monopolistic position in the market and increase the prices at a higher rate than production costs grow. On the whole, 1991 witnessed a dramatic change in profit trends. Since 1988, profits were steadily decreasing in the Russian industrial sector due to expropriatory taxation and profit restrictions. In the Far East, the situation was somewhat different, however. In 1989 and 1990, the earned profits diminished when compared to output levels of preceding years; earned profits reached eight percent in 1989, but only 0.4 percent in 1990. In 1991, though, they jumped to 168 percent. This led to a dramatic increase in liquidity among the Far East enterprises, but failed to stimulate new production or inventory gains.

In fact, the increase prompted an acceleration in the inflation rate while profits were slimmed by a rise in prices. Enterprises did not take measures to lower production costs at this time. That appeared to be one of the major factors that hindered a dramatic growth in unemployment. It was at the beginning of 1991 that a great number of jobless were expected based on problems of the previous year. These expectations were not realized due to the "wait-and-see" policy of the Far East enterprises that also resulted in a terrible shock during the first days of January 1991 as the financial situation took a radical turn.

The decline in production did not affect all branches of industry in the same manner. As a result, the structure of the Far East industrial sector changed considerably. The percentage shares of value of the machine-building and forestry industries dropped, while that of nonferrous metallurgy rose. The machine-building industry was devastated by ongoing reductions in military orders since 1989. Military machine building constitutes more than one-third of the machine-building sector in the Far East. The reduction in orders for defense products affected nearly all plants, averaging out to a 20 percent cut. Before the beginning of 1991, there was still some hope that military orders would be retained at some of the enterprises that produce expensive and high quality equipment, but these hopes were quickly dashed. Russia's defense budget has been cut and production reduced at all defense enterprises without exception. Since 1991, defense plants began to undertake efforts to manufacture civilian products, but the outcome has not been very encouraging, given the cuts in budgets and heavy expenses incurred for maintaining social infrastructure. Defense enterprises planned to increase output of civilian products in 1991 by over 20 percent, but even that increment would hardly compensate for the drop in defense production volumes.

As for the forestry industry, a steady decline in production was observed beginning in 1986. For the first time in more than a decade, the industry reached a critical stage in 1990, when the volume of wood deliveries was less than 30 million cubic meters.

Indeed, all Far East industries, except electric power and canned fish, suffered declines in per unit output since 1990. The best year overall for most products according to output indexes was 1988. During the following three years production of coal decreased by nearly 20 percent, steel by 25 percent, and lumber by 32 percent. Production of cement between 1988 and 1991 fell by 10 percent and fish by 11 percent. Nonferrous metallurgy is, perhaps, the only major industry that has not suffered, but rather increased output. To a considerable degree, this was the result of enhanced foreign economic relations in this area, particularly with the permission granted to enterprises, including those extracting precious and rare metals, to conduct foreign economic activities on their own, though within certain limits. Local authorities are today particularly interested in this branch of industry as the region now receives a certain contribution to its budget for each gram of extracted metal, for example, gold. The nonferrous metallurgy industry cannot, however, remedy the overall deteriorating economic situation of the region.

In recent years, particularly severe problems have developed within the foodstuff and commodity markets. Despite a near tripling of consumer prices (nearly 100% in 1991 alone), output value continues to fall. In current prices, the volume of production was only 100.1 percent for the first 10 months of 1991 as compared with 1990. As a result, the commodity market continued to falter in 1991 throughout the Far East region, where the population spent some six to 20 percent less of its income on goods or services.

For a number of years, great hopes were pinned on an "alternative economy" in the Far East. It was believed that changes in the form of ownership and the development of entrepreneurship would reduce monopoly control in the economy and compensate for the stagnation of the state sector. In reality, however, this was not and is not the case. The value of the nonstate sector in the Far East is still insignificant. It is largest in Khabarovsk Territory, although it does not exceed six percent there. All operations performed by nonstate industrial enterprises constitute only 0.7 percent of the volume of the entire region's industrial output. It should be noted that a great number of former state organizations are considered now to be nonstate enterprises as they are registered under various types of associations, concerns, etc. In actuality, however, they cannot be considered part of the "alternative economy" as they are part of the former state monopolistic structure with state-owned property just declared or renamed corporate.

Conditions for the creation of alternative economic structures were never favorable in 1991. With an absence of guarantees of economic stability, money was invested exclusively in the trading and financial spheres. The rapidly increasing inflation created conditions conducive to this. About 20 financial exchanges of various types and a great number of commercial banks have been created in the Far East. In the Khabarovsk Territory alone there are nearly 30 commercial banks. Many of them flourished following the deterioration of the former huge state-run banks and their regional affiliates (for example, the banks created on the basis of the territorial divisions of the Soviet Union Industrial Construction Bank, Promstroibank).

Rapid development of a commercial and financial infrastructure during a period of stagnation and declining production was also conducive to inflation. The initial rapid rise of stock market and bank values as well as prices of brokers' posts collapsed by the end of 1991. Because of a lack of goods, the exchanges trade basically in brokers' posts. Supplies have already exceeded effective demands, and expectations concerning investments in broker offices have not been fulfilled.

FIRST RESULTS OF THE SHOCK REFORM

January 1992 witnessed a great change in the economy of Russia and the Far East. The policy of price liberalization promulgated as early as November 1991 led to unexpected consequences not only for the population but also for entrepreneurs and the government. Although the unpredictability of the consequences was foreseen by analysts, a complete understanding of all possible impacts became clear only two or three months after January 2, 1992.

In actuality, the shock was not the result of price liberalization alone. The price rise and the cessation of any control over prices on the part of the government was only the first step in the chain of events triggered on the night between the first and the second of December 1991. Two most important events occurred on that night: first, a liberalization of prices of most commodities and services; and second, the abolition of budget subsidies on most commodities and virtually all foodstuffs.

The liberalization of prices began with governmental instructions prescribing their rise. True liberalization of prices suggests a competition among manufacturers in Russia; however, this is not the case as the existing manufacturing sectors have been monopolistic for decades and today still have no alternative. What really happened was a paramount and disastrous skyrocketing of prices with the reference level legalized by the government. New prices had been established for all goods and services, and overnight they became two to three times more expensive. Half of all income earned from the one-time price rise was added to circulating trading and manufacturing funds, while the other half went to central and local budgets.

In the Far East, prices increased on average nine-fold to eleven-fold during the period from January to February 1992. However, wages in January remained equal to those in December 1991, i.e., around 750 rubles per month, while the costs of living escalated to 1,500 to 2,500 rubles per month. For some time local authorities managed, by measures of compulsion, to keep the prices of foodstuffs within the bounds declared by the government, which was trying simultaneously to help citizens survive until their wages rose. Nevertheless, even with prices under control, the processing (dairy and meat-packing plants) and agricultural enterprises (especially cattle and poultry breeding complexes) found themselves in a desperate situation. Demand for their production fell by several times. In addition, concurrently with the price rise, the government stopped granting subsidies for most products, foodstuffs included. Formerly, low prices had always been facilitated by government subsidies. This time, it was the historical subsidies that led to skyrocketing prices.

The shock from higher prices was so severe that consumer demand fell dramatically, particularly for foodstuffs. Conditions that previously allowed enterprises processing agricultural goods to sustain full capacity deteriorated. Meat-packing and dairy plants were on the verge of a production standstill.

Analogous problems appeared in the manufacturing sector. A sharp rise in prices for metals and construction materials and a subsequent rise in prices for construction services (nearly 12-fold, at the requests of the construction firms) virtually prevented the signing of construction contracts for the year 1992. In fact, because customers had limited capital resources, housing and urban construction was also brought to a halt.

The price rise, naturally, was not bounded by new values specified in governmental decrees. During the first two months of 1992, the general price index for industry was 936 percent for the region. Demands for industrial and technical merchandise dropped but could not disappear completely without a full cessation of production activity. The debts owed by one enterprise to another began to increase, and by the end of March the total debt in the region neared 70 billion rubles (thus growing 12-fold in three months).

A most important and harmful side effect of the price rise on economic development was that it helped to curtail productivity. Inflation and its unpredictability leads enterprises to insure themselves and inflate prices for fear of future devaluation of their assets. In December 1991, for example, expectation of inflation made enterprises increase their stocks of goods which sharply decreased their liquidity of assets and stimulated demand for credit. In the second half of March 1992, inflation increased again in connection with the liberalization of prices of energy resources. And again, a reduction in goods for sale led to an accumulation of goods in warehouses.

As a result of the disruption of horizontal ties between industrial enterprises in the Far East, the disintegration of the centralized system of supplies, a considerable narrowing of the domestic market due to the price rise, and a decline in the effective demands since January 1992, the decline in production assumed menacing proportions. The drop in January's production value in the Far East was around 17 percent, and in February, 14 percent, while in March it rose again due to a crisis in solvency. The abrupt price rise and the decrease in effective demands did not encourage the enterprises to independently adjust their prices. Increased prices allow for higher internal wages, as well as the opportunity to become insured against unpredictable price policies of suppliers.

Another instrument that the government decided to employ to solve budget problems was a strengthening of the taxation system and a toughening of credit terms. The imposition of a value-added tax concurrently with the sharp rise in prices clearly was a fiscal measure designed to provide increased tax receipts to the budget. In accordance with the legislation, 100 percent of value-added tax returns should, after the first of April 1992, go to the federal budget. Only profit taxes and income tax returns will go to the local budget. The general rate of taxation was dramatically increased: the value-added tax rose by 28 percent (included in the price of the product), the profit tax by 32 percent (for the manufacturers of products or services, while for banks, exchanges and trading-agential enterprises this tax was hiked by 45 percent), and the taxes for social insurance and the employment fund went up by 37 percent (calculated from the paid wages sum).

Such taxes, in many cases, make investment of capital into industry senseless. Equally senseless would be the investment of capital in trading and financial operations. The total trading premium added to the selling price of a commodity is fixed at 25 percent; it does not depend on the number of sellers. Properly implemented taxation controls make the existence of exchanges and trading organizations hardly necessary. These circumstances, together with the new law on exchanges, led to a hasty conversion of numerous exchanges into trading houses.

In reality, the Russian Far East taxation system as it is structured does not stimulate productivity, but retards the production process which may ultimately lead to even further curtailing of production, in addition to a larger number of financial problems for the enterprises. It is very unlikely that this would be compensated for by a decrease in the budget deficit. Furthermore, because of the reduction of the tax base, the reduction of the budget deficit itself will become improbable in light of the production decline and bankruptcies in the financial sphere.

The curtailing of production theoretically corresponds to a general scheme of anti-inflationary measures, but it does not take into account the nature of the Russian economy, i.e., that the shortage of goods is one of the major reasons for the inflationary pressure. Under such conditions production cuts can only result in intensification of inflation.

In January 1992, in classical anti-inflationary policy tradition, a sharp rise in interest rates on bank credit was supported. The Central Bank of Russia began to sell credit at a 20 percent per annum interest rate which resulted in an increase of the bank credit interest rate to 25 percent and more. By March 1992, commercial banks were already offering credit at 40 percent and higher.

As a rule, banks operating in the Far East do not have large monetary resources. In Khabarovsk Territory, for example, the average share of commercial banks' own credit resources is around 60 percent. An increase in credit costs has created instability in the commercial banks. In addition, a crisis of solvency of the banks' clientele jeopardizes the banks' own existence. Uncertainties in legislation and conditions under which entrepreneurship is supposed to develop have made banks practically cease giving out long-term credits in the Far East. As a result, investments are being curtailed.

The rise in bank credit costs has put both entrepreneurs and the state-run sector in even more difficult positions. An abrupt slowdown of turnover owing to price increases and financial difficulties of buyers and an absence of sufficient liquid savings have made enterprises increase their demands for credit. But credits are generally not available at competitive rates because of the high value of money. Many enterprises nonetheless have to obtain credits and, thus, run into more debt problems.

In practice, the present government policy leads to industrial and commercial bankruptcies. This may lead to a rapid growth of unemployment, though it is still being held down by the rise in prices on products; accumulation of property under the major financial groups that have sufficient financial reserves to buy up the assets of insolvent enterprises and banks; a rise in speculation on the commodity market due to the further increases in commodity shortages; and, consequently, the rise in prices, another devaluation of the ruble, a slump in the ruble exchange rate with respect to the major currencies, and another inflationary spiral.

The financial system is virtually in a state of paralysis. Every commercial entity has debts. In fact, the number of potential bankruptcies at present nears the number of enterprises and firms in the region. To disentangle the knot of mutual debts it is necessary to establish an interbank debtor registry with all subsequent crediting based on overall balances. Two obstacles hinder the Russian Central Bank from implementing this system. First, it is difficult to accomplish physically, as a universal information network comprising all the banks is necessary. The absence of such a network makes it very time-consuming to attempt to establish such a system. Second, the Central Bank is frightened by the necessity of giving out additional credits to those registered. Further credits may require billions of additional rubles which could generate a new wave of cash and, consequently, create a subsequent inflationary spiral. On the other hand, bankruptcies are not controlled by legislation. Therefore, beginning a process involving an interbank debtor registry could plunge the Far East into further chaos. Thus bankruptcies may lead to exactly the opposite of what the government expects.

Mass unemployment, if it were to occur in the Russian Far East, could quickly disturb a very delicate civil peace and plunge the country into widespread disorder, which would finally result in complete disintegration of the economy. The concentration of industry in the hands of a few financial groups could be a positive factor only in the case where the new owners are competent in formulating rational market strategies and have access to financial support, which is indispensable for modernization and reorganization of assets. A question therefore arises as to whether the country has enough money to execute the transfer of the property from one owner to another. Even at a time when assets are devalued, large sums of money will be necessary. The question is: who has the money, and how much?

Privatization and development of commercial enterprises—as alternatives to governmental organizations—have not yet brought about any palpable results. First, privatization is a time-consuming process. Second, in many cases it is being carried out under direct (or covert) control of former managerial enterprises, i.e., with former owners "ex officio" becoming legal owners. This fact alone makes the privatization essentially just a matter of changing signs and courses of profit distribution. Third, human psychology changes much more slowly than the forms of ownership. In order to change ownership only a legal certificate or decision is necessary; yet it takes a long time for the new owner to act differently in the market.

Privatization is proceeding, although slowly. So far, however, no results are evident because even after some privatization, the production sector of the country will remain monopolistic in its overall structure. President Yeltsin's decree on anti-monopolistic policy specified regulatory measures to be imposed on the monopolists. These measures include a compulsory product distribution scheme and price controls—in short, they simply amount to a distribution mechanism as had existed in earlier times. But the monopolists are extremely numerous, and a strict observance of the anti-monopolistic legislation will hinder reform in a large sector of the economy and bring back former methods of controlling economic activities. In addition, regulation measures can be applied to the monopolists only after consumers seek change through the anti-monopolistic committee. Thus, the consumer finds himself in an awkward position, facing a choice that will either make his life easier in the future or will require him to prepare for a most critical situation caused by the retaliatory sanctions taken by the monopolist against him.

The decrease in the exchange value of the ruble, due to mass bankruptcies and economic shock in the domestic market, makes it imperative that the country open up to foreign investments. The flow of foreign capital, theoretically, is the most likely solution to Russia's industrial problems. However, two questions arise here. First, do we really want our economy to be transferred under the aegis of foreign capital. Second, will foreigners want to buy up our enterprises at a fair dollar exchange rate, and if not, then on what terms. The second question is easier to answer than the first. The foreign investor will desire political and economic stability and a guarantee of ownership and, particularly, laws on mortgage rights and the legalization of private property.

PACIFIC BASIN COOPERATION

An increasingly popular option to assuage the severity of the current crisis in the Far East is seen as the establishment of a self-governing economic region, which would engage in close cooperation with neighboring Pacific Basin countries, particularly Northeast Asian countries. Since 1990 some steps in this direction have already been taken. Although Moscow does not want to relinquish the right to dispose of the currency and natural resources of the Far East, in July 1991, under the pressure of local authorities, the Russian Federation government issued Special Decree No. 815, which granted the enterprises of the Far East the right to market 30 percent of their production output in both domestic and foreign markets. In addition, exports can also be licensed by local authorities, while all the foreign currency revenues will remain with the enterprises (60%) and the region (40%). However, by the end of December 1991 two other decrees issued by the Russian president had come out and reversed the situation somewhat: one decree concerned the currency adjustment policy, while the other dealt with the new rules for granting licenses, defining quotas and establishing customs duties. At the beginning of January 1992 the Customs Committee of Russia banned the passage across the borders of all goods licensed in accordance with the rules specified in the previous decrees. Thus, all Far East export cargoes were stopped at the border.

With great difficulty, a special permit to let the cargoes pass on until the first of February was granted by the Customs Committee. Simultaneously, the Russian government promised to prolong the validity of the earlier Decree No. 815. Today, however, the situation remains uncertain. If the Far East does not regain the privileges granted to it in mid-1991, its foreign economic activity will deteriorate substantially. There is a great risk that the investment agreements already prepared by foreign firms in accordance with the 1991 rules of taxation and currency regulations may never be concluded.

In fact, regulations applying to foreign economic activity and currency have changed so much that the prolongation of the privileges envisaged by Decree No. 815 appears very unlikely, as it would mean a discrediting by the government of its own policy. The latter comes down to two major points. First, the government feels that all should be equal and equally badly off. Whether this proposition is politically or economically justified is dubious. Besides, it suggests similar rules of control and operation for everyone in the country, which is absolutely unacceptable, as it is senseless to hope for a unified economic policy for such a vast and diversified country as Russia. Second, it is highly expedient to spare no efforts to settle our foreign debts. This has been recognized as reasonable by the Russian government, but the privileges granted to the Far East region never covered the whole spectrum of foreign economic operations.

It has become increasingly clear that under the present conditions, the Far East region must adopt the model of an open mixed economy that was approved by the Silayev government in May 1991 and impeded by the new unified course of the Gaidar government. What is necessary for the region is a change in the entire system of regulation of the economy, with a special emphasis on cooperation with Pacific Basin countries. The region's reorientation toward the Pacific community is already going on, independent of the central government's plans. The number of joint ventures is expected to triple in 1992; by the end of 1992 there will be as many as 700 joint ventures, according to analyst estimates. With this growth in the number of joint ventures, the currency regulations currently in force may be bypassed. Foreign capital, although still minimal, has penetrated into the region. And it is very likely that in 1993 joint ventures and joint stock companies of various types will become a noticeable force, if not in the scope of their operations, then in number.

The main reason the Far East region is interested in foreign investors is the considerable capital contribution that can be made to its key economic areas in the absence of investment funds from Russia. Among the key areas are the energy industry, seaports, production of civilian consumer goods at the defense enterprises, and processing of agricultural products.

The economic situation of the Far East region will also be a decisive factor in assessing the prospects for cooperation with Northeast Asia. Instability and a steady deterioration of the economy suggests that, as a large potential sales market and supplier of valuable raw materials, Russia has not yet been integrated economically into the Northeast Asian macro-region. At the same time, the Far East can hardly overcome its current economic crisis unassisted if it does not receive from the federal government powers to act as an economic region with foreign trade and investment contacts. So far Russia has refused to grant such powers, and the situation has deteriorated quite rapidly. It is very unlikely that the central government will be able to pull its large and inflexible economic system out of the current crisis as it hopes. Granting autonomy to its regions is an inevitable step of the federal government that is presently busy consolidating its central distribution power. And in this step we see the real start of efficient international cooperation that can be effected in Northeast Asia.

CONCLUSIONS

The Russian Far East, considered by many to be a frontier region, has suffered economic hardships in recent years as a result of ineffective government policies coinciding with a deteriorating Soviet Union. This chapter has described the many inefficiencies in the region's price, wage, and credit systems, and efforts to improve conditions through shock reform and other economic measures. Remarkably, prices in the Far East rose around ten-fold during the month-long period from January to February 1992. Inflation continued at record levels throughout the year, hampering nearly all efforts of economic recovery.

In the months ahead, Far East government officials will continue efforts at improving economic and living conditions in the area, in part by encouraging trade with neighboring Pacific Basin nations and foreign investment from the region. Joint venture opportunities are plentiful in the Far East, given its tremendous resource base, though, admittedly, obstacles such as inadequate infrastructure must be dealt with. Nonetheless, because of its strategic location, the Russian Far East holds promising opportunities for investment and business cooperation with the many economically robust countries of the Asia-Pacific region.

PART THREE: INTERNATIONAL TRADE AND RELATIONS

Chapter 13

ISSUES AFFECTING NORTHEAST ASIAN MINERALS AND ENERGY MARKETS

Terry Sheales and Vyrene Smith

INTRODUCTION

In this chapter, some of the broad issues likely to affect industry developments and trade in minerals and energy commodities in Northeast Asia in the 1990s are examined. As will become clear from the chapter, many of these issues can be expected to have a bearing on the rate and extent of development of the mineral and energy resources of the Russian Far East region.

The countries of the Northeast Asian region (which for the purposes of this chapter include Japan, the Republic of Korea, the People's Republic of China, Republic of China/Taiwan and the Russian Far East) have a common interest in developing their mineral and energy resources. In pursuing this common interest, countries should be aware of the need to adopt resource policies that will help them achieve broader economic development for the benefit of their respective populations over the longer term. Rapid economic growth in the region, the pace of economic policy reform and the increasing globalization of both capital and goods markets, coupled with concern over the environment, mean that there is greater interest in most countries in reviewing current policies towards the mining and related processing sectors. Consideration of future resource development, trade and industry policy options open to governments to enhance the contribution that these sectors make to national economies is integral to the review process.

While the Russian Far East and China are rich in resources, their future development as suppliers will be substantially affected by domestic and international economic factors. For the Northeast Asian region as a whole, important factors (other than economic growth) likely to affect the mineral and energy industries of the region include resource development policies in competing countries; global, regional and local environmental considerations; technological

J.P. Dorian et al. (eds.), CIS Energy and Minerals Development, 191–223.

changes affecting supply and demand; and trade policies of the importing and exporting countries. However, before looking at these in detail, it will be useful to examine world trade in minerals and energy commodities and how Northeast Asian countries have become increasingly important in this trade.

TRADE IN MINERALS AND ENERGY COMMODITIES

During the past three decades, changes in the distribution of world income have produced marked shifts in the pattern of world trade in minerals and energy commodities. Of greatest significance, perhaps, have been the rapid economic growth rates achieved by the economies of Northeast Asia. Thus, the main focus of the discussion in this section is how the Northeast Asian region has emerged as a major importer of resource commodities.

The recovery of the Japanese economy from the devastation of World War II provided the initial stimulus to the growth in demand for commodity imports, with further impetus received as other Northeast Asian nations rapidly industrialized. By 1989, the combined income of the region was US$3,524 billion, making it the third largest economic region in the world after the United States and the European Community.

As shown in Table 13.1, at the beginning of the 1960s the developed countries in North America and Europe accounted for the major share of world imports of minerals and energy commodities. For commodities such as bauxite and refined copper, the developed countries, as the major regions of industrial activity, accounted for over three-quarters of the world's imports. Imports of minerals and energy commodities in Asia as a whole were relatively small as industrialization was still reasonably limited. For many minerals and energy commodities, the Asian region accounted for substantially less than 10 percent of world imports.

Major changes in the patterns of trade began to occur from about the mid-1960s onwards as rapid economic growth in much of Asia resulted in the region displacing either North America or Europe as a major destination for commodity exports. By the late 1980s, Asia had supplanted North America as the second largest importer of aluminum, refined copper and lead, and copper ore, and had become the world's largest importer of iron ore. The aggregated data shown in Table 13.1 obscures the fact, however, that much of the growth in minerals and energy imports occurred in relatively few countries of Northeast Asia.

The export-oriented growth strategies for manufactures pursued by the resource-deficient countries of Northeast Asia—Japan, Taiwan and the Republic of Korea—and the consequent increase in industrial production have been the main driving forces behind growth in regional demand for minerals and energy products. The growth in these countries' demands for several resource commodities is illustrated in Table 13.2. In the table, it can be seen that from about the mid-1970s, Japan was joined by Korea as a major regional importer of these commodities.

Associated with the rise in demand for minerals and energy commodities in Northeast Asia, there has been the emergence of several Southeast Asian countries as major sources for these commodities. Increased domestic and international investment, directed toward the development of the mineral and energy reserves, has promoted the rapid development of related industries in a number of relatively resource-rich countries. For example, in Malaysia, Indonesia and the Philippines, mineral production has expanded quickly with the support of industry assistance measures and attractive foreign investment packages. The close proximity of these resource-rich Southeast Asian countries to the rapidly developing economies of Japan, Korea and Taiwan has been an important factor in stimulating intraregional trade. Australia has also been a major beneficiary of stronger demand in Northeast Asia for resource commodities.

Rapid economic change in China and economic and political restructuring in Russia means that these countries will become increasingly important suppliers of mineral and energy resources. The Russian Far East is well endowed with many minerals, including iron ore, coal, gold and a wide range of nonferrous metals (Findlay and Edwards, 1991). Significant amounts of coal, tin and gold ores are currently mined, and there are large oil and gas deposits offshore Sakhalin Island. Russia is already a major supplier of coal to Japan, accounting for 14 percent (in value) of total Japanese imports of coal in 1989 (Japan External Trade Organization, 1990).

ECONOMIC DEVELOPMENT

Probably the single most important determinant of future changes in trade in minerals and energy commodities in the Northeast Asian region will be the rate of economic growth in the principal importing countries: Japan, Korea and Taiwan. Not only will economic growth rates be the major direct determinant of demand, they will also affect the nature, rate and sources of investment in the development of new resources and infrastructure to satisfy that demand. For the newly emerging and potentially very large suppliers of resource commodities in the region—the Russian Far East and China—it is investment in their mineral and energy industries and related infrastructure which will be of greatest immediate importance.

Macroeconomic policies in various countries of the region will create the underlying environment in which the mineral and energy industries must operate. Fiscal policy, through the taxation of mineral and energy resource exploitation and through expenditures on support infrastructure, has the potential to either distort or promote the efficient allocation of resources among sectors. Monetary policy can exacerbate or smooth the cyclical nature of minerals and energy prices and investment through the discretionary use of tools such as exchange rates (where exchange rates are fixed and thought of as a discretionary tool for

Table 13.1. Imports of Selected Commodities, by Region

	Unit	1960	1965	1970	1975	1980	1985	1986	1987	1988	1989
Bauxite (aluminum content)											
World	kt	3 660	4 686	6 483	7 786	8 899	7 445	7 210	7 575	8 029	8 523
America [a]	kt	2 665	6 265	3 875	3 444	4 327	5 343	4 810	5 313	5 649	6 180
Europe [b]	kt	517	697	1 157	2 038	2 101	2 310	2 130	2 082	2 109	2 289
Asia [c]	kt	226	387	868	1 110	1 357	3 570	2 408	1 990	2 268	2 391.5
Aluminum											
World	kt	1 115	1 713	2 767	3 208	4 896	6 701	7 051	7 405	8 126	8 437
America	kt	141	485	329	434	537	928	1 411	1 299	1 086	1 004
Europe	kt	748	859	1 609	1 244	2 084	2 345	2 580	2 527	2 933	3 179
Asia	kt	80	92	360	561	1 377	2 160	1 941	2 522	3 130	3 170
Copper ore (Cu content)											
World	kt	280	301	599	1 096	1 286	1 547	1 576	1 551	1 677	1 710
America	kt	73	36	30	85	27	80	76	57	58	97
Europe	kt	76	89	146	243	253	258	288	290	337	329
Asia	kt	129	170	415	708	947	974	1 000	995	1 061	1 074
Refined copper											
World	kt	1976	2133	2 528	2 705	3 472	3 409	3 465	3 604	3 658	3 779
America	kt	130	130	133	143	474	410	528	535	398	344
Europe	kt	1426	1628	1 861	1 826	1 963	1 677	1 756	1 718	1 806	1 868
Asia	kt	108	152	219	221	383	613	601	821	1 018	1 115
Refined lead											
World	kt	962	1077	1 272	1 185	1 404	1 329	1 337	1 314	1 295	1 378
America	kt	199	200	224	94	86	139	145	201	163	137

Europe	kt	541	590	745	671	835	677	731	625	632	711
Asia	kt	61	96	78	90	213	237	230	267	271	308
Lead ore (Pb content)											
World	kt	417	481	701	603	579	693	681	904	911	992
America	kt	125	111	102	79	44	43	87	230	235	262
Europe	kt	254	315	425	323	315	386	353	403	449	497
Asia	kt	24	40	137	119	155	165	165.2	187	163	174
Iron ore (actual weight)											
World	Mt	155	215	278	375	385	373	362	364	393	413
America	Mt	40	51	48	52	31	22	22	22	25	25
Europe	Mt	80	97	128	76	77	134	125	127	139	147
Asia	Mt	15	39	102	[illegible]	152	169	139	141	154	170
Unwrought nickel											
World	kt	-	-	-	275	328	302	321	374	373	394
America	kt	-	-	-	110	113	91	94	107	109	116
Europe	kt	-	-	-	116	136	128	132	151	156	167
Asia		-	-	-	15	33	43	50	69	62	70.6
Crude petroleum											
World	Mt	-	742	1170	1412	1479	1108	1212	1230	1282	1355
America	Mt	-	96	114	267	288	191	146	271	289	332
Europe	Mt	-	365	601	565	546	386	418	410	418	427
Asia	Mt	-	156	271	332	256	286	290	285	304	329

kt = thousand tonnes   Mt = million tonnes

a. United States and Canada. **b.** Western Europe. **c.** Includes Middle East, excludes China.

Sources: UNCTAD Commodity Yearbook, various years; Yearbook of Energy Statistics, United Nations, various issues

Table 13.2. Selected Minerals and Energy Imports by Northeast Asian Countries

	Unit	1960	1965	1970	1975	1980	1985	1986	1987	1988	1989
Hard coal											
Japan	Mt	-	-	-	62	68	93	90	91	101	102
Korea	Mt	-	-	-	0.6	7.2	19.4	21.3	21.3	24.6	25.2
Taiwan	Mt	-	-	-	-	-	-	-	-	-	
China	Mt	-	-	-	-	-	-	-	-	-	-
Iron ore (actual weight)											
Japan	Mt	15	39	102	132	134	125	115	112	123	127
Korea	Mt	-	-	0.07	1	9	11	12	17	16	23
Taiwan	Mt	-	-	-	-	-	5	5	6	8	9
China	Mt	-	-	-	1.4	5.2	10	12	12	13	13
Lead ore (Pb content)											
Japan	kt	24	39	137	119	150	152	159	171	162	173
Korea	kt	-	-	-	-	-	6	-	3	1	1
Taiwan	kt	-	-	-	-	-	-	-	-	-	-
China	kt	-	-	-	-	-	-	-	-	-	-
Refined lead											
Japan	kt	23	39	2	17	91	88	70	49	76	83
Korea	kt	2	2	2	5	21	42	35	76	55	97
Taiwan	kt	-	-	-	-	-	12	16	21	19	20
China	kt	2	9	27	51	32	4	5	3	5	5
Bauxite (Al content)											
Japan	kt	220	362	838	1078	1299	756	496	402	462	488
Korea	kt	-	0.1	0.2	0.2	0.5	2	4	7	6	7

Taiwan	kt	-	-	-	-	-	5	11	8	9	8
China	kt	-	-	-	-	-	-	-	-	-	-
Aluminum											
Japan	kt	23	42	258	378	910	1575	1366	1835	2292	2363
Korea	kt	3	7	3	17	74	200	185	200	218	282
Taiwan	kt	-	-	-	-	-	147	156	195	194	223
China	kt	5	4	9	331	67	488	266	148	75	175
Copper ore (Cu content)											
Japan	kt	129	169	414	692	835	813	820	805	928	910
Korea	kt	-	-	-	7	71	106	117	144	102	125
Taiwan	kt	-	-	-	-	-	47	52	44	30	38
China	kt	-	-	-	33	16	69	75	60	52	60
Refined copper											
Japan	kt	59	84	160	168	228	358	275	352	420	487
Korea	kt	-	-	4	6	10	60	84	107	124	93
Taiwan	kt	-	-	-	-	-	48	106	161	172	274
China	kt	10	0.2	69	100	128	356	171	76	84	70
Crude petroleum											
Japan	Mt	-	85	169	228	217	165	162	153	160	170
Korea	Mt	-	2	9	16	25	27	31	30	36	40
Taiwan	Mt	-	-	-	-	-	16	17	17	18	20
China	Mt	-	0.1	0.4	0	0.5	0.3	0.4	0.3	0.5	0.5

kt = thousand tonnes Mt = million tonnes

Sources: UNCTAD Commodity Yearbook, various years; Yearbook of Energy Statistics, United Nations, various issues achieving short-run outcomes) and interest rates. It can also influence the demand for minerals and energy exports through its effect on economic growth. It is on this latter issue—economic growth—that the remainder of this section will be focused.

The Australian Bureau of Agricultural and Resource Economics' (ABARE) assessment (which is updated regularly in ABARE's *Agriculture and Resources Quarterly*) is that the short-term outlook for the world economy is somewhat mixed, with several of the world's major economies at critical turning points (ABARE, 1992). Economic activity in the United States, a major market for manufactures produced in the Northeast Asian region, is showing signs of recovery following the contraction in 1991. However, growth in the Japanese and German economies is showing signs of slowing. The performance of the latter two, the world's second and third largest economies, respectively, will be of critical importance to the overall outlook for the world economy.

In the short term, ABARE is assuming that the world economy will grow slightly in 1992 and pick up further in 1993 (Table 13.3). This follows almost no growth overall in 1991. ABARE's assessment is based on the premise that a recovery in the U.S. economy will emerge and gather momentum during the remainder of 1992 and 1993, and that the economic slowdowns in Japan and Germany will be relatively modest and temporary. Continued strong growth in Southeast Asian economies will be an important positive stimulant to world economic activity.

Over the medium term, ABARE is assuming that world economic growth will accelerate for several years, mainly reflecting relatively strong growth in many OECD countries as excess capacity generated by the downturn in the past year is absorbed (ABARE, 1991a). While inflationary pressures appear likely to remain relatively subdued in most industrialized countries over the next couple of years, some upward pressure on real world interest rates seems possible. Higher interest rates may come about partly as a result of continued strong demand for investment funds and possible declines in savings in some major countries. Reconstruction in the Commonwealth of Independent States, Eastern Europe and Latin America, along with economic recovery in the OECD countries, can be expected to result in strong demand for investment funds over the medium term. The size of the U.S. budget deficit and a possible decline in savings rates in Japan, Germany, Taiwan and Korea, partly as a result of financial market liberalization in these countries, are also likely to be critical influences.

Japan

As the major economy in Northeast Asia, developments in Japan will be of the utmost importance to the minerals and energy markets of the region. Recent developments in Japan's financial markets, particularly in relation to the sharp decline in the share and property markets, can be expected to contribute to short-term uncertainty about the Japanese economy. Economic activity in Japan is assumed to slow in 1992 after averaging over five percent growth per year between 1988 and 1990 (Table 13.3). Uncertainty about short-term economic prospects is likely to adversely affect business and consumer confidence and will be reflected in both investment and consumption demand.

However, it is anticipated that the slowdown in economic growth in Japan will be temporary and relatively mild. Improved growth rates are assumed to be in evidence by 1993 and to be maintained at around three percent a year over the medium term. Falling interest rates since mid-1991 and strengthening world economic growth could be expected to provide a stimulus to the Japanese economy. However, with the economy operating at a relatively high rate of capacity utilization, there may be some constraints on growth rates.

Other Asian economies

The Asian region contains some of the world's fastest growing economies, and this is likely to hold true for the foreseeable future. Economic growth in the countries of interest in this chapter is expected to remain high in 1992–93 (Table 13.3).

South Korea and Taiwan can be expected to achieve further gains in productivity as new technologies continue to be adopted from developed economies. Labor productivity rates in both countries are still relatively low compared with those in developed countries. While slower Japanese economic growth will mean weaker demand for these countries' exports in the short term, domestic demand should remain strong with large increases in public investment planned and private investment and consumption likely to increase steadily (ABARE, 1992).

Overall, it is expected that the dynamic expansion of the South Korean and Taiwanese economies will continue into the medium term, though at more sustainable levels as infrastructure improvements are made, new technologies are adopted and measures are taken to liberalize goods and financial markets. Successful growth in the short and medium terms will continue to be closely associated with developments in the Japanese economy, with both economies benefiting from Japanese investment.

Over the past few years, economic growth in China has accelerated, rising to 5.2 percent in 1990 and around seven percent in 1991. Similar rates of economic growth seem achievable over the medium term as China draws on the benefits of a steady increase in domestic and foreign investment, rising domestic consumption and reforms to its system of trade. The southern region of China, in particular, is likely to benefit from the development of closer trade links with Taiwan and Hong Kong.

Commonwealth of Independent States

The Commonwealth of Independent States (CIS), with its large population and manufacturing base and extensive reserves of natural resources, is a potential

Table 13.3. Key World Macroeconomic Assumptions

	Unit	1989	1990	1991	1992[a]	1993[a]	1994[a]	1995[a]	1996[a]
Economic growth[b]									
OECD	%	3.2	2.5	0.9	1.8	3.6	3.8	3.5	3.2
Japan	%	4.8	5.2	4.4	2.0	3.5	3.5	3.2	3.0
Australia[c]	%	4.0	3.8	-0.2	0.8	4.5	4.3	3.5	3.5
United States	%	2.5	1.0	-0.7	1.5	4.0	4.5	4.0	3.5
Hong Kong	%	2.3	2.4	4.0	5.8	5.0	4.0	4.0	4.0
Singapore	%	9.2	8.3	7.0	6.0	6.5	6.0	5.0	4.0
Korea, Rep. of	%	6.2	9.0	8.5	7.5	8.0	7.0	7.0	7.0
Taiwan	%	7.6	5.1	7.4	7.5	7.0	7.0	7.0	6.0
Malaysia	%	8.6	9.8	8.5	8.0	7.5	7.5	7.5	7.5
Philippines	%	5.6	3.7	0	3.0	4.0	4.0	4.0	4.0
China	%	4.0	5.2	7.2	6.0	7.0	7.0	7.0	7.0
USSR/CIS	%	2.4	-3.0	-12.0	-7.0	-4.0	-2.0	0	2.0
Inflation[b]									
OECD	%	4.7	5.2	4.2	3.5	3.5	3.5	3.5	3.5

Japan	%	2.4	3.1	3.3	2.8	2.5	2.5	2.5	2.5
Australia	%	7.3	8.0	5.3	2.5	3.2	3.5	4.0	4.0
United States	%	4.8	5.4	4.3	3.5	4.0	4.0	4.0	4.0
Hong Kong	%	10.1	9.7	12.0	10.0	10.0	10.0	10.0	10.0
Singapore	%	2.4	3.5	3.6	3.2	3.0	3.0	3.0	3.0
Korea, Rep. of	%	5.6	8.6	9.6	9.5	8.5	7.0	7.0	7.0
Taiwan	%	4.4	4.1	3.5	3.4	3.0	3.0	3.0	3.0
Malaysia	%	2.8	2.6	4.5	5.0	6.0	5.0	43.0	4.0
Philippines	%	10.6	12.6	18.0	7.5	7.0	7.0	7.0	7.0
Interest rates[d]									
Japan[e]	%	4.5	6.9	7.7	6.0	6.8	7.5	8.0	8.5
World[f]	%	9.8	10.1	9.1	7.5	8.0	8.7	9.2	9.6
Nominal exchange rates	Yen/US$	138	145	134	125	124	122	120	120

kt = thousand tonnes Mt = million tonnes

a. ABARE assumptions. **b.** Change from previous period. **c.** Year ended 30 June. **d.** Prime rates. **e.** Japanese short-term bank loans over one million yen. **f.** Weighted average of the commercial bank lending rate to prime borrowers in the United States, Japan, western Germany and the United Kingdom.

Sources: ABARE, 1992; Organization for Economic Cooperation and Development; Reserve Bank of Australia; International Monetary Fund; Australian Bureau of Statistics.

future major consumer and supplier of a range of minerals and energy commodities. However, the rate of development, particularly with respect to domestic demand, is likely to be slow while constituent republics experience severe recessions as they strive to make the transition from a centrally planned to a market-based economic system.

The economic situation in the CIS remains extremely uncertain. In the short term, output is likely to fall further in response to the disruption and uncertainty caused by economic restructuring. However, the declines in economic growth of the last few years are expected to slow and eventually be reversed to attain positive growth in the second half of the 1990s.

While these comments relate to the CIS as a whole, the situation could improve more rapidly in the Russian Far East. This will be the case especially if improved economic growth and hence demand for raw materials in Japan, Korea and Taiwan result in strong foreign investment in minerals and energy commodity production in that part of Russia.

DOMESTIC POLICIES AFFECTING RESOURCE DEVELOPMENT

The diversity of economic and political structures, stages of development, cultural values and differences between countries over national objectives means that there is no single set of policies which meets all countries' specific circumstances. The extent of development of the mining and processing sectors and the contribution of these sectors to individual national economies vary widely in the region. This reflects differences in resource endowments, stages of development and economic objectives, and implies different policy challenges for each country.

While there is a virtual plethora of domestic policies that may have a bearing on the development of minerals and energy trade, there are at least four areas which warrant closer attention because of their regionwide implications. These include recognition of the need for a clear policy distinction between mining and processing; policy instruments that can affect mining development; privatization of resource industries; and foreign investment in the resource industries.

Mining and processing links

In formulating policies relating to resource industry development, there are some important economic distinctions to be made between policies specific to mining and minerals processing and other policies that affect these activities. An obvious prerequisite for the development of mining is the availability of resources of sufficient quality to be developed. That is, as resources have to be found, a key policy focus will be to attract exploration efforts. The subsequent development of identified minerals and energy deposits requires different, but related, policy measures. These latter policies will be concerned with the legal rights to those resources (property rights) and how the wider society can obtain a reasonable return from developing the resource.

Because minerals processing is basically just a particular component of a country's manufacturing sector, policies relevant to manufacturing in general are frequently more relevant to the management of the minerals processing sector than are mining-specific policies. The successful development of a downstream minerals processing sector is frequently associated more with comparative advantages based on access to technology, availability of a range of input products, a skilled workforce, transport infrastructure, proximity to key markets and sites at which pollution standards are acceptable than to the existence of a domestic mining industry. For example, Japan has a large steel industry based on imported iron ore. Similarly, a successful metal fabrication industry does not need to be based on a domestic mineral processing sector. Again, Japan imports only very small amounts of bauxite and alumina for aluminum smelting. Instead, since the rise in energy prices in the 1970s which shifted any comparative advantage away from smelting, Japan has preferred to import the bulk of its aluminum metal requirements directly for fabrication.

When developing policies affecting primary minerals and energy production and value-adding processing, it is important to take a national perspective. The major policy objective should be to ensure a proper allocation of resources between these two types of activities and between these sectors and the rest of the economy. In broad economic terms, the guiding principle should be for a nation to exploit its comparative advantages to maximize economic welfare. By doing this, countries will be better able to avoid the creation of industries which end up requiring long-term assistance to survive, and will thus contribute to an efficient pattern of regional production and trade with benefits for all participants.

In seeking to maximize economic welfare by exploiting its comparative advantages, a country will need to be conscious of impediments to mineral and energy resource development and to the opportunities for new value-adding industries. From a policy perspective, this means reviewing such things as current taxation policies, foreign investment rules, research and development policies, industry policies, domestic preference arrangements and trade barriers on inputs. Reviews of these policy-related issues will enable a country to better assess how policy distortions might inhibit the flow of resources into activities in which it has a comparative advantage.

Policies affecting mining development

Issues of resource property rights, land access and how a society can best obtain a return from developing its mineral and energy resources will all have a bearing on the rate and extent of mining development. These issues are discussed below.

Property rights for mineral deposits and taxation of production from these sites are issues which have generated considerable debate in many societies. Changes to operating rules, particularly if these occur frequently and in non-predictable ways, add to the risks faced by potential investors and can reduce the amount of investment relative to what it might be in the absence of such risks. With respect to property rights and scarce nonrenewable resources, it can be argued that resources will be allocated efficiently in a competitive economy if the property rights are well defined and there is a full set of markets (futures, insurance, capital and equities) for risk bearing (see Hinchy, Fisher and Wallace, 1989, for more detail). Clearly, for the Russian economy, which is in transition from a centrally planned to a market-based system, it will take some time before such markets can be established and made fully functional.

The possibility of a government changing the rules part-way through the life of a project and consequently eroding the value of private property rights, commonly known as 'sovereign risk', is a major problem for the mining industry. Similarly, anticipating the reaction of mining companies to perceptions of this key area of mining risk is a major problem for national policy makers. Government-induced uncertainties about the likelihood of being able to convert a right to explore into a right to mine and unexpected changes in taxation and royalty arrangements during a project are important areas of sovereign risk and can be major deterrents to investment in the sector.

Royalties, which are payments to the owners of a resource deposit (typically a government) for the right to exploit that resource, can be a significant cost for miners. Royalties usually have been levied at a specific rate on volume of output or as a percentage of the value of output (*ad valorem*). In recent times, some governments have been moving toward profit-based royalty systems which mean that payments reflect more closely a project's capacity to pay. This latter approach has the advantage of being less likely to distort production decisions because the government shares in the risks of positive or negative returns from the projects. (For a fuller discussion of the advantages and disadvantages of alternative royalty arrangements see ABARE, 1991b).

Taxes on mining activities are often wide ranging and can be either direct or indirect. Some of these taxes may put mining at a disadvantage while others may benefit mining relative to other economic activities. Correcting these distortions would lead to more appropriate investment decisions, both within the mining sector and elsewhere in the economy. Arrangements that put mining at a disadvantage include tariffs on imported equipment, limited tax deductibility for exploration expenditures, and plant depreciation (and other deductibility issues relating to the expenditure on such things as housing and welfare facilities away from the mine site), and discriminatory local government taxes calculated on the basis of mine output or transport of that output.

The offering of tax concessions to attract foreign investment is an important issue for many countries. Tax concessions are a subsidy, so there is a fundamental question of whether the cost of the subsidy (direct and in terms of any distortions it causes to the allocation of resources among different sectors of the economy) exceeds the full benefit of awarding it. Competition between countries in offering tax concessions to attract foreign investment has the effect of increasing the size of the subsidy (and hence cost to a country's economy and taxpayers) without changing the benefits of the project.

Issues of access to land for uses such as exploration and mine development are of growing importance worldwide. Conflicts over land use are becoming more common in situations where present processes for resolving conflicts do not allow sufficiently for an objective balancing of the full social costs of conservation or development. One approach to resolving land-use conflicts is to rely as much as possible on market forces to ensure that surface and subsurface resources are put to their most valuable use. A crucial advantage of a market-based approach that is based on private property rights to land is that it would help ensure that individuals take proper account of any costs that their activities impose on others (ABARE, 1991b).

Government intervention in land-use decisions may be necessary where market approaches are not considered feasible (as is often the case for public goods such as environmental assets). Decisions in these cases will need to be based on a careful evaluation of the costs and benefits of alternative uses of public land to ensure that the community derives the maximum benefit from natural resources owned and controlled on its behalf by governments.

PRIVATIZATION

During the 1980s there were considerable changes in the world's social, political and economic structures. One consequence has been that some governments are changing the role they play in the economic life of their countries. This role change is reflected (in part) in an opening of those economies to greater international competition and a greater emphasis given to the role of the private sector as a means of stimulating export-oriented growth (Asian Development Bank, 1991).

In Northeast Asia, it is in Russia that the most significant changes are occurring. Changes to the Russian economic system in recent years have created opportunities for large-scale restructuring of ownership and management. Over time, transfers of state-owned primary industries to the private sector can be expected to have significant effects on the location and pace of resource development. Developing a full understanding of the benefits available through flexible and market-responsive private ownership and the attendant problems in ensuring that markets remain effective and undistorted will present a major and critical challenge to policy makers.

A key argument in favor of the privatization of various industries that has taken place worldwide in the past decade or so is that it will result in improved economic efficiency as market forces determine the allocation of resources between various competing end uses. However, a further important effect of privatization on trade and resource development patterns will be that it is likely to increase the potential for greater amounts of direct foreign investment in the minerals and energy sectors of the affected countries.

FOREIGN INVESTMENT

Direct foreign investment has been important in the past development of the mineral and energy resources of the region—and it is likely to remain so in the future. Within the region, countries such as Russia and China lack much of the capital necessary to rapidly develop their mineral and energy resources by themselves. International firms specializing in mineral and energy production are more likely to be able to develop resource deposits profitably because they already have considerable market knowledge, financial expertise and market access compared with domestic firms (or governments) that have little experience. Thus, with direct foreign investment, the development of the mineral and energy resources of the region would be expected to be more efficient and rapid than otherwise.

The transfer of technical knowledge, ideas, managerial talents and entrepreneurial skills is another important aspect of direct foreign investment in the development of mineral and energy resources. Without the transmission of skills from the developed countries, some of the region's countries may not have the technical knowledge or managerial skills and supply of skilled local labor for project development or to put in place and operate the infrastructure required for profitable industrial development. An important aspect of this transfer of technical expertise is that once it is imported into a particular industry, there is usually considerable potential for knowledge gained to 'slip over' into other industries.

One of the major constraints on the development of the mineral and energy resources of some countries is likely to be that of infrastructure. The development of transport infrastructure in the Russian Far East may be one of the major determinants of the extent, location and pace of mineral and energy development in that region. For example, there are large oil and gas reserves offshore Sakhalin Island, but no pipelines which could be used to transport the gas or oil produced to a major market such as Japan. To take advantage of such market opportunities, Russia will probably need the injection of considerable foreign capital and expertise through foreign direct investment to supplement and further develop its existing systems for transporting minerals and energy commodities.

As privatization of resource industries may well require at least some foreign capital input, especially in Russia and China, the privatization process will need to be accompanied by appropriate foreign investment laws. Direct foreign investment will be impeded and many of the advantages of privatization will be forgone if laws governing such investment are unduly restrictive. To a large extent, the amount of capital that a less developed country can obtain from abroad will be influenced by the internal policies in the borrowing country.

An important policy aspect of foreign direct investment is the desirability of minimizing the sovereign risk that foreign companies perceive exists when they invest in domestic industries. Given the long-term nature of resource projects, the best way for governments to minimize the sovereign risk problem is to formulate equitable and clear longer term objectives and policies for resource sector development, and to minimize changes to policy settings in response to shorter term disturbances (ABARE, 1991b). In this respect, guarantees against expropriation or nationalization of resources or infrastructure once they have been developed are likely to be integral to investor confidence.

The experience of the 1960s and 1970s, when there was a tendency among some developing countries to nationalize successful resource developments, which subsequently often became run down and uneconomic to operate, highlights the issue of sovereign risk and some critical linkages in the mining industry. These linkages of mutual dependence are mostly between those controlling access to mineral resources (countries and governments), those with access to capital (companies and financiers), and those who buy the products (markets and customers) (Salazar Sparks, 1992).

Convertibility of a country's currency and the ability to move capital freely can be expected to facilitate the introduction of investment funds as well as the development of an export trade in the raw commodity and the import of capital equipment used by the resources sector. Achieving convertibility for the ruble is likely to be one of the key factors in the restructuring of the Russian economy. Just as it needs to facilitate the easy movement of capital, a country that receives foreign investment funds needs to put in place a legislative framework to protect against the possibility of transfer pricing between related firms so as to avoid or reduce income tax obligations.

ENVIRONMENTAL CONSIDERATIONS

The challenge of integrating environmental and economic objectives to ensure sustainable economic development is increasingly occupying the attention of policy makers around the world. In the developed market economies, serious environmental problems have occurred as a result of intensive resource use, and governments are having to respond to public concern about reducing or eliminating these problems. In the less developed and former centrally planned economies, the links between poverty, economic and population growth and degradation of the environment are being increasingly recognized.

Various aspects of the environmental issue are examined in this section, starting with sustainable resource development. Other aspects examined include the role of technology in sustainable development and the role of government in achieving environmental goals. Each of these issues, albeit to varying degrees, is relevant to policy makers in Northeast Asia.

Sustainable resource development

The issue of the sustainability of resource use has been around since at least the eighteenth century, when Thomas Malthus wrote about what he perceived as being the problem of population growth outstripping subsistence or output growth. While the Malthusian arguments have been substantially refined since then, they remain basically unchanged. The basic argument is that economic growth in tandem with population growth cannot increase indefinitely. This implies that attempting to maximize economic growth now (in the absence of savings and investment) will involve a cost (in the form of lower living standards) to future generations.

Proponents of the above view argue that the preferred alternative should be a low- to no-growth society in which a reasonable standard of living can be sustained indefinitely. In considering the arguments from an economic perspective, the key factor is the nature and magnitude of market responses to the higher resource prices resulting from apparent resource depletion. Observed responses to higher real prices for resources include demand reduction, supply increase (including more intensive recycling) and increased substitution. In all of this, price-induced new technology plays a key role.

The limits to growth argument was supplemented in the late 1980s by a broader concern over the consequences of resource use for the natural and physical environment. A growing range of environmental problems was seen to be the result of economic growth. Environmental issues were brought into international forums by concerns about such problems as acid rain, tropical rainforest depletion and the thinning of the ozone layer. The prospect of global climate change because of human activities augmenting the natural greenhouse effect of the earth's atmosphere has added a new dimension to the 'limits to growth' argument.

Attempts by governments to address environmental concerns and to give policy effect to the concept of sustainable development have both direct and indirect impacts on minerals and energy markets. Examples of direct impacts include legislation to reduce the use of lead because of concerns over its health effects and possible moves to limit consumption of fossil fuels because of their role in contributing to greenhouse gas emissions. However, at this time of policy uncertainty, the overall impact of attempts to limit consumption is difficult to determine. For example, limiting use of lead and fossil fuels or raising their prices forces a search for substitutes which will stimulate demand for other mineral and energy resources. Perhaps the most important challenge for minerals and energy markets in the longer term is to draw together the concepts of sustainable development and growth in resource use through acceptance of the real costs of environmental impacts in product prices and through community acceptance of the links between improved national income and better environmental standards.

Of significance in the debate about sustainable development is the fact that there is as yet no standardized definition of what the term means. Nevertheless, it can be observed that most definitions involve in some way the concept of intergenerational equity—that is, the concept that the actions of the present generation should not diminish the prospects of future generations. Rose and Cox (1991) noted that there are both ecological and economic interpretations of what this means in practice.

In its narrowest form, the ecological view is that the present generation should pass on to future generations no fewer resources than it inherited—that is, the sustainability of an economic system depends on whether any of the resources on which it relies are being irreversibly lost. However, the ecological view ignores the role of technological progress and the fact that the available capital stock includes not only natural, but also manufactured and human capital. Deliberate substitution between these types of capital is possible. For instance, the returns from natural capital depletion can be invested in reproducible manufactured and human capital. This means that the economic view is essentially one that asserts that if technical progress is rapid enough, economic growth can be sustained indefinitely, even with a finite stock of nonrenewable resources (Hartwick and Olewiler, 1986).

Technological progress and the environment

Apart from technological change being a factor in maintaining or increasing a nation's capital stock, technical progress also results in changes over time in the uses to which individual materials are put and to the demand for those materials. The traditional theory of changes in the use of materials in the course of industrialization is that there is a life-cycle pattern in the demand for any material (Cox, Nagle and Lawson, 1990). Consumption of a newly introduced material initially increases much faster than the rate of growth of the economy as a whole. This rapid growth is associated with improvements in the technology of processing, lower prices and higher product quality. In the next phase, additional innovations make it possible to use the material more efficiently, increasing the ratio of the value added in manufacturing to the quantity of the material used. In this phase, the intensity of use of the material peaks and begins to decline. Consumption may still increase, but more slowly than economic growth. In the final phase of the cycle, the demand for the material is further curtailed as new market opportunities diminish and substitution possibilities increase. Technological change will continue to mean improvements in efficiency of use as well as in the availability and cost of substitutes. In absolute terms, consumption tends to level off and may even begin to decline.

As suggested by Labson, Jones, Gooday and Neal (1992), examples of the traditional life-cycle pattern of materials demand abound. In energy markets, for example, coal displaced renewable sources of energy in the early stages of industrial development. From the late nineteenth century, oil consumption growth became more rapid and oil eventually overtook coal as the most widely used energy source. From the 1950s, growth in uranium use was rapid as nuclear power technology was developed. From the 1960s and particularly following the oil price shocks of the 1970s, developments in natural gas transport technology saw world natural gas use rise rapidly. With the growth in use of new energy sources, the intensity of use of the older sources declined.

Technology transfer also has implications for the aggregate use of individual materials. Diffusion and adoption of new technologies and techniques in different countries are frequently observed to be more rapid the later those technologies are introduced. This usually means that the intensity of use of materials is lower the later a country is on the path of industrial growth. As a result, sustainability of resource use is probably best viewed in the context of life-cycle theory and technology transfer. The simple trend extrapolation of resource use over time, as is essentially the approach used by those with a Malthusian view of the implications of economic and population growth, is misleading.

Government intervention to achieve environmental goals

Environmental goals can be achieved through the use of market mechanisms or by regulation. While market mechanisms are more efficient in an economic sense, they may be less dependable than direct regulation in terms of achieving specific outcomes. Even when relying on market mechanisms, government action may still be required to ensure that the basic requirements for markets to function efficiently are present. These requirements include the specification of property rights, freedom of transfer of those rights and the availability of information (Rose and Cox, 1991). In many countries though, despite the economic advantages of a market-based solution, the environmental goals of society have tended to be pursued largely through regulation.

A major reason for government intervention to achieve environmental goals is the failure of prices for some goods and services to reflect fully the environmental costs associated with their production and use. The reason for this is that many environmental resources are not commonly traded in markets. For example, while there is a market for coal, there is no market for the atmosphere into which the products of coal combustion are released. There are markets for the products of a mine, but there are no or weak markets for the environmental amenity that the mine site provided before mining operations.

In the Northeast Asian region, the extent of government intervention in mineral and energy markets to achieve environmental goals ranges from stringent regulation (as in Japan) to relatively limited intervention (as appears to be the case in China and Russia). Government intervention can take many forms. For example, the use of some materials can be directly regulated, such as by quantitative restrictions. Also, emissions of pollutants from processing or manufacturing plants or end-use devices can be restricted through the determination of emission 'standards' and the issue of permits which allow these standards to be achieved over time. Alternatively, the cost or physical availability of some materials can be affected by environmental policies not directed at the material itself.

Examples of such policies include restrictions on access to land for exploration or mining, or restrictions on the siting of processing plants.

Because of its well-documented adverse effects on humans when ingested, lead provides a good example of a mineral that is subject to a wide range of environmental controls and of the paradoxical outcomes that can result from such controls. In some countries certain uses of lead, such as for water piping, in paint, or for inclusion in motor vehicle fuel additives, have been banned or restricted for years. However, while there are human health benefits to be gained by restricting lead use, there are also economic costs. Alternatives to lead are often more expensive. Also, there can be conflicts between policies to reduce lead use and other environmental goals that have the indirect effect of increasing demand for lead. For example, moves to increase the use of electric vehicles to cut petroleum fuel-related emissions are likely to result in increased demand for lead to use in lead acid batteries. Restrictions on lead use in developed countries are also likely to put downward pressure on the world price. But this in turn will likely mean increased consumption in countries with less stringent standards.

Another commodity affected by environmental policies is coal. Concern over acid rain has led to controls being introduced in many countries on emissions of sulfur dioxide from fossil fuel burning. Sulfur dioxide in the atmosphere is a major precursor of acid rain. This has the effect of increasing the relative attractiveness of alternative fuels to coal (which usually has the highest sulfur content of fossil fuels), encouraging the development of technologies to remove sulfur from emission streams and burn coal more efficiently, and promoting the use of higher cost but lower sulfur-content coals.

In the cases of both lead and sulfur, environmental restrictions mean a change in the patterns of resource use, but not necessarily reductions in consumption. Regulations in both cases have led to greater use of alternative resources and developments in technology. Environmental restrictions such as these, however, have a direct cost to society. They usually mean an increase in the price of inputs and/or outputs to manufacturing processes. These costs need to be offset against the reduction in environmental damage costs that have been achieved through the regulation. Whether or not a socially beneficial outcome results will depend on the balance between these costs. From a policy perspective, it is important to achieve environmental goals using measures that maximize the chances of such an outcome.

Climate change

The issue of climate change provides a good illustration of the problems associated with resource use, environmental quality and sustainable development. There is concern that human activities are leading to an increased concentration of greenhouse gases in the atmosphere which in turn could lead to global climatic changes, including increased average temperatures, a rise in the sea level, and greater frequency of severe weather events. A major factor in the increased concentration of greenhouse gases in the atmosphere is combustion of fossil fuels. Governments worldwide are addressing the question of whether emissions of greenhouse gases should be reduced, and if so, how this should be achieved.

Setting aside the question of the scientific legitimacy of expectations that climate change will occur because of emissions of greenhouse gases, an efficient policy for dealing with the problem of climate change would be to limit greenhouse gas emissions to the point where the cost of reducing emissions by a marginal amount is equal to the benefit obtained by so doing (Thorpe, Sterland, Jones, Wallace and Pugsley, 1991). Markets may fail to achieve this outcome because the atmosphere is an unpriced resource. However, there are considerable practical problems involved in identifying this optimal point. The costs of an uncertain amount of climate change are difficult to determine. The rate of climate change and the uncertainties surrounding the timing and extent of damage are further complicating factors. Valuing some of the environmental consequences of climate change is also conceptually as well as practically difficult.

The most straightforward way to achieve a reduction in greenhouse gas emissions is to impose a common target for emissions reductions to be achieved across countries. But this direct regulatory approach is a costly way to address the problem. Uniform restrictions do not take into account differences in consumer preferences, competitiveness and emission reduction costs between countries. Market measures, which automatically take such factors into account, are in general a more efficient means of reducing emissions.

The most commonly discussed market measures for reducing greenhouse gas emissions are taxes and tradable emission permits. A tax would force emitters to take into account the cost of damage caused by the emissions. If the tax were set at the rate where the marginal cost of emission abatement equaled the marginal cost of damage caused, the optimal level of emissions would result. Initially, however, this point would be difficult to determine and the tax rate would have to be adjusted over time. In addition to the challenge of maximizing participation in such a scheme, there are also practical problems with administering the large and complex tax system that would be required.

In principle, a tradable emissions scheme could overcome these administrative problems and still achieve an optimal outcome (Hinchy and Fisher, 1991). Because they represent very much a market-based solution, proposals for tradable emission permits may well come to the fore in policy debates concerning national emission targets following the United Nations' Conference on Environment and Development in Rio de Janeiro in June 1992.

Tradable emission permits provide a good example of the attractions of market-based policy solutions in general. One advantage of a tradable emissions scheme is that governments do not need to know the relative abatement costs of each polluter to find the least-cost way to reduce emissions. For example, within a country it is feasible that such a scheme would allow some firms to 'buy' their emission reductions from those firms for which achieving reductions is easier and cheaper. A highly energy-efficient firm may find it cheaper to purchase emission permits from a less energy-efficient firm. The less energy-efficient firm could then use these funds to, say, purchase or develop technologies to reduce its emissions.

TECHNOLOGY AND THE MINERALS AND ENERGY INDUSTRIES

Apart from the earlier discussion of the role of technological change in relation to the environment, such change is also important because of its direct impact on minerals and energy demand and supply. On the demand side, technical progress means that the uses to which individual materials are put and the demand for those materials can change markedly over time. Technological change can expand the range of uses for a material, but it can also make the use of substitutes more competitive and improve the efficiency of processes so that less of a given input is required per unit of output from a process. Increased efficiency of supply, leading to lower production costs and a decrease in price, will also have a considerable influence on materials consumption.

The mining sector

Considerable advances have been made in recent years in the area of resources exploration. Technological change in equipment designed to detect ore bodies using methods that have a low impact on the environment has been an important area of change—for example, the use of airborne surveys and improved low-cost drilling methods to discover and then determine the extent of various mineral and energy deposits.

During the past twenty years, the principal trend in mining itself has been toward increasing the scale of operations to improve productivity and reduce costs. To date, most of the technological advances in the mining industry have revolved around drilling and explosives technology. However, increases in scale, particularly for opencut mining, could be approaching practical limits in terms of the size of specialized equipment and facilities needed and the loss of capital flexibility implied with very large projects.

While the scope for mining operations to continue to grow in size appears limited, there is potential for significant productivity gains from technology associated with improvements to materials handling and processing. Although firms will still take advantage of economies of scale where available, the industry can also be expected to make greater use of smaller ore bodies with lower capital requirements.

Major advances in mine productivity in recent times have come from technological advances associated with the conveying of ore and in-pit crushing. With production cost savings of up to 50 percent having been suggested (Ericsson, 1991), the mining industry is likely to continue research into technologies associated with the transport of minerals. For example, high angle or vertical conveying of material reduces reliance on trucks, whether above or below ground, and has the advantage of reducing the amount of energy expended on moving empty vehicles around the mine site. However, whether or not this is done depends on a range of factors, including the expected life of the mine.

Technologies for increased grinding of minerals on site will promote more efficient processing of ore because of easier conveyance from the mine and because of the opportunity provided for better blending. The latter will assist the efficiency of downstream processing activities that require feedstock material of constant grade for optimal processing efficiency. High pressure grinding of bauxite, for example, allows alumina refineries to operate with a less than one percent variation in the alumina content of bauxite, thus increasing the efficiency with which alumina is extracted from bauxite.

With ore grades tending to diminish in quality over time, grinding mills at the mine site can be expected to increase in size due to the rising volume of material that must be processed in order to achieve a given amount of output. In this area, it appears that there is scope for significant energy savings if only the mineral or beneficiated ore is brought to the surface. McMahon (1992) has suggested that systems involving automated extraction, grinding and blending of minerals underground and continuous feed via conveyers to downstream processing plants may soon be possible.

Minerals processing. One of the major advances in minerals processing in recent decades has been the improvement in processors' ability to economically recover minerals from ore bodies with ever decreasing concentrations of the target mineral. For example, with gold, copper and zinc it has become profitable to rework tailings dumps at some old mine sites. Aside from this type of technological development in minerals recovery, changes in the iron and steel industry provide a good all-round case study of how technological change can give rise to substantial productivity improvements, especially with respect to the amounts of capital, labor, raw materials and energy required in processing.

In the past, the iron and steel industry has responded to rising labor and energy costs mainly with increased mechanization and automation. The result has been to produce an industry with enormous capital requirements. Pressure to reduce the capital intensity of this industry has led to the development of novel production techniques with smaller capital costs and simpler raw material needs (Labson et al., 1992).

The traditional method of producing steel using a blast furnace involves large-scale plants to achieve the level of efficiency necessary to be competitive. An efficient modern blast furnace requires an annual output of about three million tonnes to achieve the necessary economies of scale. A sinter plant and coking oven is required to produce the raw material inputs to the blast furnace, making the integrated blast furnace route extremely capital-intensive, with an investment of about US$4.5 billion for the construction of a grassroots plant (Cusack, 1992). Some advances have been made in reducing coke requirements by injecting pulverized coal. While consumption of pulverized coal in blast furnaces is increasing, there are two other steel-producing techniques which may offer greater potential to reduce the need for specialized raw materials and the facilities to produce them.

The first of these is the electric arc furnace method. The quantity of steel produced by electric arc furnaces has grown threefold in the past ten years. This method uses steel scrap and, in some cases, a small amount of pig iron to produce low-value steels. The plants are much less capital-intensive than the traditional blast furnace, with annual output rarely exceeding 500,000 tonnes. On the other hand, the main disadvantage of the electric arc furnace method of steel making is the limited range of products and the need for quality scrap.

The second alternative method for producing steel is by directly reduced iron (DRI) making technology. Direct reduction negates the need for the coking ovens, sinter plants and blast furnaces associated with traditional steel making. Iron ore is mixed with coal in a reactor to directly reduce iron ore to pig iron and semi-steels. A second generation of the technology is being developed to enable the production of high-quality steel from variable grade iron ores.

Direct smelting technologies also offer similar benefits for base metal smelting, although technological advances in this area are less developed than for the iron ore and steel industry, particularly where the feedstock contains several base metals such as lead zinc or lead copper. Nevertheless, new Sirosmelt/Isasmelt technology, which results in improved performance of smelters in the recovery of base metals and other metals, including nickel and tin, is now being used commercially in Australia and elsewhere.

The energy sector

In contrast to the above discussion of technological change in mining and minerals processing, it is useful to see how technological change has affected the types of inputs (primary fuels) used in the production of a secondary energy source—electricity. The electricity industry has experienced considerable technological change affecting both demand and supply. The proliferation of electrical appliances and increased electrification of developing countries has meant that electricity consumption growth has been particularly rapid compared with other forms of energy. For example, over the period 1977–87, world commercial energy consumption increased by 20 percent, while electricity consumption increased by 43 percent (World Resources Institute, 1990). In both developed and developing countries, growth in electricity consumption far outstripped that for energy as a whole.

Accompanying this rapid growth in consumption have been substantial changes in the fuel mix used in electricity generation. While changes in energy prices and perceptions of stability in energy markets were major factors in these fuel mix changes, technology has also played an important part. In Japan, for example, coal accounted for about 80 percent of fossil fuels burnt in electricity generation in 1960. During the 1960s, as real oil prices fell, both the share of oil and the absolute amount of oil used for electricity generation increased rapidly. By 1973, the amount of coal used was only half that in 1960, and two-thirds of Japan's total primary energy requirements for electricity generation were met by imported oil. The first oil price shock in 1973 triggered a major shift away from oil, and this was reinforced by the second oil price shock in 1979. The use of coal began to increase again, but more significant was the rapid rise in the use of natural gas and nuclear power. By 1990, the shares of energy sources in Japanese electricity generation were oil, 27 percent; nuclear, 26 percent; gas, 22 percent; hydroelectricity, 13 percent; coal, nine percent; and others, three percent (Labson et al., 1992).

In the 1980s, however, concern over the safety of existing nuclear technologies and the lack of commercial technology for long-term storage or disposal of nuclear wastes has slowed considerably the rate of growth in nuclear power use. It seems that further technological developments will be required before the rate of growth in nuclear energy can increase. The rapid growth in the use of natural gas as a fuel for electricity generation reflects the development of new technology that has enabled natural gas to be liquefied for transport in ships. This opened up previously isolated markets, such as Japan, to natural gas trade. Liquefied natural gas trade has increased from 1.5 million tonnes in 1970 to 44 million tonnes in 1990, and this trade is expected to increase substantially in the next ten to fifteen years (Gillan, 1992).

With prices of primary energy changing rapidly and often unpredictably, there have been significant changes in another aspect of electricity generation. Increasingly, in countries such as Japan, individual plants are being built or modified to use more than one fuel. Multifuel power stations offer short-term flexibility in fuel choice and longer term security against unanticipated energy price increases or supply interruptions.

Environmental considerations will continue to force change in the structure of the energy sector, with technological development a major factor in this change. For example, considerable research continues into improving the safety of nuclear power technologies and into the safe disposal and storage of nuclear wastes. Research is also continuing into renewable energy technologies. On this front, the efficiencies of solar thermal and solar photovoltaic devices are improving steadily over time, while the unit costs are decreasing. While still some way from widespread commercialization, these technologies are finding niche applications and should increase in importance over time (Labson et al., 1992). Efforts to reduce greenhouse gas emissions can be expected to increase the interest in renewable energy technologies because of their low or negligible levels of greenhouse gas emissions. Such efforts will also stimulate research into more thermally efficient fossil fuel technologies.

TRADE POLICY ISSUES

Trade policy encompasses a range of tools that can be used by a government to affect the involvement of its country in the international economy. The ways in which trade policy measures are employed will reflect the views of a government about the nature of the international trade environment and the country's place in it. Economists have traditionally advocated a free-trade policy position because of the substantial long-term efficiency gains derived from such a policy stance. However, in pursuing strategic medium-term objectives, government policy makers have often sought to achieve domestic gains through interventionist trade policies which distort the quantities and prices of goods traded in the international economy.

The expected continuation of relatively high growth in demand for minerals and energy commodities in the Northeast Asian region means that there are considerable opportunities for regional exporters to also expand their activities. The relatively limited availability of domestic mineral and energy resources in Japan, Korea and Taiwan indicates that there are significant market opportunities for established major resource exporters such as Australia, as well as for those with less well-developed resource exporting sectors such as the Russian Far East and China. In taking advantage of the growing international demand, exporters will benefit from a reduction in barriers to trade and from the development of policies within their own countries aimed at improving sectoral competitiveness.

There is a range of trade and industrial policies employed by countries in the region which can adversely affect the free flow of trade in accordance with the principles of comparative advantage. Such measures are typically used to target imported goods and affect their import price or restrict the quantities imported. Among them are ad valorem tariffs, specific duties, import quotas and voluntary export restraints. Policies aimed at assisting domestic industries directly to enhance their international competitiveness include local content requirements, export subsidies, export credit subsidies and national procurement policies (Krugman and Obstfeld, 1988). Quite often, accompanying such protective measures is a bureaucratic structure which impedes the flow of imports of commodities. This form of protection is difficult to detect and its effects are also difficult to quantify compared with tariffs. Examples of these policies include inadequate staffing of import clearance agencies, import licensing requirements, irrelevant health and safety regulations and foreign exchange controls.

Tariff barriers

A tariff on an imported good raises the price which competing domestic producers can charge and consumers must pay for a product. This has the effect of raising domestic production in the industry in question and reduces domestic demand for that product, thus reducing imports. The increased domestic production in the protected industry draws capital and labor away from other more efficient sectors of the economy which receive less or no assistance or protection. Consequently, production in these sectors declines. Additional impacts will be felt by other domestic firms that use the protected industries' output because they will face higher costs. If such industries are export-oriented, their international competitiveness will be eroded because of the relatively higher input costs they must pay compared with those paid by international competitors.

The effect of a tariff on economic efficiency depends on whether an economy is operating at or close to full capacity. When most factors of production in an economy are fully employed in their most economically efficient roles, the imposition of a tariff will draw resources away from their most efficient uses and into the protected sector. At less than full employment, the loss of efficiency from protection may not be as great, as it uses some idle capacity. Nevertheless, there will still be economic costs associated with competition for some scarce resources and higher prices.

The net effect of tariffs is a loss of world and national welfare. This occurs as resources are not allocated to maximize gains from comparative advantages and lower levels of international trade result. The benefits of specialization (a larger consumption set and access to a greater variety of goods) and trade are therefore diminished. These losses are compounded if protected industries (as a result of reduced competition) fail to adopt new technologies and invest less in research and development.

With respect to the Northeast Asian region, tariffs on resource imports typically increase with the degree of processing (Glance et al., 1991). This type of escalating tariff structure can act as an impediment to increased value-added trade on an interregional and intraregional basis. The rate at which tariffs escalate varies from country to country. As can be seen from Figure 13.1, tariff escalation is particularly rapid in China and more gradual in Japan, Taiwan and Korea. This type of tariff structure creates a bias toward imports of unprocessed raw materials at the expense of value-added imports.

A recent development in relation to tariff protection is the pressure in some parts of the developed world for trade restrictions (taxes or bans) and other punitive actions against countries whose environmental regulations fall below the standards of those imposing the restrictions. However, the World Bank (1992) notes that studies in several developing countries show the claims that free trade leads to environmental degradation and that industries migrate to such countries seeking comparative advantage from their lower environmental standards cannot be substantiated.

Nontariff barriers

While tariffs are probably the most common trade protection measure employed, nontariff barriers to trade have grown in importance as tariffs on many items have been slowly reduced, largely through the GATT process. Nontariff barriers to trade include all other forms of government intervention applied at either the national border or through domestic assistance policies. Nontariff barriers such as import quotas and foreign exchange controls are applied at national borders. These nontariff barriers have similar effects to tariffs in that they act to restrict imports and raise the domestic price of the good in question.

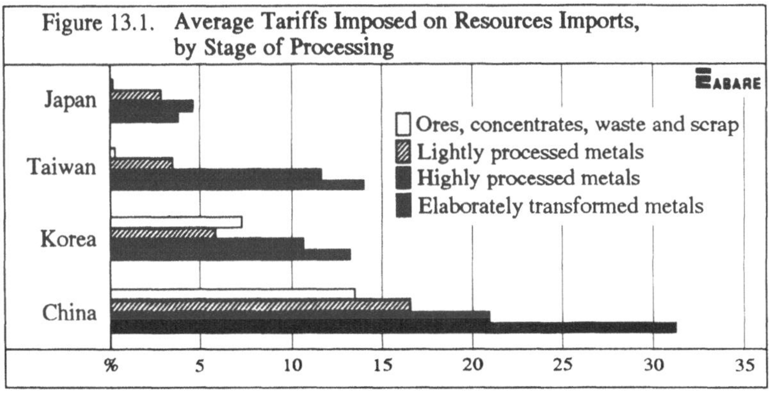

Source: Glance, 1991

Figure 13.1. Average Tariffs Imposed on Resources Imports, by Stage of Processing

There are, however, important differences between these types of nontariff barriers to trade and tariffs on imports. First, the price increases caused by nontariff barriers are often less visible and more difficult to measure. Second, while tariffs drive a wedge between world and domestic prices, they do not prevent domestic prices from reflecting movements in world prices. Nontariff barriers, on the other hand, often break this link, leaving little incentive for domestic producers and consumers to adjust to changes in world market conditions. Third, tariffs generate government revenues which can be used to fund programs or reduce other forms of taxation. In contrast, most nontariff barriers do not generate government revenue.

Glance et al. (1991) have noted that, in the Asian region, the uses of nontariff barriers to trade tends to mirror those of tariffs in that they escalate with the degree of processing. Thus, nontariff barriers increase the overall level of protection faced by mineral exports beyond that posed by tariffs. It was also found that nontariff barriers tend to be commodity specific. For example, in Japan, Taiwan and China, 29–40 percent of all resource tariff line items are affected by nontariff barriers. In Korea, nontariff barriers to resource imports affect around three to five percent of tariff line items.

Perhaps the most common form of nontariff barrier in Northeast Asia is the practice of nonautomatic licensing of imports. Licenses are required so that imports can gain access to domestic markets. The discretionary ways in which governments employ licensing arrangements and the variable nature of the requirements to satisfy these arrangements can be particularly restrictive to trade. In Taiwan, the categorization of imports into different levels of desirability acts as a de facto quantitative restriction on trade. Licensing arrangements in Japan have similar results but are achieved by requiring exacting and copious informational material.

Environmental, health and safety requirements are prominent nontariff barriers to trade, especially in Japan and Taiwan (Glance et al., 1991). The degree of subjectivity which surrounds these standards has led to different governments imposing different standards on imports of similar goods. Such differences in standards have the potential to bias trade and development toward nations with less stringent health, safety and environmental regulations. In many cases these are the developing nations. As with many other nontariff barriers, the effect of these regulations is hard to evaluate as it is difficult to apply a cost to such normative subjects as health, safety and the environment.

Domestic assistance policies

Domestic policies which can adversely affect trade in resources include a wide range of government intervention measures such as production subsidies, product standards and discriminatory tax schemes. Such measures encourage domestic production and/or restrict trade relative to what would otherwise be the case. Determining the distorting effect of these interventions is often difficult, especially where the measures may have some justification on other grounds, such as health and safety standards.

Examples of domestic protective measures include discriminatory tax schemes, production subsidies, market structures and particular institutional and commercial trading arrangements, public procurement policies, subsidies to inputs of production, targeted credit allocations, and measures to assure security of resource supplies.

A nation may impede the free flow of international trade through institutional arrangements which govern business behavioral codes and market structures and commercial trading arrangements as well as government procurement policies. On the latter, governments have the potential to obscure trade flows because of the enormous amounts of goods and services they purchase. Governments tend to buy domestically for reasons such as prestige and security. If these purchases are from protected industries with higher than world prices, inefficiencies are introduced into the system as more could be purchased overseas for the same price.

Regional trading blocs

One potential threat to the multilateral trading systems encompassed in the GATT is the emergence of regional trading blocs like the European Community and the North American Free Trade Agreement (NAFTA). These trading blocs have been formed to serve several purposes: to improve national welfare through income and efficiency gains of reducing intraregional barriers; to increase regional negotiating leverage; and to promote regional political cooperation and security (Schott, 1991).

Regional trading blocs represent very much a 'second best' solution to the problem of reducing barriers to trade in order to improve economic welfare in member countries. For example, such agreements do not necessarily result in the emergence of trade patterns reflecting comparative advantages. They may well serve to perpetuate external trade barriers already in existence within the member countries and reduce market access for more efficient producers.

In overall terms, the elimination of trade barriers, including border measures (such as tariffs, quotas and import licensing arrangements) and domestic assistance measures which protect industries from import competition, can be expected to result in expanded trade and stronger economic growth in Northeast Asia. Given the economic complementarities that exist within the region, the interests of exporting and importing countries would be served by removing arrangements that restrict opportunities for competitive and efficient exporting countries to improve their standards of living and that prevent processing and manufacturing industries in importing countries from obtaining their input requirements on the best possible commercial terms.

CONCLUSIONS

Northeast Asia is an area which is likely to experience continued strong growth in demand for resource commodities into the twenty-first century. Following the major political and economic changes of recent years, the Far East region of Russia can be expected to play an increasingly significant role in minerals and energy trade. In fact, it is likely that over the next one or two decades, the region could emerge as one of the major suppliers to the relatively resource poor but large consumers of resource commodities—Japan, Korea and Taiwan. China can also be expected to develop its resource supply industries to take increasing advantage of demand growth in the region. However, the resource industries in the region face complex and dynamic production and market circumstances in the 1990s.

Economic growth will be the engine driving both import demand and, hence, investment in resources development. While the short-term outlook for regional economies is somewhat patchy (especially for Japan), longer term prospects for the region are for solid growth for much of the decade. The rate at which resources are developed in the region and in other major supplying areas, such as Australia and Southeast Asia, will also be affected by domestic policies relating to resource development and by environmental concerns.

Concerns over global warming, if translated into policy actions, may have substantial effects on the energy-producing industries in general and the coal industry in particular because of its major role in electricity generation. There would also be significant implications for the size and location of the major energy-using industries of the region, such as aluminum. The development and adoption of new technologies relating to electricity generation and use will be an important element in any response to the concerns about climate change. Similarly, technological change will contribute to the conservation of scarce mineral resources and the pursuit of sustainable development ideals. New ways to increase recycling and use less materials and energy in production will be increasingly developed and adopted for sound economic as well as environmental reasons. Research efforts to secure technological change and boost the productivity of capital, labor and raw material inputs in mining will, however, require appropriate market signals and policy settings by governments.

If the countries of Northeast Asia are to maximize their development potential, each will have to pay particular attention to the avoidance and/or removal of impediments to trade in minerals and energy commodities. Tariff and non-tariff barriers to trade result in reduced national incomes for the countries imposing them, as well as reduced incomes for potential external suppliers of the product. The net result is reduced economic activity and a less prosperous regional economy.

Finally, while the various issues discussed in this chapter will have different implications for different countries, they are nonetheless of overriding importance in considering the future of the minerals and energy sectors of the region. The degree to which they can be dealt with effectively within each country's policy processes will largely determine not only the future of these industries but also the extent to which Northeast Asia can continue to grow in importance in the global economy. Given the substantial base of human skills, manufacturing capacity, and abundance of natural resources in areas such as the Russian Far East and China, the future economic prosperity of the region should be assured.

REFERENCES

ABARE, 1991a, "Macroeconomic Setting," *Agriculture and Resources Quarterly,* Vol. 3, No. 4, pp. 450–455.

______, 1991b, *Minerals and Metals Trade in the Asia Pacific Region,* Report for the Economic and Social Commission for Asia and the Pacific, Bangkok, Thailand.

______, 1992, "Macroeconomic Setting," *Agriculture and Resources Quarterly,* Vol. 4, No. 1, pp. 10–13.

Asian Development Bank, 1991, *Asian Development Outlook,* Manila.

Cox, A.J., B. Nagle, and K. Lawson, 1990, *Factors Influencing World Demand for Metals,* ABARE Discussion Paper 90.71, AGPS, Canberra.

Cusack, B.L., 1992, "Hismelt—New Generation Ironmaking," paper presented at the National Agricultural and Resources Outlook Conference, Canberra, 4–6 February.

Ericsson, M., 1991, "Minerals and Metals Production Technology, A Survey of Recent Developments," *Resources Policy,* Vol. 17, No. 4.

Findlay, C., and A. Edwards, 1991, "Economic Relations between the Soviet Union and the Pacific: An Assessment of Opportunities," in Drysdale, P., and M. O'Hare (ed.), *The Soviets and the Pacific Challenge,* The Australia-Japan Research Centre, Australian National University, Canberra.

Gillan, P., 1992, "Opportunities for Expanded Trade in Liquefied Natural Gas in the Asia Pacific Region," *Agriculture and Resources Quarterly,* Vol. 4, No. 1, pp. 66–78.

Glance, S.H., G. Murtough, B.G. Johnston, and J. Winton, 1991, "Barriers to Resource Commodity Trade with Asia," *Agriculture and Resources Quarterly,* Vol. 3, No. 2, pp. 241–254.

Hartwick, J.M., and N.D. Olewiler, 1986, *The Economics of Natural Resource Use,* Harper and Row, New York.

Hinchy, M.D., and B.S. Fisher, 1991, "Global Emission Trading for Greenhouse Gases: Possibilities and Constraints," ABARE paper delivered at the Conference on Environmental Strategies for Asia Pacific Oil and Gas, IBC Asia (Conferences) Ltd., Kuala Lumpur, 26–27 August.

Hinchy, M.D., B.S. Fisher, and B. Bowen, 1991, "Global Carbon Dioxide Emission Trading: A Possible Way to Encourage the Use of Clean Coal Technologies?" ABARE paper presented at the International Conference on Policy Issues and Options for Environmentally Sound Coal Technologies, Beijing, 1-6 December.

Hinchy, M.D., B.S. Fisher, and N.A. Wallace, 1989, *Mineral Taxation and Risk in Australia,* ABARE Discussion Paper 89.8, AGPS, Canberra.

Japan External Trade Organization, 1990, *White Paper on International Trade—Japan 1990,* Tokyo.

Krugman, P.R., and M. Obstfeld, 1988, *International Economics: Theory and Policy,* Scott, Foresman and Company, Illinois.

Labson, B.S., B. Jones, P. Gooday, and M.J. Neck, May 1992, "Minerals and Energy to the 21st Century," ABARE paper contributed to the Pacific Economic Cooperation Council's Fifth Minerals and Energy Forum, Sydney, 13–15 May.

McMahon, D., 1992, "Escaping from Drill and Blast Tyranny—A Path to Automated Mining," *Mining Review,* Vol. 16, No. 1, pp. 12–16.

Rose, R., and A. Cox, 1991, *Australia's Natural Resources: Optimizing Present and Future Use,* ABARE Discussion Paper 91.5, AGPS, Canberra.

Salazar Sparks, J., 1992, "International Opportunities in the Mining Industry: An Australia-Chile Partnership," paper presented at the Australian Mining Industry Council, 1992 Minerals Industry Seminar, Canberra, 7 May.

Schott, J. J., 1991, "Trading Blocs and the World Trading System." *The World Economy,* Vol. 14, No. 1, pp 1–17.

Thorpe, S., B. Sterland, B.P. Jones, N.A. Wallace, and S. Pugsley, 1991, *World Energy Markets and Uncertainty to the Year 2100: Implications for Greenhouse Policy,* ABARE Discussion Paper 91.9, AGPS, Canberra.

World Bank, May 1992, *International Trade and the Environment,* edited by Patrick Low, Washington, D.C.

World Resources Institute, 1990, *World Resources 1990–91,* Oxford University Press, New York.

Chapter 14

EXPLORATION AND MINING CONDITIONS IN THE ASIA-PACIFIC REGION: AN INDUSTRY PERSPECTIVE

Douglas Ritchie

INTRODUCTION

This chapter reviews some of the issues that a mineral explorer and miner must take into account when evaluating the desirability of conducting operations in various countries. The principal issues examined are the foreign investment, taxation, local equity and certainty of tenure policies adopted by host governments necessary to attract and retain investment by exploration and mining interests.

For host governments there is a justifiable need to obtain a fair reward for the use of its assets. The chapter highlights some of the wider economic benefits accruing to a nation from exploration and mining investment.

THE EXPLORATION/MINING INDUSTRY

Geologists, by nature, tend to be optimistic and generally see development potential for a range of minerals in any country in the world. While geological development potential is an obvious factor in an explorer/miner's choice of where to invest, there is a range of economic, commercial and political factors which must be considered when selecting a country in which to invest the scarce and high-risk capital required in this industry.

The minerals industry, throughout the chain from exploration through construction and development, processing, marketing and trade in derived products, is a high-risk business with many unique characteristics. A government wishing to attract and optimize the benefits of mining to its national economy should provide an investment climate and pass legislation that firmly recognize the industry's distinctive characteristics. Following are some of the conditions intrinsic to the industry:

J.P. Dorian et al. (eds.), CIS Energy and Minerals Development, 225–242.

Exploration is an extremely high-risk activity. While technical successes are not rare, the chances of making a commercial discovery are low, perhaps as low as 1:1,000 and each viable discovery must, therefore, be capable of supporting a considerable amount of unsuccessful work.

A study of U.S. exploration between 1942 and 1967 showed that the chances of an economic discovery resulting from any one program was 1:300. In a study of Canadian exploration the odds have been estimated at between 1:100 and 1:1,500. A major study of Australian exploration discovery statistics for the period 1955–1978 by McKenzie and Bilodeau showed that of 100 mineral deposits discovered, only 43 could be considered potentially economic before taxes. After taking the tax regime into account, the number of economic deposits was reduced to 33.

Exploration is expensive. Exploration is frequently conducted in remote areas with little or no existing infrastructure or established lines of supply. The sophisticated equipment, the experts required, and the means to house and sustain them must often be imported to the host country and transported over long distances. It is estimated that the industry spends US$35 to US$40 million on exploration for every economic mineral discovery. Over the seventeen year period, 1974–1991, 61 significant mineral discoveries were made in Australia (Blain-Chris, 1992). To make these discoveries, a massive A$13.7 billion (in current dollars) was spent on exploration. Of these discoveries, only 45 have yet become operating mines.

Exploration is essential for sustained development. As ore reserves from existing mines are depleted, replacement can be achieved only through exploration for and development of new mines either nearby or elsewhere. Profits and cash flows must be high enough during the limited life of a mine to fund further exploration and closing expenditures (rehabilitation), and retain earnings for further developments so that the assets can be replaced with new discoveries.

Each mine is unique. The mine and associated processing facilities must be designed to cater to the unique spatial, physical and chemical parameters of each ore body. No two ore bodies are identical. In addition, geographical location may impose special considerations. In framing legislative regimes, it is often the case that governments give little thought to retaining the flexibility required to deal with mining development on a case-by-case basis.

Mineral exploration and mine development is long term, not a short-term, quick payback exercise. Large mines of a company the size of CRA Limited, for example, represent long-term commitments of large capital amounts. For example, the new Kaltim Prima coal mine developed by CRA and BP in Kalimantan, Indonesia, required over five years to explore and more than two years to develop and commission. The total commitment was about US$600 million with an expected mine life of around 20 years for currently discovered resources; it is likely that this mine life will be capable of being extended beyond the 20-year period. The Broken Hill mines in Australia have been operating for around 100 years and other mines developed in Australia (e.g., Hamersley iron ore, Weipa bauxite and Argyle diamonds) have been designed for a life of 20 years or more. The long time frame of world-class mineral developments (see Figure 14.1) means that mines need to be able to withstand the vagaries of the international markets and that, at the outset, the explorer/miner has to consider a broader range of risk factors than investors in other sectors whose level of risk is substantially lower (Mackenzie and Bilodeau, May 1982).

The mining industry must compete in international markets. With few exceptions, domestic consumption of minerals falls well short of production (especially when considering the output from world-class mines) and the bulk of production must be exported. As is well known, considerable fluctuations in metals prices have occurred over the last decade. Because export contracts are typically priced in U.S. dollars, the industry is exposed to currency fluctuations, which can be quite severe. Desirably, a mine will be profitable in all years, but in most instances some losses occur or at least very marginal returns on capital invested are experienced, and these must be sustained by the very much higher profits in periods of higher prices. The industry's profile is very low in these lean years compared with the prominence given it by the media during times of higher profits.

Most responsible mining companies endeavor to sustain their exploration efforts at constant levels throughout the good times and the bad. A principal reason for this is that if the highly trained expert staff required to run quality exploration efforts are laid off in the bad times, they are very difficult to replace. Another reason is that exploration acreage usually carries with it minimum (though often high) expenditure commitments which must be maintained if the ground is to be retained.

A number of mining operations (principally for gold) which have been developed in recent years are essentially short term in nature, designed to quickly exploit a known resource during a period of high prices and then to cease involvement in the mine. While such mining operations can generate considerable benefits, these investments are made to take advantage of high price periods and do not necessarily take into account longer-term cash flows. It is therefore possible that some of these mines can be subject to conditions (e.g., noncontributing national equity, higher royalty rates, etc.) which could not be sustained in the long term by a mining operation which has to weather significant periods of low prices. Governments need to differentiate between the two types of investment.

Mining companies are often required to provide social, transportation and industrial infrastructure which would normally be provided by the government for other industries. The provision of infrastructure adds significantly to the

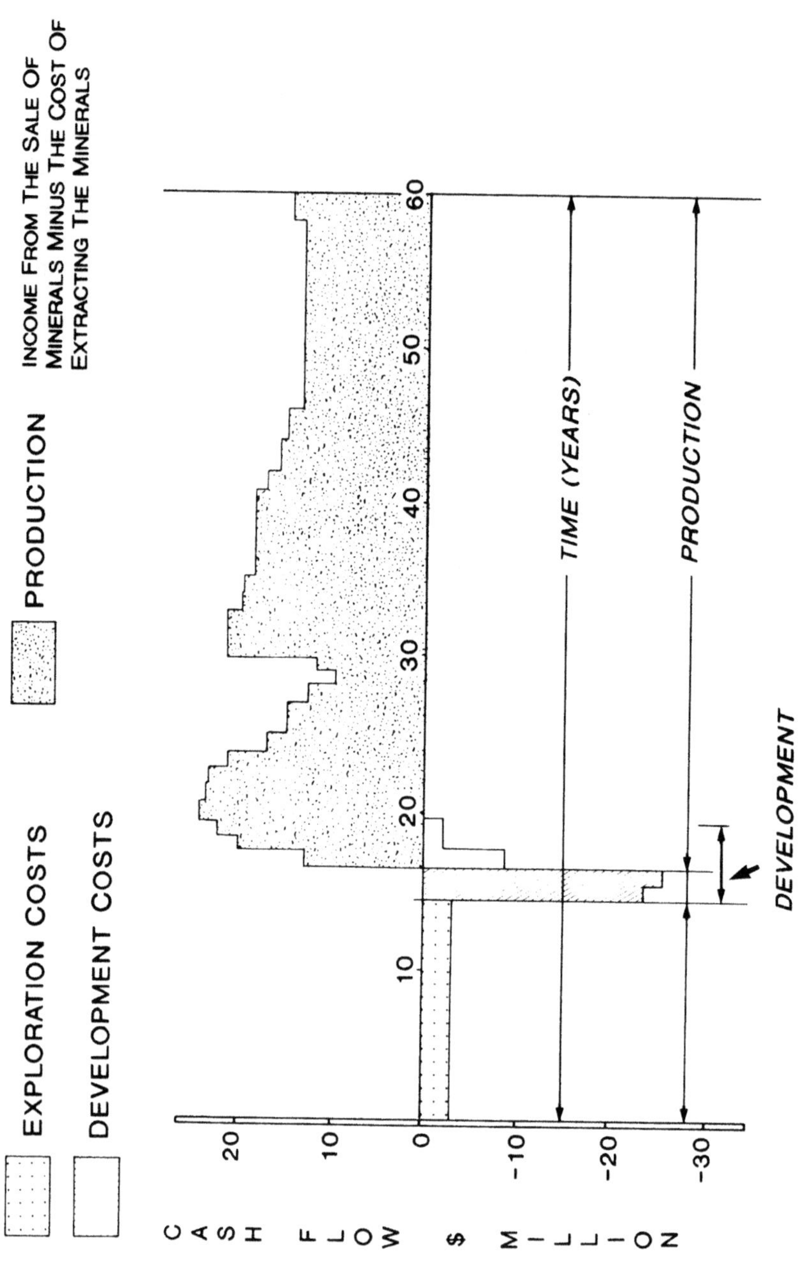

Source: McKenzie and Bilodeau, May 1982.

Figure 14.1. Mine Life Cash Flow

development costs of the miner and often benefits the host government by opening up otherwise remote and inaccessible parts of the country to potential development of other economic activities unrelated to mining.

Mining operations are capital intensive. This partly explains the industry's high productivity, but carries with it the disadvantage of heavy interest and loan repayment commitments. High cash flows are usually necessary in the early years of a mine to fund high loan repayments because lenders perceive the higher risk in mining and lend on short-term bases. Wear and tear on equipment in mining is heavy, and both replacement costs and incremental investments are high in order to maintain production levels after the early years because of declining head grades and deeper mining levels.

BENEFITS PROVIDED BY THE EXPLORATION/MINING INDUSTRY

Before turning to a more detailed review of the criteria that a minerals exploration and mining company will apply in an evaluation of where to invest its resources, it is useful to consider some of the (not always readily apparent) benefits which will accrue to those countries which actively seek and encourage investment in our sector.

Minerals are generally owned by the State, which regulates when they may be explored and mined. In return for permitting the extraction of its resources, the State is entitled to an appropriate reward. The rewards to the State are many, including the inflow of investment monies, the creation of new jobs and training of nationals in new areas of expertise, and development of previously remote areas by provision of access roads, educational and medical facilities, railways, ports and the like. The costs to the company of providing these indirect benefits to the country are often not fully appreciated by governments when calculating the level of direct charges to be imposed on a development. The State also receives considerable direct monetary benefits, including taxes on the profit of the mining company, taxes on the wages of the company's employees, royalties from the sale of production and transport levies on the use of such State-owned facilities as roads, wharves and ports.

In Australia, the mining industry contributes about 6.5 percent of GDP and about 51 percent of exports (see Figure 14.2). It directly employs about 105,000 people, and it has been determined that each job in the Australian mining industry creates 2.7 jobs in other industries.

Individual projects can be highly significant for employment. For example, development of the new Hamersley iron mine at Marandoo is expected to employ about 800 workers during construction and about 350 during operation.

The economic output multiplier for the minerals sector in Australia has been calculated at 1.7. For each A$1.00 of mineral sales (whether domestic or export), an economy-wide output of A$1.70 is generated. In terms of capital growth, for

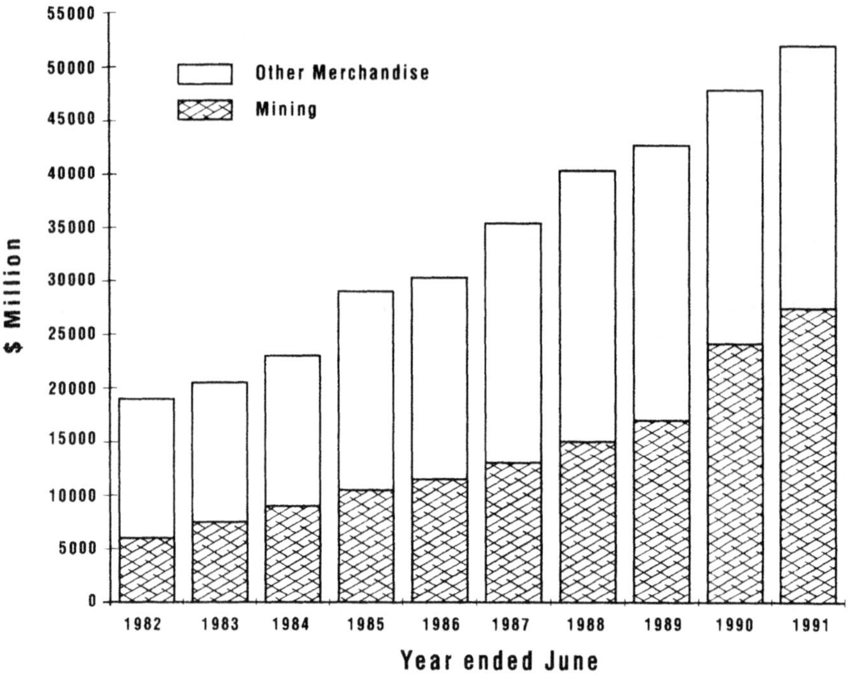

Figure 14.2. Australian Merchandise Exports, 1982–1991

each dollar spent in the minerals sector, an additional dollar of nonminerals sector capital investment is generated (provided, of course, that the resources in the economy are not already fully committed).

Another interesting example is the impact of the Bougainville copper project on the Papua New Guinea (PNG) economy. The Bougainville deposit cost A$20 million (1971 dollars) to find, define and evaluate, and the mine cost a further A$400 million to develop. About 66 percent of the mine's profits in the 20 years of its operation went to the PNG or provincial governments. Total receipts paid to the PNG government averaged 16 percent of total PNG government domestic revenues for the years 1975 to 1985. Bougainville's net sales revenues averaged approximately 19 percent of PNG's gross national product for the years 1980 to 1985. The closure of this mine resulted in a 10 percent devaluation of the Kina and a drastic pruning of government expenditure. Fortunately for PNG, other large mines have come on stream, reducing somewhat the impact of the Bougainville closure.

Freeport's operations in Indonesia provide another example of the impact of mineral projects. Associated with its operation in Irian Jaya, the company has constructed a seaport, airport, power plant and a complex road system. Freeport was Indonesia's largest taxpayer in 1988 and 1989 and annually purchases around Rp. 90 billion in domestically produced goods and services.

While these statistics will vary from country to country and depend on specific circumstances, they illustrate the potential for large indirect benefits provided by the mining industry by increasing the wealth of the country and its people. Direct benefits from taxation and employment are generally on a large scale.

What then are the issues which an explorer will address in determining whether to invest the large sums required to find, delineate and extract a resource?

PRINCIPAL INVESTMENT CRITERIA FOR THE MINERALS SECTOR

The purpose of this chapter is to examine and discuss the primary criteria that will govern the investment decisions of exploration/mining companies in the Asia-Pacific region. In practice, the criteria used apply worldwide, as there are no unique criteria to be applied in considering the Asia-Pacific region per se.

This chapter has attempted to set the scene for the discussion by highlighting some of the unique aspects of the exploration and mining industries as compared to other economic enterprises and some of the benefits accruing to host countries that provide a regime attractive to investment by our industry. While the exploration and mining industries each have their own special aspects and requirements, a decision to embark on exploration will necessarily require a prior examination of the host country's ability to satisfy the investor's criteria in providing a setting favorable to the development and operation of a mine.

As an example of perceived risk to investment, a recent *Euromoney* country risk analysis follows, with risk plotted against land area for selected countries with mineral development potential (Figure 14.3).

Political Factors

While the political system that operates in a country with geological mineral development potential does not necessarily affect its attractiveness as an exploration site, a number of political factors do.

Political stability. A stable, continuous regime, free of turmoil and dramatic change, is an important consideration for foreign investors. This does not necessarily mean the favoring of one political party over another to retain government, but rather the seeking by the investor of a host country which provides a setting where successive governments of perhaps different political persuasions have a record of encouraging foreign investment and, in particular, promoting the development of the country's minerals sector.

Law and order. A potential foreign investor will favor a country where its staff, its investment and its plant are unlikely to be placed at risk of physical attack.

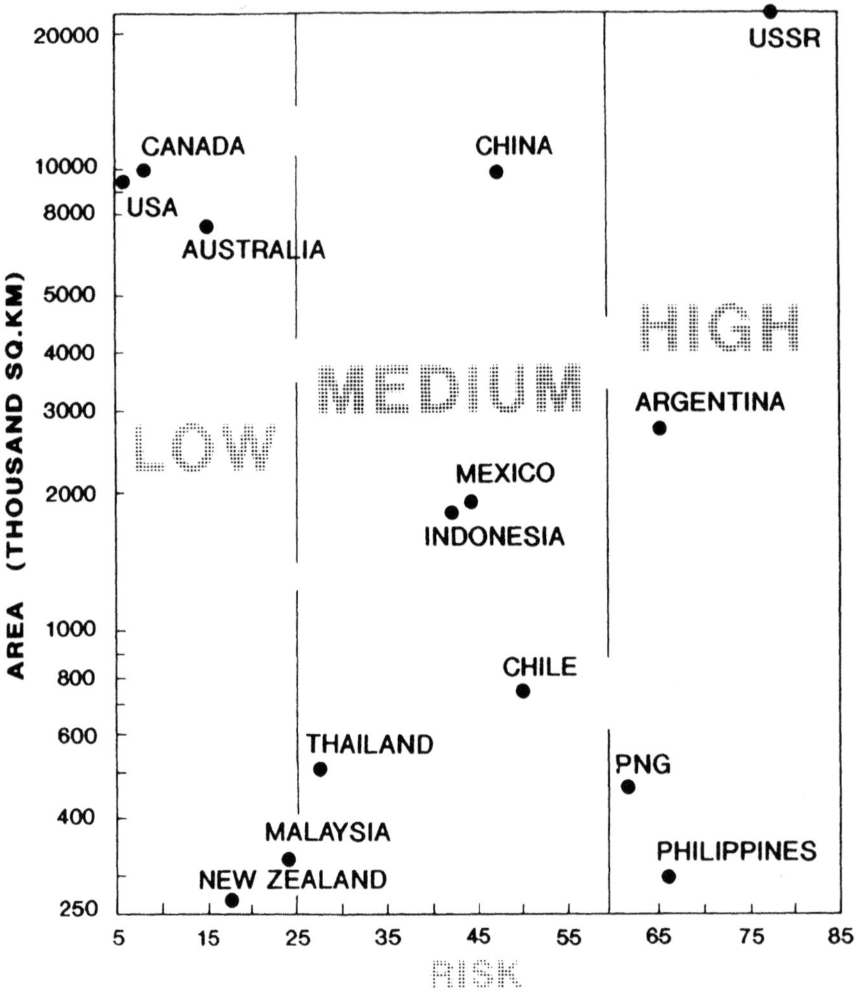

Source: *Euromoney*, September 1991.

Figure 14.3. Euromoney Country Risk for Selected Countries

Attitude towards foreign investors. It is important to an investor's perception of the potential for successful operations in a country whether or not the regime maintains open and nondiscriminatory attitudes and policies towards foreign investors.

Bureaucratic efficiency. A government bureaucracy free of corruption and inefficiency will be instrumental in creating the flexibility and freedom necessary for productive exploration work.

Political risk is obviously an important factor in investment decisions of any kind, but more so in an industry that has such a high capital investment up front as the minerals industry. If minerals development can occur in a country that is not resource rich but has less political risk, some cost penalties can be worth carrying. Clearly, however, there are limits to this and the outcome may be that a resource remains undeveloped until a more opportune time.

It should not go unremarked that assessment of political risk is subjective and difficult to quantify compared with the assessment of defined economic factors such as taxes, royalties and the like. Political risk, however, must always be an important element in the investment decision equation.

Foreign exchange and expatriation of capital

A very important factor in determining the desirability of a country for investment is its foreign exchange regulations and the ability for a foreign investor to repatriate profits. In a study undertaken in the United States to determine the importance of different foreign risk factors to large U.S.-based multinational companies (not mineral specific), 84 percent of respondents indicated that the issue of repatriation of profits was highly important—more important in relative terms than all other factors (see Table 14.1).

The institution of a regime which allows investors to consistently and freely repatriate profits as well as freedom to market its products is essential to most investors. It is important that any rules relating to capital, including repatriation of foreign exchange, local pricing policy, and withholding taxes, are fair, clearly defined and not subject to sudden change or redefinition. The more certain and defined these matters are from the outset of exploration, the more attractive will be foreign investment in the resources sector of the host country.

While many countries are cautious about control over their hard-earned foreign currency, it is essential that companies be allowed to repatriate earnings. Profits, growth and survival are fundamental goals for all companies. The mining company's exploration strategy is guided by both its profit expectations and its survival considerations. Governments should be sensitive to these issues.

Equity participation

In the case of high-risk, high capital cost investments which typify the mining industry, company aims are likely to include a desire for majority equity ownership and a requirement for all minority participants (including the government) to bear their pro rata shares of the risk from the outset. This aim must often be balanced with the desire of the government to impose local participation on a venture either by way of a direct government-held or private sector-held equity by nationals from the outset or by way of a post-discovery divestment in favor of nationals at a more advanced stage.

In a survey of more than 50 major international mining companies representing 60–80 percent of worldwide exploration expenditure conducted by the World

Table 14.1. Importance of Country Evaluation Factors

	Least Important (%)			Most Important (%)		
	1	2	3	4	5	N
Repatriation of profits	3	5	7	40	44	95
Price controls	5	16	27	37	15	93
Restriction on foreign equity participation	5	6	22	38	28	95
Local content requirements	7	19	15	49	10	94
Restrictions on employment of expatriate managers	14	29	26	22	8	94

Note: Column N is the actual number of respondents to each question.
Source: World Bank, 1990.

Bank in 1990, mandatory majority government or local private participation was regarded as a major disincentive or barrier to investment in the minerals sector. However, it should also be noted that minority private local participation was regarded by many as a positive factor (Table 14.2).

It makes some sense, especially in developing countries, for governments to encourage local participation in mining ventures where this is sensibly and rationally administered. Among other benefits, it provides a means for a country to assimilate expertise in the minerals sector.

If this is the case, then the local partner should be capable of funding its proportionate share of exploration and development expenditure (and thus, also take a share of the risk) or at least provide expertise or other services to the value of its share in the project. If a government is fully aware that local partners are not in a position to pay their way, then insistence on local participation is counterproductive to exploration and mining.

In the long term, few explorers can afford to assume 100 percent of the risk in return for a smaller proportion of the rewards. A requirement to fund 100 percent of the capital development cost of a new mine while holding less than 100 percent of the equity means that only mines with very high projected profits are attractive. Lesser mines, which would otherwise be profitable, are not capable of being developed and, thus, opportunities are lost to both the explorer/miner and the host country.

If a government does wish to impose local equity participation on a largely foreign-funded venture, it is likely the foreign investor will feel more comfortable with a private sector partner rather than a public sector partner. Briefly, the main reasons for this are:

- diversion of public funds to high-risk areas from traditional areas such as the provision of roads, hospitals, schools, etc., may color the general public's feeling against the development;

Table 14.2. Major Disincentives to Mining Development Related to Government Policy

	Prohibitive	Major Disincentive	Minor Disincentive	Not a Disincentive
	(------ % Responding ------)			
Mandatory majority government participation	58	37	2	3
Mandatory majority local private participation	45	34	16	5
Mandatory provisions of social services (schools, hospitals)	9	26	47	18
Restrictions on negotiating wages	—	42	45	13
Expatriate personnel limitations	5	34	45	16
Mandatory minority government participation	—	21	45	34
Mandatory minority local private participation	—	13	45	42
Mandatory training of nationals	—	—	29	71

Source: World Bank, 1990.

- limited expertise within the government to protect its investment;
- potential for conflicts of interest between the government's role as both regulator and promoter of company interest; and
- reduction in the incentive of the private sector to invest and assume risk.

Taxation regime

As previously noted, compensation to a government for the use of its mineral assets is justified. However, governments must be aware that their policies, particularly in the taxation area, have a critical bearing on the ability of explorers/miners to develop internationally competitive mines and thus create wealth for the host country. In other words, governments must caution against the imposition of taxes that remove incentives for explorers to take the very high risks involved in exploration work. In designing taxation policies, consideration should be given to the high-risk/high-cost nature of exploration and the great potential for sustained development and wealth created through the development of economic mineral deposits.

Mining companies can be expected to consider any reduction in their tax burden desirable. By reverse logic, one might assume that an increase in the tax burden is always beneficial to the state; however, this is demonstrably not the case.

The "total tax" system, which includes company taxes, additional profits taxes, royalties, withholding taxes, depreciation rates and in some instances various regional taxes, represents costs and, as such, reduces the value of a resource to the developer. The total tax system, therefore, should seek to minimize any distorting effect it has on investment decisions and must take into account its long-term effect on the mining industry. A high total tax may initially boost government revenues, but in the long term it will reduce receipts and the prospects of sustained development through exploration. Investors will choose to place their funds where there is an investment climate more favorable to the special needs of the mining industry. There are well-documented cases where this has happened, notably in Canada in the 1970s.

It should be noted that government revenues associated with various tax burden levels are not realized all at once. A relatively high tax burden may discourage new mine development and further exploration but not force existing mines to close prematurely.

Alternatively, a lower tax burden may collect only modest tax revenues from existing mines but more than make up the difference by stimulating exploration and sustaining development through the creation of new mines. However, most of the revenues associated with the latter tax regime will only be realized in the future. Most governments prefer to earn revenue now rather than later and so do not take much cognizance of this concept.

A major constraint in many mining development negotiations lies in the fact that the host country's fiscal regime is set down in the country's legislation, and there often is little or no flexibility on the part of government negotiators in adjusting the fiscal package to the requirements of a particular project (or indeed, to the special needs of the mining industry as a whole). Governments should be aware that high-risk investments require a different fiscal structure than low-risk investments. The lower the probability of success (the higher the risk) of an investment, the lower will be the probability-adjusted net present value of the project—prospective investors generally take probability factors into account in estimating their minimum acceptable after-tax rate of return. Countries not currently enjoying the fruits of major mining development should also note that once one or two large-scale projects are initiated and exploration efforts are successful, other companies are often willing to negotiate contracts to explore an area of similar geologic structure that provide for less generous financial terms than were accorded the initial investors, for example, the successive "generations" of minerals Contracts of Work in Indonesia.

Flexibility, not uniformity, is required in a tax system if the full potential value of a country's mineral endowment is to be realized. Such flexibility may, for instance, discount normal tax rates (perhaps for royalties and during the initial years of post-commercial production) and accelerate depreciation for a particular development where unusually high infrastructure costs are to be borne by the developer.

Security

The term security is used to mean the ability to explore within defined parameters and utilize the successes of exploration by mining and marketing the resource under a tenure guaranteed by the government. Commitment by explorers of large amounts of capital requires that a tenure should be certain and, if exploration is successful, a right to mine should be guaranteed. A 1989 survey of 32 international mining companies by the East-West Center of Hawaii disclosed that the "right to mine" was considered critical by 97 percent of respondents (Figure 14.4). A similar conclusion was reached by the 1990 World Bank survey of international mining companies (Table 14.3).

Flexibility needs to be built into budgets and unnecessary bureaucratic procedures avoided so that the exploration team has the freedom to move and adapt quickly to its new operating environment and exploration results. This flexibility can be provided within set limits and normal discipline and accountability. Overly tight control will demotivate and reduce chances of success.

The ability of the exploration industry to respond to the demands of the host societies for growth and prosperity may be more dependent on the policies of governments than on the evolving skills, ideas and technology of exploration science. Governments, therefore, have a responsibility to provide a stable and efficient operating regime for the industry as well as access to land and data.

Those nations that do not offer security, stability and appropriate policies for the longer term will not attract exploration and will forfeit opportunities to discover and develop their mineral resources. Most mining investors will favor

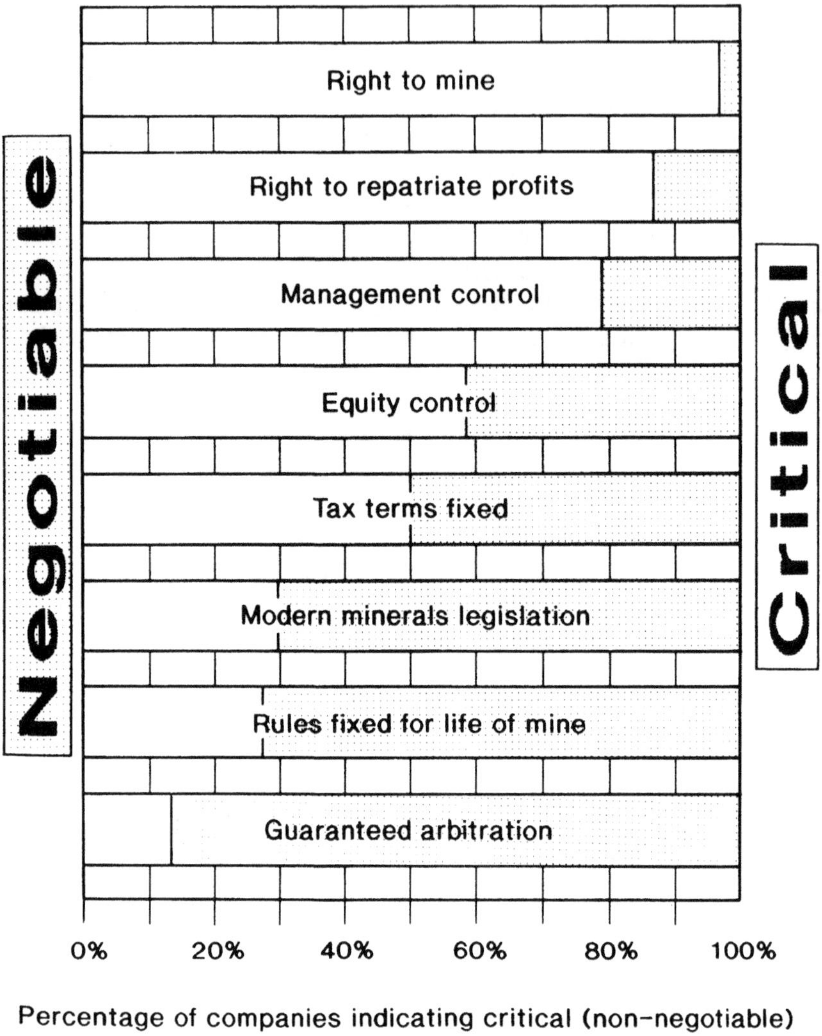

Figure 14.4. Critical and Negotiable Factors for Major Mineral Exploration in a Country

countries with stable fiscal regimes and reasonable incentives for investment over those with a higher mineral potential and changeable or discriminatory rules.

CONCLUSIONS

The considerations which explorers make before committing to exploration go well beyond the simple issue of mineral endowment. By nature, exploration carries high costs and high risks and the exploration environment itself is constrained by numerous factors beyond the control of the explorer.

Table 14.3. Major Incentives to Mining Development Related to Government Policy, Economic Environment and Mining Legislation

	Essential	Major Incentive	Minor Incentive	Not an Incentive
	(------	------ % Responding	------	------)
Guarantee of mining rights before starting exploration	50	50	—	—
Guaranteed tax regime	52	45	3	—
Convertibility guarantee	47	53	—	—
Established mining code	46	46	8	—
Permitted external accounts	41	56	3	—
Realistic exchange rate	29	63	6	2
Accelerated amortization	17	63	15	5

Source: World Bank, 1990.

Mineral resources are indeed finite, but evidence suggests that these limits are not yet being approached. Given this situation and the great potential benefits that exploration can contribute to mining companies, governments and societies, there is a strong argument for countries to provide a fair and stable system within which explorers may operate—that is "bona fide" explorers, explorers committed not to short-term fund raising, but to the development of economic, cost-competitive mines and ultimately, prosperity and wealth for host countries.

If one looks at the global context (see Figure 14.5) it can be seen that in most of the stable regions of the world with mineral development potential, considerable sums are invested in mineral exploration. It is also worthy of comment that the Asia-Pacific region in the years 1988–89 accounted for about one-third of total global exploration expenditure (A$915 million). However, about 85 percent of exploration expenditure in the Asia-Pacific region was spent in Australia alone.

It should also be noted that the bulk of the exploration expenditure in the region, excluding Australia, was made by foreign companies and that it was almost exclusively confined to Papua New Guinea and Indonesia, countries which have well-developed mining laws and firm foreign investment regimes which specifically address the exploration/mining sector.

These investments did not occur by chance, but were encouraged by firm and supportive government policies. That is not to imply that Indonesia and Papua New Guinea provide ideal investment climates. Indeed, it may be concluded that the lack of proper regimes in other countries in the region have, by default, assisted Indonesia and Papua New Guinea to reach this position.

It is suggested that other host governments that seek to benefit from mining investment in their countries should consider these facts and conclude that the provision of a regime attractive to exploration/mining is, to a large degree, dependent upon their own actions. We are, after all, in an international industry and all countries must compete for investment monies to find and procure mineral wealth and sustain economic and social development.

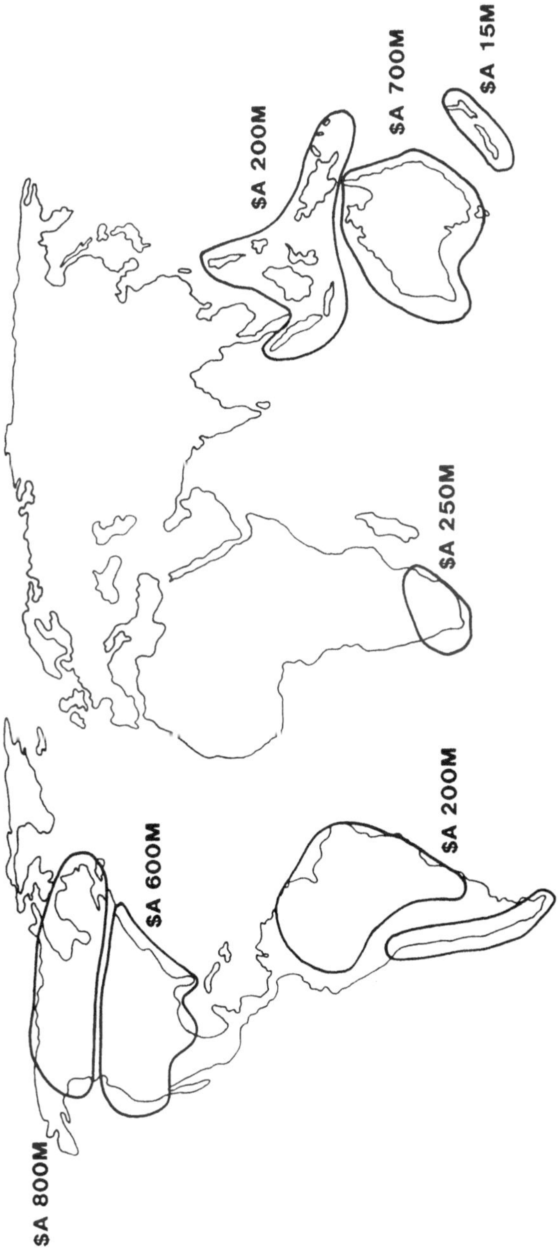

Figure 14.5. Global Mineral Exploration Activity, 1988–1989

REFERENCES

Blain-Chris, 1992, ''Is Exploration Becoming More Cost-Effective?'' ABARE, Canberra, Australia.

Cook, D.R., 1986, ''Analysis of Significant Mineral Discoveries in the Last Forty Years and Future Trends,'' *Mining Engineering* (February).

Eggert, Rod, 1987, ''Metallic Mineral Exploration: An Economic Analysis,'' *Resources for the Future,* Johns Hopkins Univ. Press.

Johnson, Charles J., February 1990, ''Ranking Countries for Mineral Exploration,'' East-West Center, Hawaii.

McKenzie, Brian W., ''Looking for the Improbable Needle in a Haystack: The Economics of Base Metal Exploration in Canada,'' (Working Paper No. 19). Centre for Resource Studies, Queens University, Kingston, Ontario.

Mckenzie, B.W. and M.L. Bilodeau, May 1982, ''A Study of the Economics of Mineral Exploration in Australia,'' paper presented at AMIC Seminar July 1983, Canberra.

Preble, J.F., P.A. Rau, and A. Reichel, 1988, ''The Environmental Scanning Practices of U.S. Multinationals in the Late 1980s,'' MIR Vol. 28. 1988/4.

World Bank, July 1990, ''Results of a Survey of International Mining Companies Regarding Determinants of Mineral Exploration and Investment in Developing Countries,'' Mining Unit, Africa Technical Department.

Chapter 15

SOVIET ASIA-PACIFIC MINERALS TRADE: PAST PERFORMANCE AND FUTURE OUTLOOK

Ping-Sun Leung, John F. Yanagida, and Vitaly T. Borisovich

INTRODUCTION

The former Soviet Union has the largest mineral industry in the world. In 1983, the latest year for which data was available, mineral production was valued at US$221.3 billion, which was about 24 percent of global minerals output value. The market value of minerals which can be sold on the domestic and world markets was estimated to be about 31 to 35 trillion rubles in 1991. In 1990, minerals and mineral products accounted for 51.6 percent of the total exports, while mineral imports accounted for only eight percent of total imports.

Although the total Soviet Asia-Pacific trade volume has been rather small, it is anticipated to increase in the future because of the new foreign policy initiatives. Japan and India are the major Asia-Pacific trading partners. Soviet mineral exports to Japan and India amounted to about 850 and 675 million rubles, respectively, in 1990. This chapter documents the regional mineral trade flows between the Soviet Union and major Asia-Pacific countries for the period 1980 to 1990. Production and consumption trends of the mineral sector of the former Soviet Union and major Asia-Pacific countries are also analyzed. Prospects for expanded mineral trade between the new Commonwealth of Independent States (CIS) and Asia-Pacific countries are assessed.

The Asia-Pacific region is defined in this chapter as all countries in Asia and Oceania, but excludes countries in the Americas. Minerals, as used in this chapter, include energy minerals, ferrous and nonferrous metals, precious metals, and nonmetallic minerals. However, only a selected set of nonferrous metals and energy minerals for the major Asia-Pacific countries is analyzed in detail. Analyses of past consumption and production trends in the Asia-Pacific region and the former Soviet Union are used to infer possible future increases in mineral trade with this region.

J.P. Dorian et al. (eds.), CIS Energy and Minerals Development, 243–292.

THE SOVIET MINERAL INDUSTRY

The former Soviet Union has the largest mineral industry in the world. It leads the world in the production of iron ore, manganese ore, nickel, cadmium, nitrogen, potash, natural gas and oil. In addition, it ranks as the second-leading producer of aluminum, lead, chromite, zinc, magnesium, gold, phosphate and sulfur. Table 15.1 presents the 1990 or 1988 production levels of selected minerals and mineral products in the Soviet Union as well as their shares of world production. In 1988, the contribution of mining and manufacturing amounted to 43 percent of the Soviet Net Materials Product (The Economist Intelligence Unit, 1990). The mineral sector was also the major source of export earnings, contributing to 51.6 percent of total export value in 1990.

SOVIET FOREIGN TRADE: 1980–1990

The former Soviet Union was active in international trade in the past decade. Total value of Soviet foreign trade increased from 94.1 billion rubles in 1980 to 131.6 billion rubles in 1990, with an average annual growth rate of 3.4 percent. However, the share of exports as a percentage of Soviet gross national product (GNP) was still rather small, estimated at 4.5 percent in 1989. Between 1980 and 1990, total exports grew at an annual rate of 2.1 percent, while total imports grew 4.7 percent annually (see Table 15.2). The difference in the growth rates of imports and exports generated a trade deficit of 9.8 billion rubles in 1990.

The major Soviet trading partners were the Council for Mutual Economic Assistance (CMEA) members, developed market economies, developing countries, and other socialist countries. In 1990, the percentages of total Soviet exports which were destined for the CMEA members, developed market economies, developing countries, and other socialist countries were 43.2, 36.1, 14.0 and 6.7, respectively, while the shares of imports from these countries were 44.4, 39.7, 9.5 and 6.4 percent, respectively. Table 15.2 shows the geographical distribution of Soviet foreign trade for selected years from 1980 to 1990. The values of Soviet exports increased during the first half of the past decade and decreased in the latter half in all geographical areas. This was primarily due to the decline in world oil prices in late 1985. As the Soviet Union was heavily reliant on oil exports as a source of revenue, this drop in price had a significant effect on export values. However, the value of Soviet exports to all geographical areas was higher in 1990 than it was in 1980. The overall rate of increase in the value of exports was highest for exports to all other socialist countries at an annual rate of 4.7 percent, followed by the three percent increase for CMEA members, 2.3 percent for developing countries, and 0.6 percent for developed market economies.

Table 15.1. Production of Selected Minerals and Mineral Products in the Former Soviet Union

Commodity	Unit (tonnes)	1990 Production[a]	Percent Share of World Production	World Rank
Ferrous metals				
Chromite	1000 t	3,240[a]	27.8	2
Iron ore	1000 t	251,000[a]	27.4	1
Crude steel	1000 t	163,000[a]	21.0	1
Manganese ore	1000 t	9,200[a]	38.5	1
Nonferrous metals				
Bauxite	1000 t	5,350	4.7	5
Alumina	1000 t	4,000	9.6	3
Aluminum	1000 t	2,200	12.2	2
Mine lead	1000 t	490	14.8	2
Mine copper	1000 t	900	10.0	3
Mine zinc	1000 t	870	11.9	2
Mine tin	1000 t	13	6.2	7
Mine antimony	1000 t	5,400	10.1	3
Cadmium	1000 t	2,400	12.0	1
Magnesium	1000 t	80	22.4	
Mine nickel	1000 t	212	24.2	1
Mercury	1000 t	1,400	27.4	1
Precious metals				
Mine silver	1000 t	1,380	9.4	5
Gold	1000 troy ounces	9,000[a]	15.4	2
Diamond	1000 carats	11,000[a]	11.7	4
Nonmetallic minerals				
Nitrogen	1000 t	20,500[a]	20.7	1
Phosphate	1000 t	38,820[a]	23.7	2
Potash	1000 t	11,000[a]	35.0	1
Salt	1000 t	15,500[a]	8.4	3
Sulfur	1000 t	10,700[a]	18.3	2
Energy minerals				
Coal	million t	772[a]	16.2	3
Natural gas	billion cu. ft.	27,200[a]	40.0	1
Crude oil	million 42-gallon barrels	4,586[a]	21.4	1

a. The figures are for the year 1988 instead of 1990.

Sources: *World Economy Minerals Yearbook* 1988 and *Metallstatistik 1980–1990*, 78th Edition, 1991.

Table 15.2. Soviet Foreign Trade, 1980–1990

	In Billion Rubles					Annual % Change		
	1980	1985	1988	1989	1990	1980–1985	1985–1990	1980–1990
TOTAL								
Exports	49.6	72.7	67.1	68.8	60.9	7.9	-3.5	2.1
Imports	44.5	69.4	65.0	72.1	70.7	9.3	0.4	4.7
Balance	5.1	3.3	2.1	-3.3	-9.8			
CMEA								
Exports	19.5	32.5	31.8	31.8	26.3	10.8	-4.1	3.0
Imports	17.1	30.3	32.8	33.4	31.4	12.1	0.7	6.3
Balance	2.4	2.2	-1.0	-2.1	-5.1			
Other socialist countries								
Exports	2.6	4.3	3.8	4.3	4.1	10.6	-0.9	4.7
Imports	2.2	4.6	3.6	4.1	4.5	15.9	-0.4	7.4
Balance	0.4	-0.3	0.2	0.2	-0.4			
Developed market economies								
Exports	20.7	26.3	21.9	23.1	22.0	4.9	-3.5	0.6
Imports	20.1	26.8	23.3	27.6	28.1	5.9	1.0	3.4
Balance	0.6	-0.5	-1.4	-4.5	-6.1			
Developing countries								
Exports	6.8	9.6	9.6	10.1	8.5	7.1	-2.4	2.3
Imports	5.1	7.7	5.3	7.0	6.7	8.6	-2.7	2.8
Balance	1.7	1.9	4.3	3.1	1.8			

Source: Soviet Foreign Trade Yearbooks.

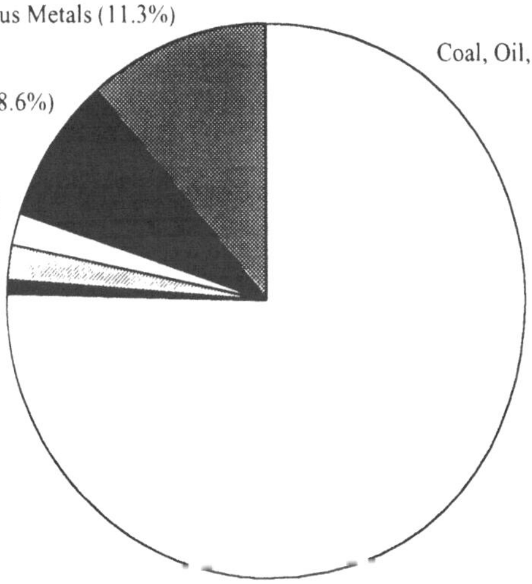

Figure 15.1. Composition of Soviet Mineral Exports, 1990

SOVIET MINERAL TRADE: 1986-1990

The commodity structure of Soviet foreign trade reflected a desire to use exports of natural resources to finance purchases of agricultural products and machinery and equipment (Bradshaw, 1990). In 1990, mineral exports accounted for 51.6 percent of total Soviet exports while mineral imports accounted for only eight percent of imports. Energy minerals led all exports, making up 38.9 percent of total exports in 1990. Figure 15.1 displays the composition of Soviet mineral exports in 1990, while Table 15.3 shows Soviet mineral trade of major mineral groups for the years 1986 to 1990. The value of Soviet mineral exports decreased from 37.9 billion rubles in 1986 to 31.4 billion rubles in 1990, declining at an average annual rate of 4.6 percent. The largest decrease during that same period came from energy minerals, whose export value declined at an annual rate of 6.8 percent, followed by drops in value for nonmetallic mineral exports at 5.6 percent and ferrous metals at 2.8 percent.

In contrast, the largest increase in exports was recorded for nonferrous metals which grew in value from 1.4 billion rubles in 1986 to 2.7 billion rubles in 1990 at an annual growth rate of 17.3 percent. Exports of precious metals, consisting primarily of gold, silver and platinum, increased at a rapid annual rate of 16.0 percent during 1986 to 1990. Exports of diamonds grew 7.7 percent during the same period. With respect to mineral imports, except for nonferrous and precious metals, all other mineral groups have shown decreases. Imports of nonferrous metals increased at a mere 0.8 percent rate, while imports of precious metals grew 5.1 percent annually for the period 1986 to 1990. Overall, mineral imports declined at an annual rate of 9.1 percent.

Tables 15.4 and 15.5, respectively, show the net exports and imports of selected minerals and metals as a percentage of consumption in 1988 and the principal

Table 15.3. Soviet Mineral Trade, 1986–1990

	In Million Rubles					Annual % Change
	1986	1987	1988	1989	1990	1986–1990
Exports						
Total minerals	37,946	37,293	34,412	34,475	31,406	-4.6
Energy minerals (coal, oil, gas & products)	31,436	30,780	27,164	26,327	23,687	-6.8
Ferrous metals	3,986	3,848	3,812	4,086	3,558	-2.8
Nonferrous metals	1,428	1,461	2,201	2,612	2,706	17.3
Precious metals (gold, silver, platinum)	322	456	347	488	584	16.0
Diamonds	469	462	618	698	630	7.7
Nonmetallic minerals	303	285	272	264	241	-5.6
Imports						
Total minerals	8,284	7,444	8,207	7,595	5,644	-9.1
Energy minerals (coal, oil, gas & products)	2,840	2,329	2,853	2,168	1,830	-10.4
Ferrous metals	3,711	3,467	3,777	3,422	2,116	-13.1
Nonferrous metals	1,457	1,449	1,408	1,805	1,507	0.8
Precious metals (gold, silver, platinum)	9	7	11	11	11	5.1
Diamonds						
Nonmetallic minerals	267	192	159	189	180	-9.4

Source: Soviet Union Ministry of Geology.

Table 15.4. Soviet Estimated Net Exports of Selected Minerals and Metals as a Percentage of Consumption, 1988

Commodity	Percent of Consumption	Principal Destinations
Aluminum	33	Japan
Asbestos	18	Japan
Cadmium	3	
Chromium ore	21	
Diamond, gem	400	
Gas, natural	113	
Gold		
Iron ore and concentrate	21	Japan, China
Magnesium	2	
Nickel	3	
Nitrogen	31	China
Perlite	31	
Petroleum		
Crude	25	
Refined product	14	
Platinum	111	Japan
Potash	27	
Titanium	15	

Source: *USSR Mineral Yearbook,* 1988.

Asia-Pacific destinations and sources. Japan and China were both major importers of Soviet iron ore and concentrates in 1988, while Japan was the principal destination of Soviet aluminum, asbestos, and platinum exports, and China was the largest market for Soviet nitrogen. The Soviet Union relied heavily on barite from North Korea, bauxite and alumina from India and Japan, tin from Indonesia, Malaysia and Singapore, and tungsten from China and Mongolia as well as North Korea. However, the Asia-Pacific region has played only a modest role in total Soviet foreign trade which will be further discussed in the next section.

SOVIET ASIA-PACIFIC TRADE[1]

Following Bradshaw (1988), Soviet trade with the Asia-Pacific region can be classified into three trading groups: the socialist nations of Vietnam, Kampuchea, Laos, Mongolia, North Korea and China; the industrially developed market economies of Japan, Australia and New Zealand; and the member states of the Association of Southeast Asian Nations (ASEAN) including Indonesia, Malaysia, the Philippines, Singapore, Thailand and Brunei. Table 15.6 presents Soviet trade with each of these three Asia-Pacific trading groups for the years 1980 and 1986.

Table 15.5. Soviet Estimated Net Import Reliance of Selected Minerals and Metals as a Percentage of Consumption, 1988

Commodity	Percent of Consumption	Principal Sources
Barite	43	North Korea
Bauxite and alumina	60	India, Japan
Bismuth	70	
Cobalt	44	
Fluorspar	67	
Iron and steel		
High quality product	7	Japan
Magnesite	28	North Korea
Lead	3	
Mica	9	India
Molybdenum	9	Mongolia
Sulfur	6	
Tin	38	Indonesia,Malaysia,Singapore
Tungsten	43	China,Mongolia,North Korea
Zinc	6	

Source: *USSR Mineral Yearbook,* 1988.

Total Soviet trade with this region increased from 6.7 billion rubles in 1980 to 10.5 billion rubles in 1986 at an annual growth rate of 7.6 percent. Although there was an increase in Soviet trade with this region, its share of total Soviet trade was still rather small, accounting for 7.2 percent in 1980 and eight percent in 1986. Soviet exports to this region as a share of total exports increased from 5.3 percent in 1980 to 7.8 percent in 1990, while the proportion of Soviet imports from this region to total imports decreased from 9.2 percent in 1980 to 8.2 percent in 1986. During the period from 1980 to 1986, Soviet exports to this region increased faster than Soviet imports at an annual growth rate of 12.4 percent versus 3.8 percent.

The difference in the rates of growth of Soviet exports and imports moved the former Soviet Union from a sizable trade deficit with this region in 1980 to a fairly balanced trading position in 1986. However, due to the different financial arrangements associated with the trading groups, the Soviet trade balance requires further explanation. Soviet trade with the socialist nations was conducted using a central clearing system, while trade with the developed market economies and ASEAN was conducted using convertible currency. As shown in Table 15.6, the Soviet Union had a large trade surplus with the socialist nations and a sizable trade deficit with the developed market economies. Since the trade surplus is in soft currency while the trade deficit is in hard currency, the Soviet Union could not balance its trade with the Asia-Pacific region.

Table 15.6. Soviet Asia-Pacific Trade, 1980–1986

Trading Group	In Million Rubles		Annual % Change	As a % of Total	
	1980	1986	1980–1986	1980	1986
Socialist nations					
Turnover	2,425.2	6,380.0	17.5	35.9	60.8
Exports	1,627.3	4,299.6	17.5	61.5	80.4
Imports	797.9	2,080.4	17.3	19.4	40.4
Balance	829.4	2,219.2			
Developed market economies					
Turnover	3,673.8	3,796.3	0.5	54.4	36.2
Exports	959.3	993.3	0.6	36.2	18.6
Imports	2,714.5	2803	0.5	66.1	54.4
Balance	-1,755.2	-1,809.7			
ASEAN					
Turnover	658.0	320.5	-11.3	9.7	3.1
Exports	61.2	55.0	-1.8	2.3	1.0
Imports	596.8	265.5	-12.6	14.5	5.2
Balance	-535.6	-210.5			
Total Asia-Pacific region					
Turnover	6,757.0	10,496.8	7.6	100.0	100.0
Exports	2,647.8	5,347.9	12.4	100.0	100.0
Imports	4,109.2	5,148.9	3.8	100.0	100.0
Balance	-1,461.4	199.0			

Source: Adapted from Bradshaw (1988).

As shown in Table 15.6, over 60 percent of Soviet trade with the Asia-Pacific region in 1986 was with the socialist nations, while 36 percent was with developed market economies and three percent with ASEAN. Most of the Soviet exports to this region were destined for the socialist nations (80.4 percent in 1986), while the developed market economies and ASEAN provided almost 60 percent of all Soviet imports from this region in 1986. During the period from 1980 to 1986, Soviet trade with the socialist nations grew at an annual rate of 17.5 percent, while trade with the developed market economies remained relatively stable and trade with ASEAN declined at an annual rate of 11.3 percent. As pointed out by Bradshaw (1988), "the terms of trade between the Soviet Union and the individual nations within the region are explained by the balance of political and economic factors which either promote or hinder trade at a particular moment in time."

Table 15.7 outlines the general commodity structure of Soviet trade with its major Asia-Pacific partners in 1986. A more detailed overview of Soviet mineral trade with this region for selected minerals is presented in the next section. Soviet exports to socialist nations in this region consisted primarily of machinery and equipment and oil and oil products, while imports from these countries were comprised of agricultural products and industrial raw materials. Soviet trade with socialist nations other than China and North Korea was generally in the form of foreign economic and technical assistance.

Japan has been the single most important Soviet trading partner in the region. In 1986, for example, 40 percent of the Soviet imports from the region came from Japan. The value of Soviet exports to Japan increased from 929 million rubles in 1985 to 1.184 billion rubles in 1988, growing at an annual rate of 8.4 percent. Soviet imports from Japan declined in value from 2.287 billion rubles in 1985 to 1.951 billion rubles in 1988 at an annual rate of 5.2 percent. Bradshaw (1988) reported that metals and metal products accounted for 34.1 percent of Soviet imports from Japan in 1985, while mineral fuels, nonferrous metals and gold accounted for 24.6, 16.1, and 10.8 percent, respectively, of total Soviet exports to Japan. It is important to note that the Soviet trade deficit with Japan is in hard currency which has to be financed through hard currency exports to other countries, primarily in Western Europe.

SOVIET ASIA-PACIFIC TRADE OF SELECTED MINERALS

Tables 15.8 through 15.12 present Soviet Asia-Pacific trade of the world's principal nonferrous metals—aluminum, lead, copper, zinc, and tin—for the years 1980, 1985 and 1990. Exports and imports of the refined metals are also expressed as a percentage of total Soviet consumption of the refined metals. Imports of unprocessed minerals are expressed as a percentage of total Soviet mine production of each metal. These percentages are presented in Tables 15.8 through 15.12 in parentheses. The percentage of exports/imports of refined metals to total consumption provides an indication of the relative significance of the exports/

Table 15.7. Commodity Structure of Soviet Foreign Trade with Selected Asia-Pacific Countries, 1986

Country	Total Trade Turnover in 1986 (million rubles)	Major Commodities[a]	
		Exports	Imports
China	1,822.0	Machinery and equipment, **fertilizers, ferrous metals,** sawnwood	Food
Mongolia	1,547.4	Machinery and equipment, **oil and products**, food, clothing	**Metallic ores and concentrates,** livestock and meat
Vietnam	1,612.7	Machinery and equipment, medicines, **oil and products, metals, fertilizers,** cotton and food	Rubber, flooring, carpets, fruit and liquor
North Korea	1,207.9	Machinery and equipment, **oil and products**	Machinery and equipment, **minerals,** clothing
Kampuchea	122.7	Machinery and equipment, **oil and products**	Rubber
Laos	63.7	Machinery and equipment	Ores and concentrates
Japan	3,185.3	**Coal, oil,** roundwood, sawnwood, wood chips, **iron scrap, potassium salts, nonferrous metals**	Machinery and equipment, **rolled metal, pipe, chemical products**
Australia	517.3	**Oil and products**	Wool, grain
New Zealand	93.7	Fish, **potassium salts**	Wool, **metal**
Malaysia	104.2	**Fertilizers,** medicines, cotton	Rubber, **tin**

a. Mineral and mineral products are bold-faced.
Source: Adapted from Bradshaw (1988).

Table 15.8. Soviet Asia-Pacific Trade of Aluminum, 1980–1990

	Exports of Aluminum (in tonnes)			Imports of Bauxite (in tonnes)			Imports of Alumina (in tonnes)		
Country	1980	1985	1990	1980	1985	1990	1980	1985	1990
China, PR			30,910 (1.9)						
Hong Kong	2,545 (0.1)								
India		10,000 (0.6)		78,383 (1.2)			76,927 (2.4)	38,036 (0.9)	
Indonesia	225 (0.0)								
Japan	42,136 (2.3)	78,214 (4.5)	111,537 (7.0)				60,858 (1.9)	43 (0.0)	1 (0.0)
Pakistan		390 (0.0)							
Singapore			1,105 (0.1)						
Taiwan			200 (0.0)						
Australia									71,000 (1.8)
Soviet Union									
Consumption	1,850,000	1,750,000	1,600,000						
Production	2,420,000	2,300,000	2,200,000	6,400,000	6,400,000	5,350,000	3,250,000	4,350,000	4,000,000

Note: Numbers in parentheses are % of Soviet consumption.
Source: *Metallstatistik 1980–1990,* 78th Edition, 1991.

imports to domestic consumption, while the percentage of imports to total production shows the level of Soviet reliance on imported unprocessed minerals.

Aluminum

As shown in Table 15.8, the former Soviet Union produced more aluminum than it consumed domestically during the past decade. During the same period, exports of aluminum were destined for the following Asia-Pacific nations: the People's Republic of China, Hong Kong, India, Indonesia, Japan, Pakistan, Singapore and Taiwan. Except for shipments to Japan, Soviet export volumes to these countries were small. Exports of Soviet aluminum to Japan grew from 42,136 tonnes in 1980 to 111,537 tonnes in 1990 at an annual rate of 10.2 percent. In 1990, exports of Soviet aluminum to Japan comprised seven percent of total Soviet aluminum consumption. During the past decade, the former Soviet Union has imported bauxite and alumina from Japan, India and Australia. However, the level of imports from these countries has been rather small.

Lead

In 1980, Soviet consumption of refined lead exceeded domestic lead production and a small amount of refined lead was imported from Japan (Table 15.9). Since 1984, sufficient amounts of refined lead were produced to satisfy domestic consumption as well as export needs. In 1990, small amounts of Soviet refined lead were exported to South Korea, Japan, Kampuchea, Pakistan and Vietnam. Japan has been the major importer of Soviet refined lead, amounting to about 6.6 percent of total Soviet refined lead exports. Also, small amounts of lead ores and concentrates were imported from Japan.

Copper

As shown in Table 15.10, small amounts of Soviet refined copper were exported to the People's Republic of China, India and Japan in 1990. In 1980, when Soviet production of refined copper barely covered domestic consumption, a small amount of refined copper was imported from Japan. The Soviet Union relied on Mongolia for supplies of copper ores. Imports of Mongolian copper ores grew from 44,000 tonnes in 1980 to 136,600 tonnes in 1990. In 1990, imports of Mongolian copper ores amounted to 15.2 percent of total Soviet mine production of copper. A limited amount of copper ore was also imported from the Philippines in 1980.

Table 15.9. Soviet Asia-Pacific Trade of Lead, 1980–1990

	Exports of Refined Lead (in tonnes)			Imports of Refined Lead (in tonnes)			Imports of Lead Ores and Concentrates (in tonnes)		
Country	1980	1985	1990	1980	1985	1990	1980	1985	1990
South Korea			1,100						
			(0.2)						
Japan			4,600	1,255			3,683		
			(0.7)	(0.2)			(0.6)		
Kampuchea			30						
			(0.0)						
Pakistan		851							
		(0.1)							
Vietnam			1,250						
			(0.2)						
Soviet Union									
Consumption	800,000	780,000	650,000	800,000	780,000	650,000			
Production	780,000	810,000	730,000	780,000	810,000	730,000	580,000	580,000	490,000

Note: Numbers in parentheses are % of Soviet consumption.
Source: *Metallstatistik 1980–1990,* 78th Edition, 1991.

Table 15.10. Soviet Asia-Pacific Trade of Copper, 1980–1990

	Exports of Refined Copper (in tonnes)			Imports of Refined Copper (in tonnes)			Imports of Copper Ores (in tonnes)		
Country	1980	1985	1990	1980	1985	1990	1980	1985	1990
China, PR			5,663 (0.6)						
India			7,169 (0.7)						
Japan			5,337 (0.5)	1,945 (0.1)					
Mongolia							44,000 (4.5)	98,000 (9.5)	136,600 (15.2)
Philippines							8,071 (0.8)		
Soviet Union									
Consumption	1,300,000	1,305,000	1,000,000	1,300,000	1,305,000	1,000,000			
Production	1,300,000	1,400,000	1,260,000	1,300,000	1,400,000	1,260,000	980,000	1,030,000	900,000

Note: Numbers in parentheses are % of Soviet consumption.
Source: ***Metallstatistik*** *1980–1990*, 78th Edition, 1991.

Table 15.11. Soviet Asia-Pacific Trade of Zinc, 1980–1990

	Exports of Slab Zinc (in tonnes)			Imports of Slab Zinc (in tonnes)			Imports of Zinc Ores and Concentrates (in tonnes)		
Country	1980	1985	1990	1980	1985	1990	1980	1985	1990
Hong Kong				2,003 (0.2)					
Japan				4,497 (0.4)					
India	7,542 (0.7)	16,700 (1.7)	16,700 (1.8)						
Kampuchea			30 (0.0)						
North Korea						1,700 (0.2)			
Pakistan		708 (0.1)	500 (0.1)						
Vietnam			470 (0.1)						
Australia								3,484 (0.3)	
Soviet Union									
Consumption	1,030,000	1,000,000	920,000	1,030,000	1,000,000	920,000			
Production	1,060,000	1,050,000	920,000	1,060,000	1,050,000	920,000	1,000,000	1,000,000	870,000

Note: Numbers in parentheses are % of Soviet consumption.
Source: *Metallstatistik 1980–1990,* 78th Edition, 1991.

Table 15.12. Soviet Asia-Pacific Trade of Tin, 1980–1990

Country	Exports of Tin (in tonnes)			Imports of Tin (in tonnes)			Imports of Tin Concentrates (in tonnes)		
	1980	1985	1990	1980	1985	1990	1980	1985	1990
China, PR				50 (0.2)					
Indonesia				125 (0.5)	500 (1.6)				
Malaysia				7,357 (29.4)	2,722 (8.6)				
Singapore				2,613 (10.5)	3,741 (11.9)		1,370 (8.6)	1,635 (10.2)	
Soviet Union									
Consumption	25,000	31,500	20,000	25,000	31,500	20,000			
Production	17,000	18,500	14,000	17,000	18,500	14,000	16,000	16,000	13,000

Note: Numbers in parentheses are % of Soviet consumption.
Source: *Metallstatistik 1980–1990*, 78th Edition, 1991.

Zinc

The Soviet Union exported slab zinc to India, Kampuchea, Pakistan and Vietnam during the past decade (Table 15.11). India was the major importer of Soviet slab zinc, with imports growing from 7,542 tonnes in 1980 to 16,700 tonnes in 1990. In 1990, exports of slab zinc to India comprised 56.1 percent of total exports of Soviet slab zinc. During the past decade, some amounts of slab zinc were imported from Hong Kong, Japan and North Korea. Australia was the only country in the region which has exported small amounts of zinc ore to the Soviet Union.

Tin

Soviet production of tin was not able to keep pace with domestic consumption in the 1980s. Much of the tin consumed in the Soviet Union came from the Asia-Pacific region. A large amount of tin was imported from Malaysia and Singapore, while smaller amounts were imported from the People's Republic of China and Indonesia. In addition, a large amount of tin concentrates was imported from Singapore. Table 15.12 shows the imports of tin and tin concentrates from the Asia-Pacific region.

OVERVIEW OF THE MINERAL INDUSTRY IN THE ASIA-PACIFIC REGION

The Asia-Pacific Region (APR) is composed of a diverse set of countries whose endowments in minerals range from the very resource-rich to the very resource-poor. The APR, as used in the following analysis, refers to countries in Asia and Oceania, but excludes the Americas. Following the convention used in *Metallstatistik* (1991), Asia refers to all Asian countries excluding socialist countries. Table 15.13 illustrates 1990 shares of world minerals production and consumption by a select group of countries in the APR. With the exception of tin, Asia had larger shares of world consumption of these major minerals than corresponding shares of world production. On the other hand, the Soviet Union, China and Oceania generally had larger production shares than consumption shares in 1990.

Consumption trends

Over the data period 1980–1989, two sets of consumption trends are analyzed. First, data showing the annual growth rates of mineral consumption per million people (c/mp) are presented in Table 15.14. For the world, mineral consumption (c/mp) either experienced minimal annual growth of less than one percent,

Table 15.13. Production and Consumption Shares of Major Minerals by Selected Countries/Groups in the Asia-Pacific Region, 1990[a]

Mineral	Country	Production Share (%)	Consumption Share (%)
Aluminum	Asia	6.5	23.6
	Japan	0.2	13.5
	S. Korea	0.0	2.0
	India	2.4	2.3
	China	4.7	3.6
	Soviet Union	12.2	8.9
	Oceania	8.3	1.8
Lead	Asia	9.9	16.2
	Japan	5.8	7.4
	S. Korea	1.4	2.7
	India	0.7	1.4
	China	5.1	4.5
	Soviet Union	12.9	11.6
	Oceania	4.0	1.2
Copper	Asia	14.2	25.2
	Japan	9.4	14.6
	S. Korea	1.7	3.0
	India	0.4	1.2
	China	4.6	4.7
	Soviet Union	11.7	9.2
	Oceania	2.6	1.2
Zinc	Asia	15.6	22.7
	Japan	9.7	11.7
	S. Korea	3.5	3.3
	India	1.1	1.9
	China	7.4	7.2
	Soviet Union	13.0	13.2
	Oceania	4.2	1.5
Tin	Asia	45.2	28.2
	Japan	0.4	14.5
	S. Korea	1.1	3.4
	India	0.1	1.3
	China	12.4	7.7
	Soviet Union	6.2	8.6
	Oceania	0.3	0.6
Cadmium	Asia	16.8	29.0
	Japan	12.3	24.9
	S. Korea	2.8	2.0
	India	1.4	1.5
	China	5.0	2.3
	Soviet Union	12.0	10.3
	Oceania	3.2	0.3

Table 15.13. *(continued)*

Mineral	Country	Production Share (%)	Consumption Share (%)
Magnesium	Asia	3.9	10.2
	Japan	3.6	7.4
	S. Korea	0.0	0.4
	India	0.3	0.4
	China	4.5	4.8
	Soviet Union	22.4	22.9
	Oceania	0.0	1.1
Nickel	Asia	14.4	26.4
	Japan	11.7	19.3
	S. Korea	1.0	2.8
	India	0.0	1.4
	China	3.2	3.2
	Soviet Union	26.7	13.5
	Oceania	9.2	0.2

a. See Annex 1 for mineral production and consumption data, 1980–1990.

no growth, or slight declines of less than one percent. World consumption figures generally show declining consumption (c/mp) for the period 1980–1985, but increases in consumption in the latter period, 1985–1989.

For Asia, Table 15.14 shows that cadmium had the largest consumption (c/mp) growth for the 1980–1989 period, with an average annual growth rate of 13.6 percent. All of the selected minerals did experience some consumption growth in Asia. Of the major countries in Asia, South Korea showed the largest percentage increases in mineral consumption (c/mp). China continued to experience consumption (c/mp) growth for most minerals over the period 1980–1989. The Soviet Union, however, exhibited a declining consumption (c/mp) trend.

The second set of consumption trend data is shown in Table 15.15. Intensity of use (IU) is defined as the ratio of mineral consumption in tonnes to real gross domestic product (RGDP) in billions of U.S. dollars. With the exception of cadmium, Japan experienced negative IU growth rates for 1980–1989. Similarly, the Soviet Union had negative IU growth rates, except for magnesium. China, South Korea, and India generated positive IU growth rates for all selected minerals over the period 1980–1989. As suggested by Clark and Jeon (1990) and Jeon (1989), RGDP increases in many developed countries will be accompanied by stable (constant) or declining consumption of minerals because of technological change and mineral substitution effects occurring in the production process.

Table 15.16 shows net production data (production minus consumption) for the primary metals in 1980, 1985, and 1990. It is illustrated clearly that Asia is a net consuming region (negative net producer) for aluminum, lead, copper, zinc, magnesium and nickel. For cadmium, Asia was a net consuming region in 1990. It is also evident from these data that Asia's negative net consumption

Table 15.14. Annual Growth Rates of Per Capita Consumption of Major Minerals[a]

Mineral	Country	Annual Percentage Change		
		1980–1985	1985–1989	1980–1989
Aluminum	World	-1.1	1.4	0.0
	Asia	1.1	6.4	3.4
	Japan	0.0	6.3	2.7
	S. Korea	15.0	17.4	16.1
	India	2.7	6.9	4.6
	China	1.8	0.8	1.4
	Soviet Union	-1.1	0.0	-0.6
	Oceania	4.9	2.4	3.8
Lead	World	-1.5	-0.1	-0.9
	Asia	2.4	3.7	3.0
	Japan	-0.4	0.0	-0.2
	S. Korea	18.0	20.3	19.0
	India	3.7	0.7	2.3
	China	2.3	-0.7	0.9
	Soviet Union	-0.5	-2.7	-1.5
	Oceania	-3.3	0.6	-1.6
Copper	World	-0.8	0.8	-0.1
	Asia	2.0	6.7	4.1
	Japan	0.6	3.7	2.0
	S. Korea	18.2	3.3	11.3
	India	-0.8	9.9	3.8
	China	1.3	4.4	2.7
	Soviet Union	0.1	-3.3	-1.4
	Oceania	-0.7	1.7	0.4
Zinc	World	-1.2	0.3	-0.5
	Asia	1.1	0.5	0.8
	Japan	0.1	-0.9	-0.3
	S. Korea	10.5	10.9	10.7
	India	4.8	-1.7	1.8
	China	4.8	5.1	5.0
	Soviet Union	-0.6	0.5	-0.1
	Oceania	-1.7	2.5	0.2
Tin	World	-2.5	1.1	-0.9
	Asia	-0.4	7.1	2.8
	Japan	0.2	1.0	0.3
	S. Korea	6.2	45.8	22.2
	India	-2.1	4.8	0.9
	China	1.6	10.3	5.3
	Soviet Union	4.7	-6.6	-0.5
	Oceania	-3.8	0.9	-1.8

Table 15.14. *(continued)*

Mineral	Country	Annual Percentage Change 1980–1985	1985–1989	1980–1989
Cadmium	World	-0.5	1.4	0.4
	Asia	7.6	21.6	13.6
	Japan	10.6	27.0	17.6
	S. Korea	13.7	7.3	10.8
	India	23.2	-2.3	11.1
	China	5.6	3.2	4.5
	Soviet Union	3.9	-5.6	-0.5
	Oceania	-21.4	-9.8	-16.4
Magnesium	World	-0.7	1.2	0.1
	Asia	4.6	-1.6	1.8
	Japan	2.7	0.9	1.9
	S. Korea	27.5	-12.4	7.9
	India	-11.6	25.3	3.2
	China	5.6	2.3	4.1
	Soviet Union	4.8	1.3	3.2
	Oceania	1.3	3.6	2.3
Nickel	World	-0.1	0.4	0.1
	Asia	1.3	5.7	3.2
	Japan	1.2	4.1	2.5
	S. Korea	1.7	49.8	20.8
	India	12.4	-0.6	6.5
	China	1.9	4.7	3.1
	Soviet Union	1.2	-3.4	-0.9
	Oceania	-5.1	-1.4	-3.4

a. Mineral consumption in tonnes per million people (c/mp); calculated from Annex 1.

Table 15.15. Annual Growth Rates of Mineral Consumption Per Real Gross Domestic Product (RGDP)[a]

Mineral	Country	Annual Percentage Change 1980–1985	1985–1990	1980–1990
Aluminum	Japan	-2.1	-10.7	-6.0
	S. Korea	15.5	0.7	8.7
	India	8.6	11.2	9.8
	China	7.5	1.0	4.6
	Soviet Union	-1.1	0.0	-0.6
	Oceania	4.9	2.4	3.8
Lead	Japan	-2.5	-15.9	-8.7
	S. Korea	18.5	3.2	11.4
	India	9.6	4.8	7.5

Table 15.15. *(continued)*

Mineral	Country	Annual Percentage Change 1980–1985	Annual Percentage Change 1985–1990	Annual Percentage Change 1980–1990
Lead *(cont'd)*	China	8.0	-0.5	4.1
	Soviet Union	-0.5	-2.7	-1.5
	Oceania	-3.3	0.6	-1.6
Copper	Japan	-1.5	-12.8	-6.7
	S. Korea	18.7	-11.5	4.2
	India	4.9	14.3	9.0
	China	7.0	4.6	5.9
	Soviet Union	0.1	-3.3	-1.4
	Oceania	-0.7	1.7	0.4
Zinc	Japan	-2.0	16.7	-8.8
	S. Korea	10.9	-4.9	3.6
	India	10.8	2.2	6.9
	China	10.7	5.3	8.3
	Soviet Union	-0.6	0.5	-0.1
	Oceania	-1.7	2.5	0.2
Tin	Japan	-2.3	-15.1	-8.2
	S. Korea	6.6	25.0	14.4
	India	3.5	9.1	5.9
	China	7.2	10.5	8.7
	Soviet Union	4.7	-6.6	-0.5
	Oceania	-3.8	0.9	-1.8
Cadmium	Japan	8.3	6.7	7.6
	S. Korea	14.2	-8.0	3.7
	India	30.3	1.7	16.7
	China	11.5	3.4	7.8
	Soviet Union	3.9	-5.6	-0.5
	Oceania	-21.4	-9.8	-16.4
Magnesium	Japan	0.5	-15.2	-6.8
	S. Korea	28.0	-24.9	1.0
	India	-6.5	30.4	8.4
	China	11.5	2.5	7.4
	Soviet Union	4.8	1.3	3.2
	Oceania	1.3	3.6	2.3
Nickel	Japan	-0.9	-12.5	-6.2
	S. Korea	2.2	28.4	13.1
	India	18.9	3.5	11.8
	China	7.5	4.9	6.4
	Soviet Union	1.2	-3.4	-0.9
	Oceania	-5.1	-1.4	-3.4

a. Mineral consumption in tonnes per billion U.S. dollars; calculated from Annex 1.

Table 15.16. Net Production (Production Minus Consumption) of Selected Metals for Various Countries and Regions, 1980, 1985, and 1990[a]

		Net Production ('000 tonnes)		
Mineral	Country	1980	1985	1990
Aluminum	World	774.8	-271.6	145.5
	Asia	-759.8	-1,564.5	-3,054.3
	Japan	-547.5	-1,468.1	-2,380.1
	S. Korea	-49.7	-128.1	-356.0
	India	-49.1	-31.1	13.2
	China	-200.0	-140.0	200.0
	Soviet Union	570.0	550.0	600.0
	Oceania	205.3	772.1	1,177.4
Lead	World	99.1	188.7	65.1
	Asia	-204.9	-199.7	-347.6
	Japan	-87.5	-30.1	-87.9
	S. Korea	-18.0	-44.7	-70.0
	India	-28.5	-48.0	-38.3
	China	-26.8	-20.5	36.8
	Soviet Union	-20.0	30.0	80.0
	Oceania	159.1	152.9	164.4
Copper	World	-94.1	-136.2	-87.6
	Asia	-303.1	-364.4	-1,204.4
	Japan	-138.3	-290.2	-569.5
	S. Korea	-11.1	-57.3	-140.2
	India	-54.0	-53.6	-96.3
	China	-75.0	-80.0	-22.0
	Soviet Union	0.0	95.0	260.0
	Oceania	52.3	68.5	149.2
Zinc	World	-34.2	369.2	117.8
	Asia	-271.9	-324.8	-480.7
	Japan	-17.1	-40.5	-126.8
	S. Korea	7.7	-10.8	21.2
	India	-51.7	-63.1	-55.9
	China	-31.6	-42.8	26.3
	Soviet Union	30.0	50.0	0.0
	Oceania	184.8	181.8	199.0
Tin	World	22.2	1.9	-7.1
	Asia	100.0	45.8	36.2
	Japan	-29.6	-30.2	-33.0
	S. Korea	-1.4	-1.0	-5.3
	India	-2.3	-2.3	-2.7
	China	5.0	13.1	10.0
	Soviet Union	-8.0	-13.0	-6.0
	Oceania	1.9	0.3	-0.8

Table 15.16. *(continued)*

Mineral	Country	Net Production ('000 tonnes) 1980	1985	1990
Cadmium	World	1.8	1.7	0.6
	Asia	1.0	0.7	-2.3
	Japan	1.0	0.6	-2.4
	S. Korea	0.2	0.1	0.2
	India	0.0	-0.1	0.0
	China	0.0	0.2	0.6
	Soviet Union	0.5	-0.2	0.4
	Oceania	0.7	0.8	0.6
Magnesium	World	35.1	20.6	21.7
	Asia	-11.2	-19.9	20.3
	Japan	-9.0	-13.1	-12.2
	S. Korea	-0.5	-1.8	-1.5
	India	-0.4	-0.2	-0.5
	China	-3.0	-5.0	0.0
	Soviet Union	15.0	9.0	3.0
	Oceania	-3.1	-3.3	-3.8
Nickel	World	27.1	9.1	10.8
	Asia	-2.1	-46.0	-100.8
	Japan	-14.5	-40.9	-63.9
	S. Korea	-3.0	-3.5	-15.8
	India	-7.0	-14.0	-12.2
	China	-7.0	3.6	0.0
	Soviet Union	35.0	60.0	115.0
	Oceania	63.1	73.2	77.3

a. Calculated from Annex 1.

increased (in absolute terms) over the 1980–1990 period. Japan, by far, is the principal net consuming country in Asia. South Korea and India, like Japan, experienced trends similar to Asia as a whole. China was able to reduce its net consumption of metals and in several cases (e.g., aluminum, lead, and zinc) became a net producing country. With the exception of tin and magnesium, both Oceania and the Soviet Union were net producers of metals. Also, over the time period considered, net metal production of both Oceania and the Soviet Union increased.

FUTURE OUTLOOK OF COMMONWEALTH MINERAL TRADE WITH THE ASIA-PACIFIC REGION

Figures 15.2–15.4 show the consumption and production trends of major metals for the period 1980 to 1990 in Asia, Japan, and the Soviet Union,

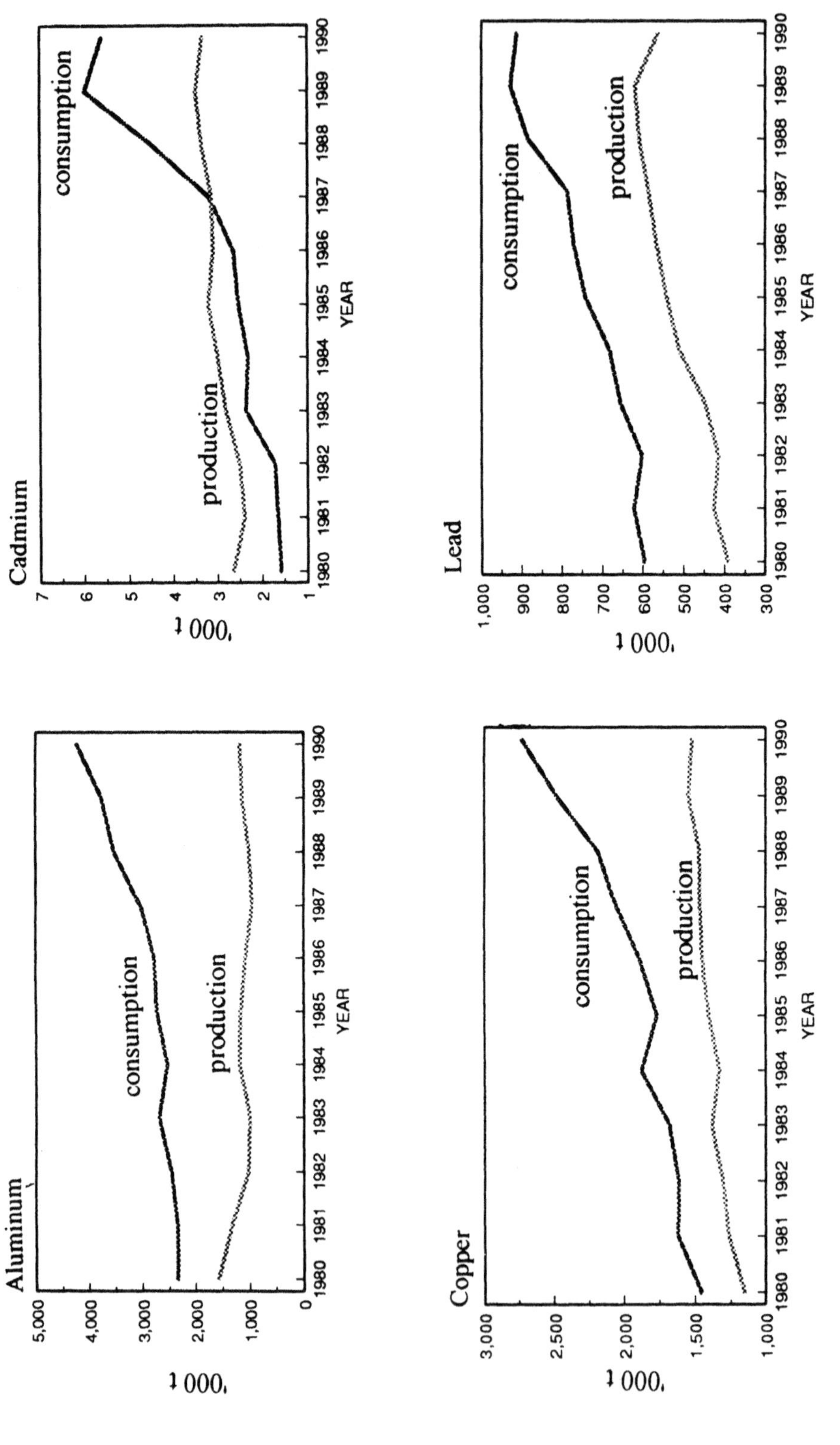
Aluminum
consumption
production
YEAR
1 000 t
Cadmium
consumption
production
YEAR
1 000 t
Copper
consumption
production
YEAR
1 000 t
Lead
consumption
production
YEAR
1 000 t

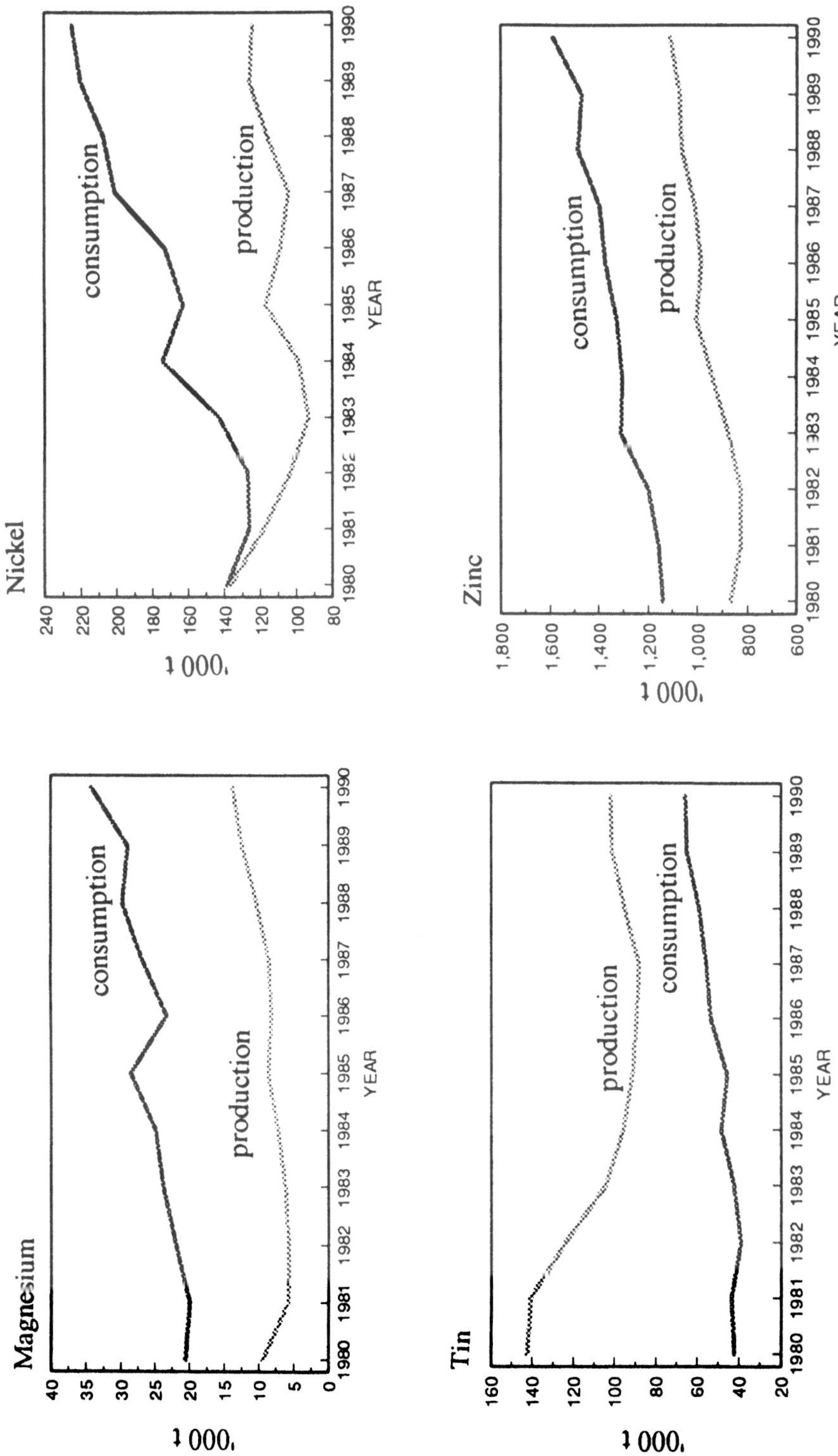

Figure 15.2. Consumption and Production Trends of Major Nonferrous Metals in Asia, 1980–1990

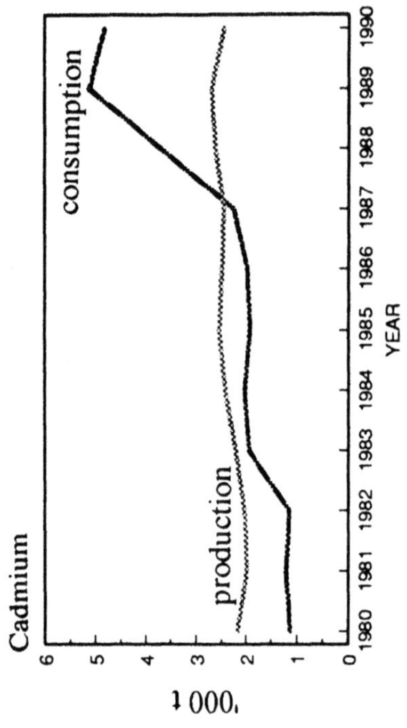
Cadmium
consumption
production
YEAR
t 000

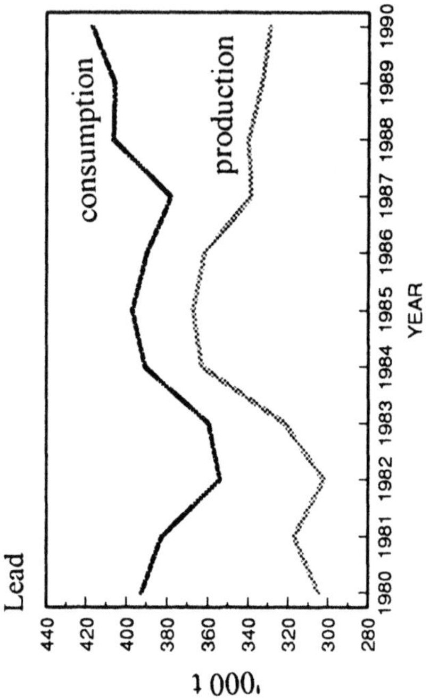
Lead
consumption
production
YEAR
t 000

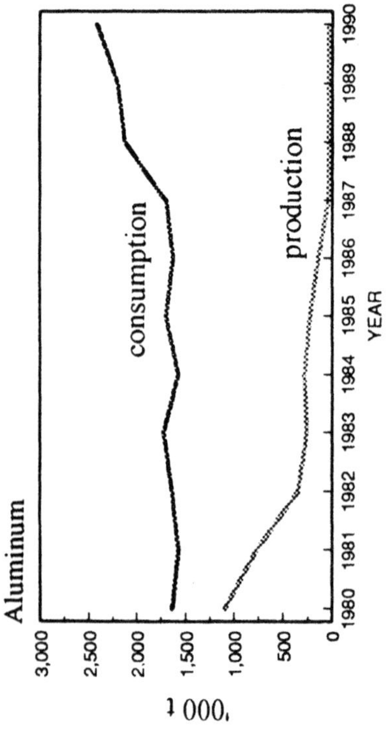
Aluminum
consumption
production
YEAR
t 000

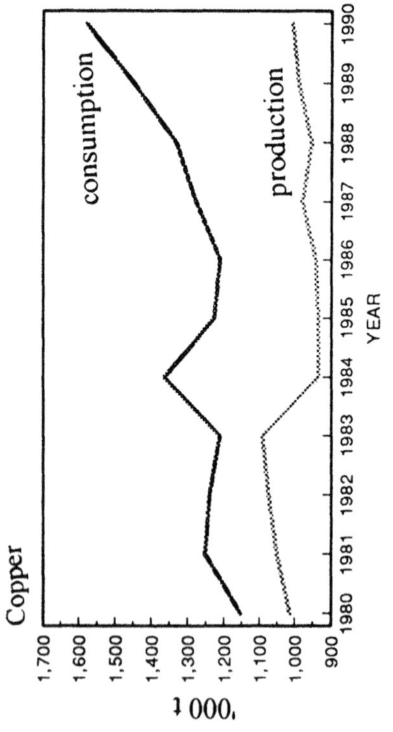
Copper
consumption
production
YEAR
t 000

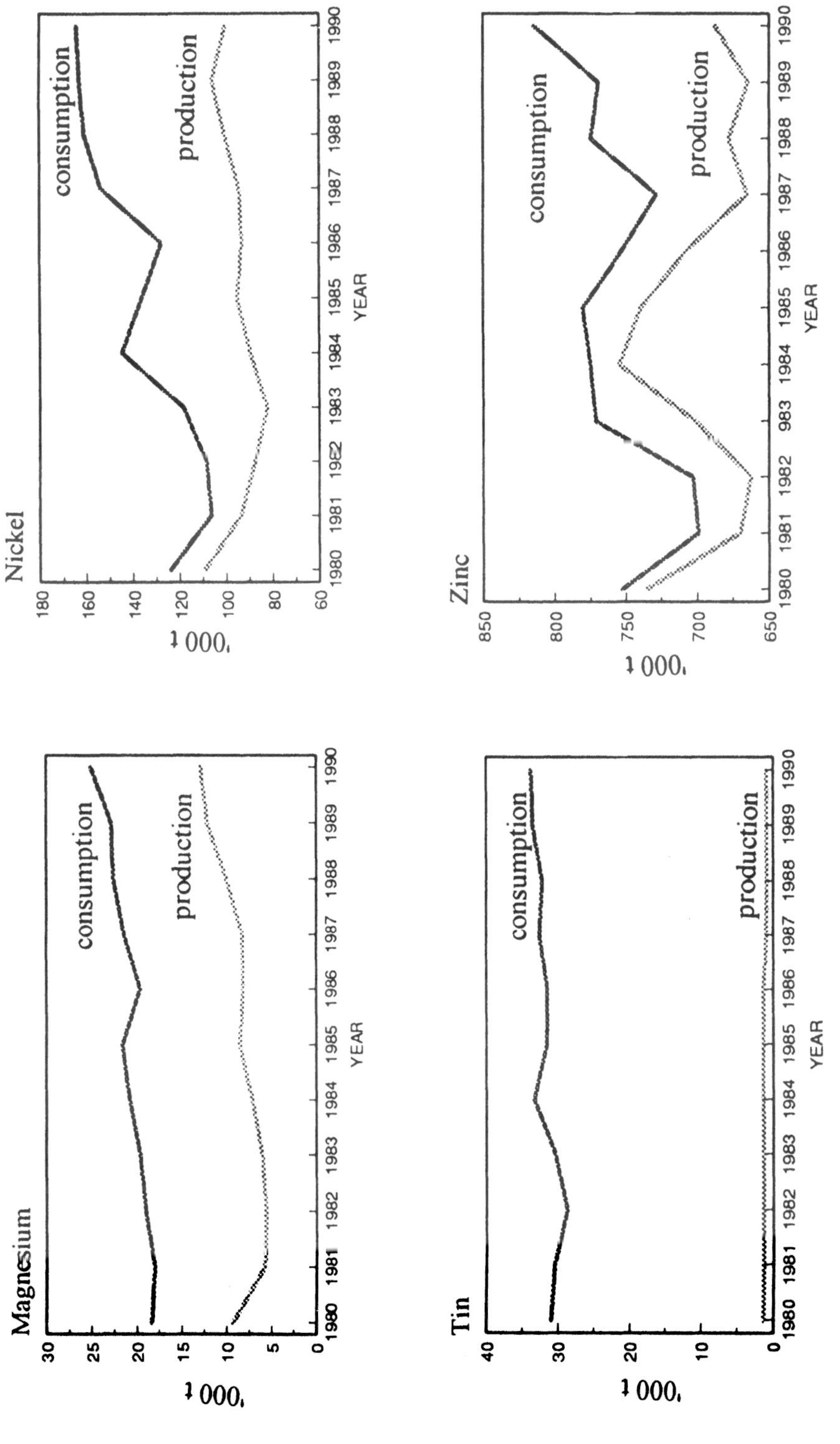

Figure 15.3. Consumption and Production Trends of Major Nonferrous Metals in Japan, 1980–1990

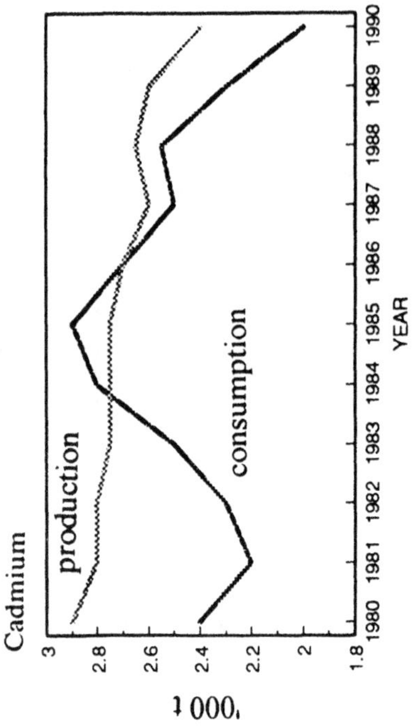
Cadmium
production
consumption
YEAR

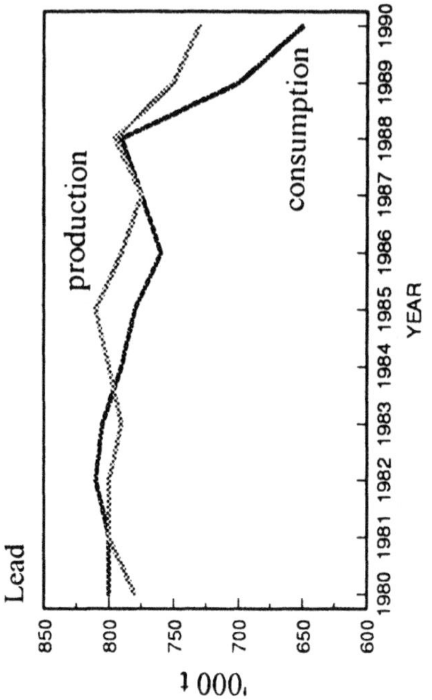
Lead
production
consumption
YEAR

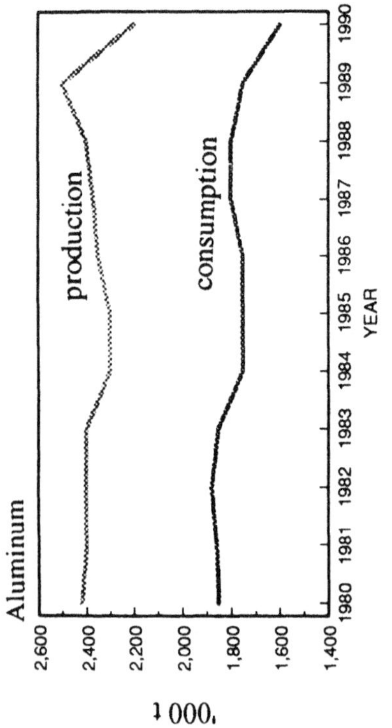
Aluminum
production
consumption
YEAR

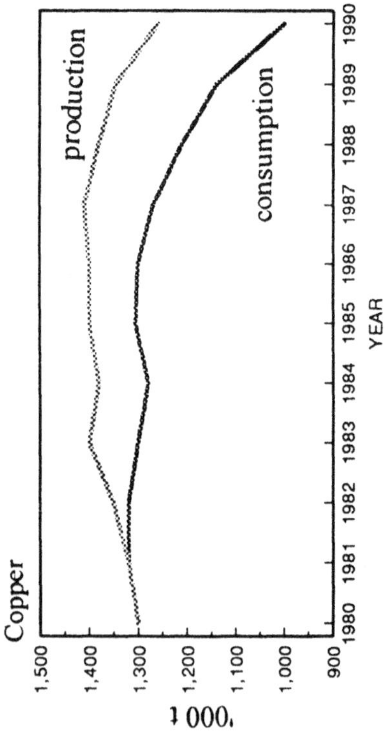
Copper
production
consumption
YEAR

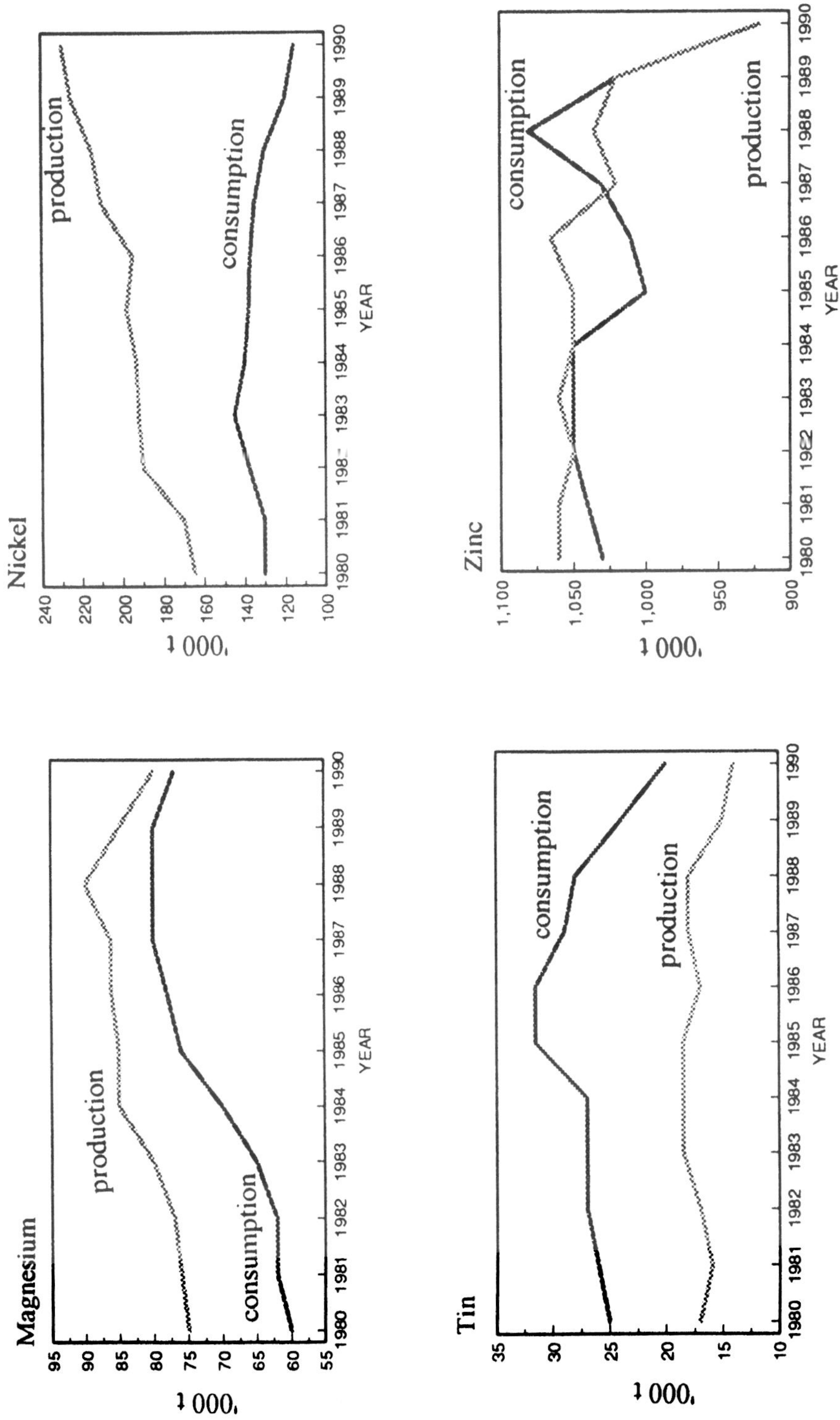

Figure 15.4. Consumption and Production Trends of Major Nonferrous Metals in the Soviet Union, 1980–1990

respectively. The increasing consumption of aluminum in Asia, and particularly Japan, together with its rather stable level of production suggest that there is potential for increased Commonwealth exports of aluminum to Asia. Although Japan shows a decreasing IU for aluminum, its high GDP growth has sustained increasing total consumption in the past decade. Japan was the major importer of Soviet aluminum, and given the trend over the past decade, it will most likely continue to be the major consumer of Commonwealth aluminum. Over the period, the Soviet Union maintained relatively stable net production of aluminum for export. However, in order to increase its aluminum exports to the Asia-Pacific region, it had to compete with countries, particularly in Oceania (Australia), where there has been a significant increase in net production in the past decade. Of special importance is the competitiveness of its product with respect to international standards of quality.

The potential for increased exports of Commonwealth refined copper to Asia is very promising. Like aluminum, consumption of copper in Asia, particularly Japan, has increased much faster than production during the past decade. On the other hand, Soviet domestic consumption of copper declined more than production, creating a large surplus for exports. However, as with aluminum, increased Soviet exports of copper had to compete with copper from Oceania (Australia and Papua New Guinea) which had shown a significant increase in net production in the past decade.

The Commonwealth is likely to be a future supplier of magnesium to the Asia-Pacific region, an area where net production of the metal has always been negative. However, Soviet net production of magnesium in the past had been small and declining.

The former Soviet Union has always been a net consumer of tin. A significant amount of tin was indeed imported from the Asia-Pacific region. This trend is expected to continue, except that the level of imports may be reduced, as the IU for tin in the Soviet Union decreased at an annual rate of 0.5 percent and the total consumption of tin declined 2.2 percent annually between 1980 and 1990.

Cadmium consumption in Asia, particularly Japan, increased rapidly in the past decade. The IU for cadmium in the Asia-Pacific region also increased. On the other hand, Soviet cadmium consumption and production declined in recent years. However, Soviet cadmium consumption decreased faster than production, generating an increase in net production. Again, any future increases in Commonwealth cadmium exports to Asia will have to compete with cadmium from Oceania which has also shown an increase in net production over the past decade.

Lead consumption in Asia has increased rapidly in the past decade. Although Japan has shown a decrease in its IU for lead, it total lead consumption has been increasing. The Soviet Union, on the other hand, had exhibited increasing net production as consumption declined faster than production in recent years. Similar to aluminum, copper and cadmium, further increases in exports of Commonwealth lead will have to compete with lead from Oceania which has a sizable net production.

The Asia-Pacific region requires a large amount of nickel to offset its production shortfall. Consumption of nickel in the region increased much faster than production. On the other hand, domestic Soviet nickel consumption declined while production has expanded. There appears to be potential for increased Commonwealth nickel exports to the region.

Zinc consumption and production in Asia increased steadily in the last decade. However, Soviet zinc production barely covered domestic consumption. Based on this past trend, there does not appear to be much promise for increased Commonwealth zinc exports to Asia, particularly because they will have to compete with the large and increasing net supplies of zinc from Oceania.

ENERGY SITUATION AND OUTLOOK FOR THE ASIA-PACIFIC REGION

The latest available data on world fuel production and consumption (see Table 15.17) indicate that in 1989, Asia (refers to all Asian countries in this section, including socialist states, due to the different data sources for metals and energy minerals) was an important producer and consumer of all forms of fuel, especially solid fuels. It is also evident that China (for solid fuels) and the Soviet Union (for liquid fuels and gas) were key players in the energy market. In this section, total energy includes solid fuels, liquid fuels, and electricity and other forms of energy. The common unit of measurement is 1000 tonnes of coal equivalent.

The growth rates shown in Table 15.18 illustrate that for liquid fuels and gas, growth in per capita consumption generally increased after 1985. However for solid fuels, growth in per capita consumption declined after 1985. These differences in growth patterns may reflect movements in relative prices for these energy forms.

The growth rates in fuel consumption per RGDP reflect a declining intensity of use for major Asian countries such as Japan and South Korea (see Table 15.19). Developing countries, like India and China, have maintained growing intensities of use for fuels as their production processes for the agricultural and manufacturing sectors become more capital intensive.

Table 15.20 indicates clearly that Asia has been a negative consuming region (negative net producer) for solid fuels. Recently, Asia has also become a negative consuming region for gas. The Soviet Union and China remained net producers of energy over the period 1980–1989. In most cases, net energy production in these countries increased. Increasing net production is particularly evident for gas in the Soviet Union.

Figures 15.5–15.7 show the consumption and production trends of total fuels, solid fuels, liquid fuels, and gas for the period 1980 to 1989 in Asia, Japan, and the Soviet Union, respectively. These trends clearly show that there is a great potential for increased Commonwealth exports of gas to the Asia-Pacific region. Asian production of liquid fuels has been more than sufficient to cover its

Table 15.17. Production and Consumption Shares of Energy by Selected Countries/Groups in the Asia-Pacific Region, 1989[a]

Fuel	Country	Production Share (%)	Consumption Share (%)
Total	Asia	27.4	23.7
	Japan	0.4	4.9
	S. Korea	0.2	0.9
	India	2.2	2.5
	China	9.2	8.8
	Soviet Union	22.2	18.4
	Oceania	2.0	1.4
Solids	Asia	31.1	35.4
	Japan	0.3	3.4
	S. Korea	0.4	1.1
	India	5.1	5.2
	China	22.7	22.1
	Soviet Union	0.2	14.9
	Oceania	0.0	1.7
Liquids	Asia	34.8	22.3
	Japan	0.0	7.1
	S. Korea	0.0	1.2
	India	1.1	1.6
	China	4.4	3.2
	Soviet Union	19.5	13.8
	Oceania	0.9	1.4
Gas	Asia	10.9	11.4
	Japan	0.1	2.7
	S. Korea	0.0	0.2
	India	0.4	0.5
	China	0.8	0.8
	Soviet Union	38.2	32.7
	Oceania	1.1	1.2

a. See Annex 2 for energy production and consumption data, 1980–1989.

consumption. With respect to solid fuels, potential exists for increased Soviet exports to Asia, except that these exports will have to compete with the expanding net supplies of solid fuels from Oceania.

CONCLUSIONS

Although mineral exports accounted for over 50 percent of Soviet exports, exports to the Asia-Pacific region, including minerals, occupied a rather modest share of total exports, estimated at about eight percent in 1986. The major player

Table 15.18. Annual Growth Rates of Per Capita Fuel Consumption[a]

Fuel	Country	1980–1985 (%)	1985–1989 (%)	1980–1989 (%)
Total	World	-0.51	0.75	0.05
	Asia	2.01	3.75	2.78
	Japan	0.47	1.39	0.88
	S. Korea	3.16	8.19	5.36
	India	4.37	5.47	4.86
	China	3.41	4.45	3.87
	Soviet Union	2.19	1.25	1.78
	Oceania	1.94	1.92	1.93
Solids	World	0.95	0.31	0.66
	Asia	4.56	3.01	3.87
	Japan	5.70	0.28	3.26
	S. Korea	9.15	2.32	6.06
	India	4.12	5.30	4.64
	China	5.03	4.14	4.63
	Soviet Union	-0.95	-0.67	-0.82
	Oceania	2.05	2.24	2.13
Liquids	World	-2.62	0.34	-1.32
	Asia	-1.32	2.45	0.33
	Japan	-2.96	1.07	-1.18
	S. Korea	-2.03	9.53	2.95
	India	4.56	4.42	4.50
	China	-3.74	6.94	0.87
	Soviet Union	-0.67	0.85	0.00
	Oceania	-0.76	1.01	0.02
Gas	World	0.42	1.89	1.07
	Asia	3.47	13.63	7.87
	Japan	9.17	4.32	6.99
	S. Korea	—	—	—
	India	21.07	18.86	20.8
	China	-3.13	2.37	-0.73
	Soviet Union	7.44	2.83	5.37
	Oceania	9.32	3.69	6.78

a. Energy consumption in tonnes per million people; calculated from Annex 2.

in Soviet Asia-Pacific minerals trade has been Japan, which imported an estimated 850 million rubles of Soviet minerals in 1990. Japan was the major importer in the region of Soviet aluminum, refined lead, refined copper, and iron ores. In return, the Soviet Union relied on Japan for high quality iron and steel products. India has also been an active trading partner, with the Soviet Union importing a large amount of slab zinc and small amounts of aluminum and refined copper valued at about 675 million rubles in 1990. The People's Republic

Table 15.19. Annual Growth Rates of Fuel Consumption Per Real Gross Domestic Product[a]

Fuel	Country	1980–1985 (%)	1985–1989 (%)	1980–1989 (%)
Total	Asia	—	—	—
	Japan	-1.62	-14.78	-7.70
	S. Korea	3.59	-7.26	-1.38
	India	10.36	9.76	10.09
	China	9.18	4.69	7.16
	Soviet Union	0.34	n.a.	n.a.
	Oceania	7.12	-0.35	3.74
Solids	Asia	—	—	—
	Japan	3.51	-15.71	-5.52
	S. Korea	9.61	-12.29	-0.72
	India	10.09	9.58	9.86
	China	10.89	4.37	7.94
	Soviet Union	-2.74	n.a.	n.a.
	Oceania	7.24	-0.04	3.94
Liquids	Asia	—	—	—
	Japan	-4.97	-15.04	-9.59
	S. Korea	-1.61	-6.11	-3.63
	India	10.56	8.67	9.71
	China	1.62	7.18	4.06
	Soviet Union	-2.47	n.a.	n.a.
	Oceania	4.29	-1.24	1.79
Gas	Asia	—	—	—
	Japan	6.90	-12.31	-2.11
	S. Korea	—	—	—
	India	28.01	23.69	26.07
	China	2.27	2.59	2.41
	Soviet Union	5.49	n.a.	n.a.
	Oceania	14.88	1.38	8.67

a. Energy consumption in tonnes per 1000 U.S. dollars; calculated from Annex 2.

of China imported modest amounts of Soviet aluminum, refined copper, and iron ores. The Soviet Union has relied on India, Japan, and Australia for its bauxite and alumina imports; Malaysia and Singapore for its tin and tin concentrates imports; and Mongolia and the Philippines for its copper ores. The Soviet Union exported oil and oil products to Mongolia, Vietnam, North Korea, Laos, Kampuchea, and Australia. It also supplied Japan with coal and oil.

Analysis of past consumption and production trends of major nonferrous metals in the Asia-Pacific region and the former Soviet Union suggests that potential exists for increased exports of Commonwealth aluminum, refined copper, magnesium, cadmium, refined lead and nickel into the region. However, further

Table 15.20. Net Energy Production (Production Minus Consumption) for Various Countries and Regions, 1980, 1985, and 1989[a]

		Net Energy Production ('000 tonnes of coal equivalent)		
Fuel	Country	1980	1985	1989
Total	World	720,281	312,837	434,440
	Asia	884,943	245,658	494,246
	Japan	-392,566	-411,167	-448,781
	S. Korea	-39,433	-48,119	-73,371
	India	-25,855	-10,755	-20,755
	China	52,336	90,179	79,432
	Soviet Union	462,776	443,645	481,533
	Oceania	13,554	54,326	65,797
Solids	World	-6,137	-63,806	-45,670
	Asia	-76,801	-133,869	-154,787
	Japan	-63,414	-95,371	-104,302
	S. Korea	-6,565	-16,364	-22,137
	India	-3,055	-2,826	-4,179
	China	2,766	3,736	10,396
	Soviet Union	19,605	16,348	19,658
	Oceania	30,757	62,680	81,967
Liquids	World	719,350	380,019	470,013
	Asia	959,792	379,144	659,595
	Japan	-298,811	-265,373	-282,816
	S. Korea	-32,868	-31,755	-47,477
	India	-22,804	-7,933	-16,448
	China	49,608	86,565	69,221
	Soviet Union	374,631	342,291	321,391
	Oceania	-17,271	-8,355	-16,177
Gas	World	6,683	-3,507	12,070
	Asia	2,140	619	-10,508
	Japan	-30,341	-50,066	-61,664
	S. Korea	0	0	-3,757
	India	0	0	0
	China	0	1	0
	Soviet Union	66,203	81,481	135,708
	Oceania	67	0	7

a. Calculated from Annex 2.

increases in Commonwealth exports of aluminum, refined copper, cadmium, and refined lead will compete with exports from Oceania, which has shown significant increases in net production of these metals in the past decade. In other words, in order to be absorbed by Asia-Pacific markets, Commonwealth products will have to be competitive in the international metal market in price and

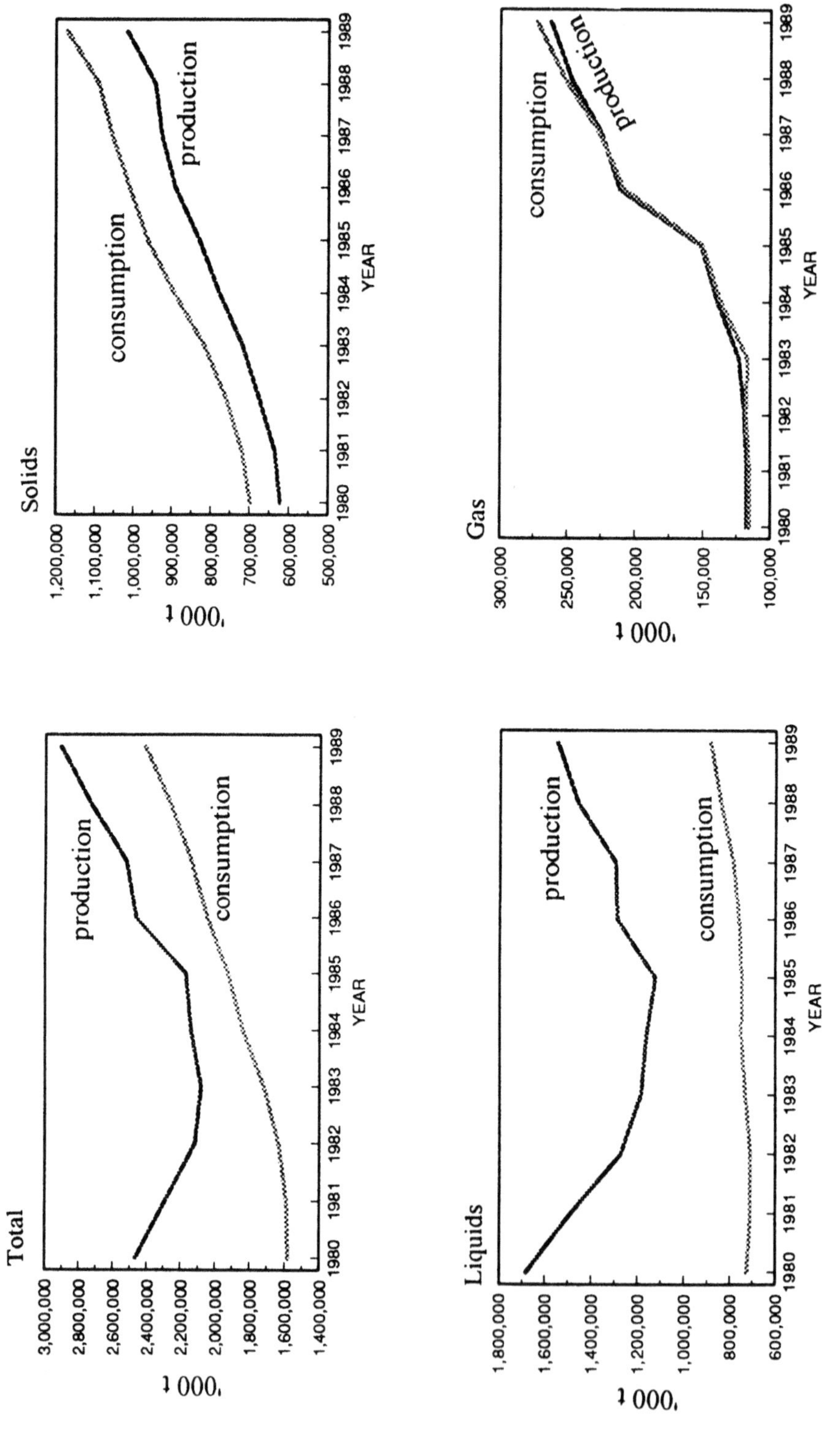

Figure 15.5. Consumption and Production Trends of Major Energy Fuels in Asia, 1980–1989

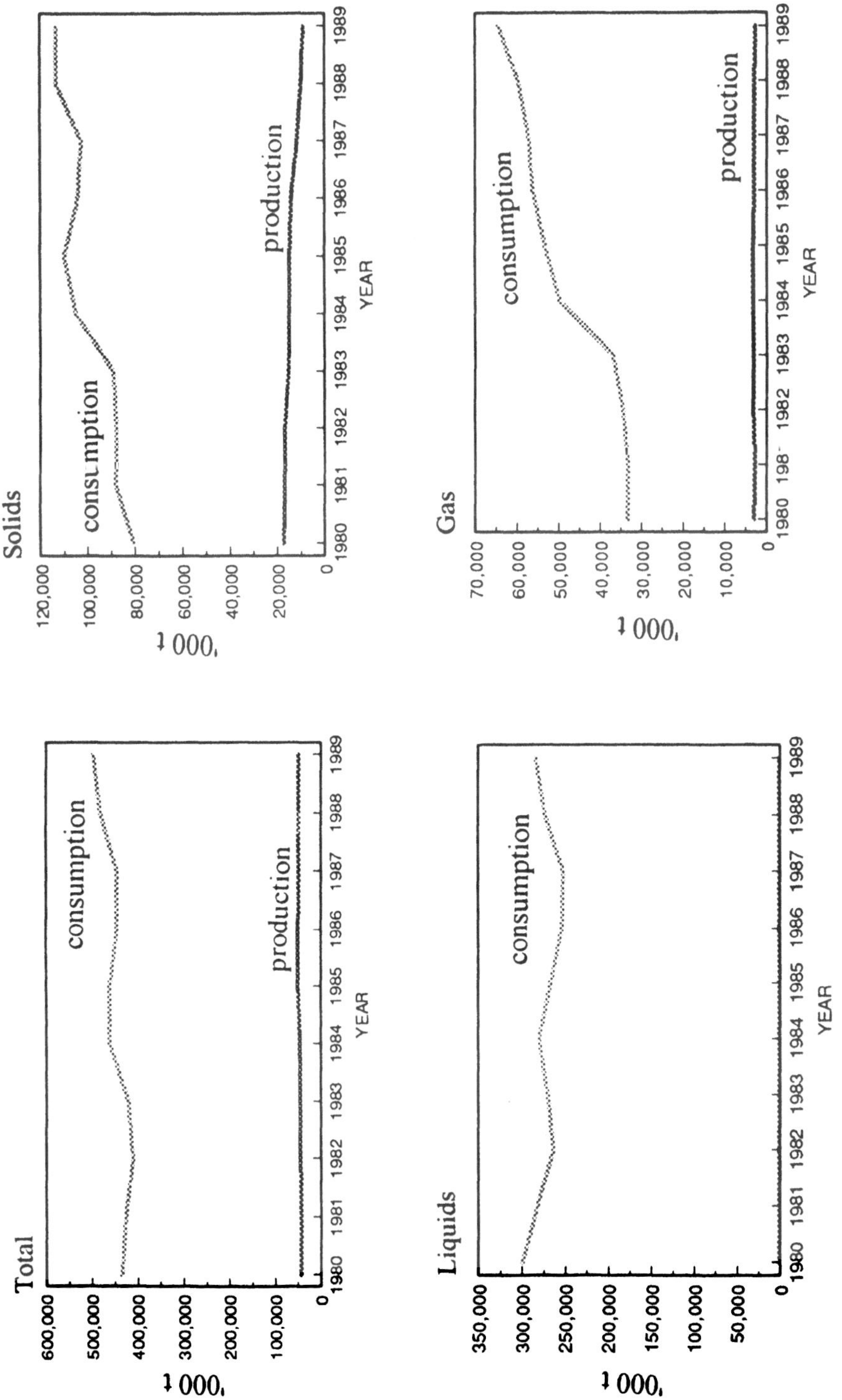

Figure 15.6. Consumption and Production Trends of Major Energy Fuels in Japan, 1980–1989

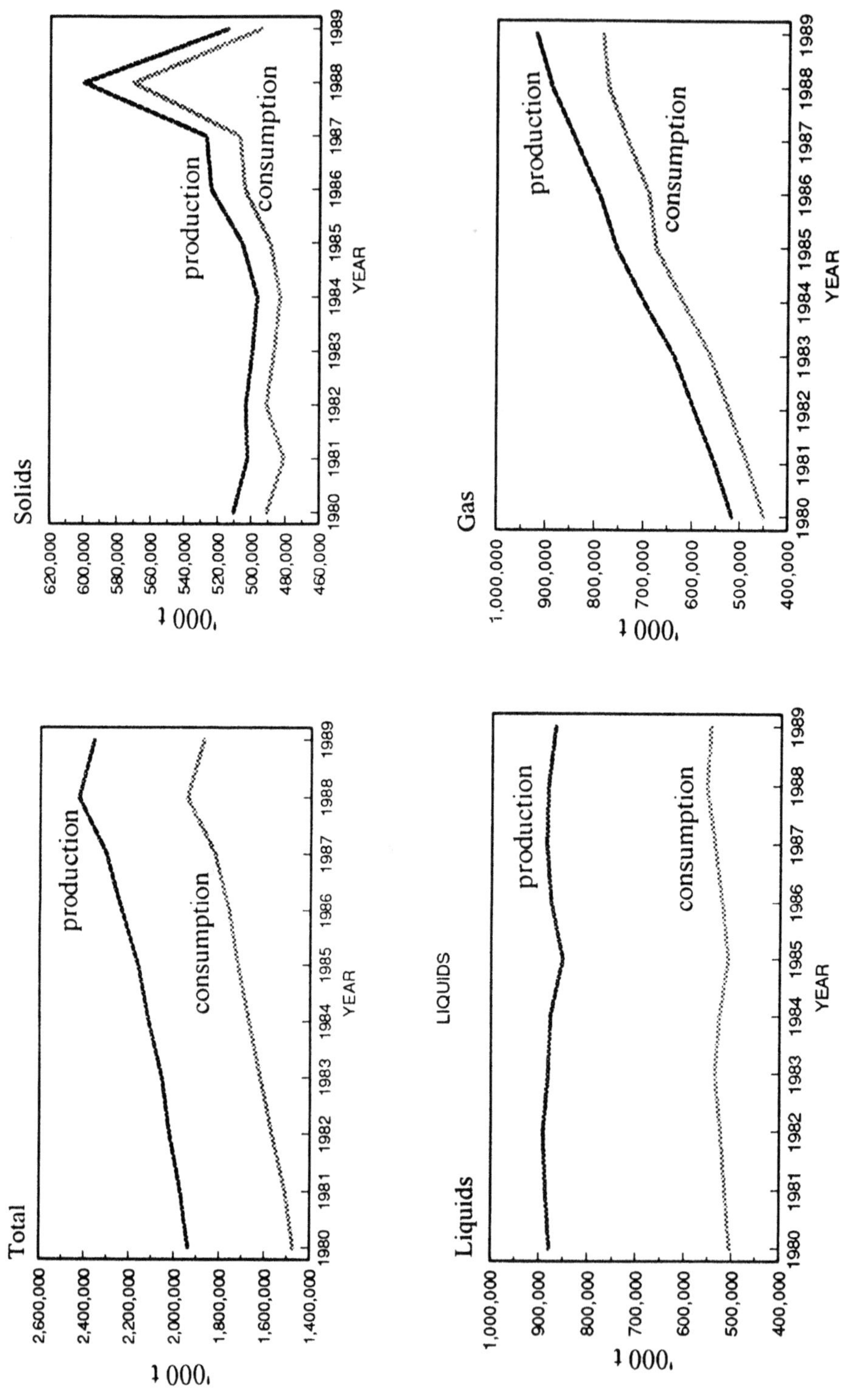

Figure 15.7. Consumption and Production Trends of Major Energy Fuels in the Soviet Union, 1980–1989

quality. The analysis also suggests that the new Commonwealth will continue to rely on tin imports from the Asia-Pacific region, and that there exists a great potential for increased Commonwealth exports of natural gas to the region. Potential also exists for increased exports of solid fuels to the area, except that they will have to compete with the expanding net production of solid fuels from Oceania.

Finally, it should be pointed out that the recent price reforms in the Commonwealth of Independent States will bring the prices of most minerals and mineral products closer to international market prices. This might lead to more efficient use of these minerals in the domestic markets and more efficient allocation of resources in the domestic production of these minerals. As a result, domestic consumption of some minerals might decrease, while production might increase. This will further enhance the opportunities for increased exports of some of these minerals to the Asia-Pacific region. However, the increase cannot be realized unless the new Commonwealth produces minerals and mineral products which are competitive in quality with respect to international standards. Furthermore, in order to address the current crisis of raw material and food needs in the new Commonwealth, additional hard currency receipts must be generated by the mineral sector, primarily through exports of energy and nonferrous and precious metals. In order to capture higher export receipts, one option is to increase the value added to these minerals which are subjected to rather low degrees of primary raw material processing at this time. This is only possible through imports of foreign capital and transfer of appropriate processing technology. Another option for increasing hard currency exports is to redirect the existing large mineral exports from the former CMEA members to countries where hard currency can be earned.

ENDNOTE

1. This section draws heavily from Bradshaw (1988).

Annex 1. Mineral Consumption and Production by Selected Countries/Regions

('000 tonnes)

		1980	1981	1982	1983	1984	1985	1986	1987	1988	1989	1990
Lead consumption	World	5,329.3	5,240.3	5,206.6	5,264.9	5,393.1	5,429.0	5,530.4	5,647.2	5,765.4	5,844.3	5,609.3
	Asia	595.9	621.7	602.6	654.7	682.0	741.0	769.5	787.8	884.3	927.5	911.2
	Japan	392.5	382.5	354.0	359.6	390.7	397.1	389.6	378.0	406.5	405.7	416.9
	S. Korea	33.0	60.0	41.7	57.5	57.4	80.8	88.3	112.4	146.0	176.0	150.0
	India	54.0	56.8	67.5	60.0	62.0	72.0	77.0	67.5	75.0	80.0	80.0
	China	204.0	171.0	191.0	210.0	224.0	243.0	249.0	256.0	250.0	250.0	250.0
	Soviet Union	800.0	800.0	810.0	805.0	790.0	780.0	760.0	775.0	790.0	700.0	650.0
	Oceania	81.5	79.9	68.3	67.6	66.9	68.8	67.4	70.3	70.0	70.6	64.8
Copper production	World	9,270.5	9,579.6	9,383.8	9,662.1	9,528.3	9,723.5	9,874.4	10,158.9	10,535.6	10,809.0	10,733.3
	Asia	1,149.7	1,262.8	1,300.4	1,379.5	1,328.0	1,401.8	1,445.6	1,464.2	1,468.5	1,546.1	1,521.0
	Japan	1,014.3	1,050.1	1,075.0	1,091.9	935.2	936.0	943.0	980.3	955.1	989.6	1,008.0
	S. Korea	72.9	113.0	115.8	134.8	140.6	150.0	165.0	157.9	170.5	178.7	184.0
	India	23.2	23.1	27.0	29.3	33.6	28.9	36.9	30.8	40.1	41.8	38.7
	China	295.0	300.0	300.0	310.0	320.0	340.0	350.0	450.0	460.0	470.0	490.0
	Soviet Union	1,300.0	1,320.0	1,350.0	1,400.0	1,380.0	1,400.0	1,400.0	1,410.0	1,380.0	1,345.0	1,260.0
	Oceania	182.4	191.0	178.1	202.6	197.2	194.3	185.1	207.8	222.7	255.0	274.0
Copper consumption	World	9,364.6	9,753.0	9,131.2	9,324.6	9,876.0	9,859.7	10,106.0	10,404.8	10,537.4	11,018.1	10,820.9
	Asia	1,452.8	1,620.0	1,613.9	1,679.8	1,878.2	1,766.2	1,886.1	2,049.9	2,189.2	2,481.3	2,725.4
	Japan	1,152.6	1,250.3	1,239.3	1,212.2	1,364.3	1,226.2	1,210.5	1,276.6	1,330.7	1,446.6	1,577.5
	S. Korea	84.0	139.8	136.7	152.3	188.0	207.3	262.3	259.0	266.3	244.8	324.2
	India	77.2	75.1	83.2	96.0	81.2	82.5	95.0	100.0	113.0	130.0	135.0
	China	370.0	370.0	380.0	380.0	390.0	420.0	450.0	470.0	465.0	528.0	512.0
	Soviet Union	1,300.0	1,320.0	1,320.0	1,300.0	1,280.0	1,305.0	1,300.0	1,270.0	1,210.0	1,140.0	1,000.0
	Oceania	130.1	144.3	132.1	128.3	125.3	125.8	118.1	125.4	130.7	134.4	124.8

(Continued)

Annex 1. Mineral Consumption and Production by Selected Countries/Regions

('000 tonnes)

		1980	1981	1982	1983	1984	1985	1986	1987	1988	1989	1990
Zinc production	World	6,256.8	6,292.0	6,061.6	6,382.1	6,665.3	6,869.6	6,825.6	7,023.6	7,245.6	7,253.7	7,096.9
	Asia	867.1	826.7	826.1	875.4	939.7	1,003.7	981.8	1,008.0	1,061.1	1,069.0	1,105.9
	Japan	735.2	670.2	662.4	701.3	754.4	739.6	708.0	665.6	678.2	664.5	687.5
	S. Korea	75.7	91.0	96.2	105.7	106.0	108.9	125.0	186.4	223.2	240.2	248.4
	India	43.6	57.4	52.6	54.1	55.4	70.9	73.8	68.9	68.9	71.4	79.1
	China	227.4	232.9	237.9	233.6	270.4	306.2	336.2	383.1	425.4	450.9	526.3
	Soviet Union	1,060.0	1,060.0	1,050.0	1,060.0	1,050.0	1,050.0	1,065.0	1,020.0	1,035.0	1,020.0	920.0
	Oceania	301.0	295.9	291.4	298.5	301.9	288.7	303.1	310.2	302.5	294.2	301.3
Zinc consumption	World	6,291.0	6,169.7	5,973.6	6,351.3	6,515.4	6,500.4	6,707.4	6,903.9	7,163.8	7,102.1	6,979.1
	Asia	1,139.0	1,155.2	1,198.0	1,310.4	1,304.7	1,328.5	1,372.0	1,396.1	1,484.1	1,467.7	1,586.6
	Japan	752.3	699.1	703.1	770.8	774.6	780.1	753.0	728.7	774.2	768.7	814.3
	S. Korea	68.0	74.4	94.2	109.0	123.8	119.7	140.9	178.7	173.0	188.0	227.2
	India	95.3	108.0	131.0	125.0	124.3	134.0	135.0	130.5	142.0	135.0	135.0
	China	259.0	278.0	304.0	322.0	337.0	349.0	382.0	409.0	385.0	450.9	500.0
	Soviet Union	1,030.0	1,040.0	1,050.0	1,050.0	1,050.0	1,000.0	1,010.0	1,030.0	1,080.0	1,020.0	920.0
	Oceania	116.2	111.9	98.1	95.4	99.1	106.9	99.4	94.2	104.1	118.1	102.3
Tin production	World	243.6	243.3	226.8	207.8	224.0	216.2	203.5	201.4	219.1	230.0	225.6
	Asia	142.2	140.9	123.9	104.2	95.9	91.4	89.3	88.1	94.8	101.0	101.9
	Japan	1.3	1.3	1.3	1.3	1.4	1.4	1.3	0.9	0.8	0.8	0.8
	S. Korea	0.4	0.2	0	0.4	1.2	1.6	1.3	1.8	2.5	2.4	2.5
	India	0	0	0	0	0	0	0	0	0	0	0.3
	China	15.0	16.5	16.5	16.5	29.9	24.6	26.0	25.0	24.0	28.3	28.0
	Soviet Union	17.0	16.0	17.0	18.5	18.5	18.5	17.0	18.0	18.0	15.0	14.0
	Oceania	5.3	4.6	3.6	3.4	3.4	3.1	1.7	0.8	0.7	0.7	0.6

(Continued)

Annex 1. Mineral Consumption and Production by Selected Countries/Regions

('000 tonnes)

		1980	1981	1982	1983	1984	1985	1986	1987	1988	1989	1990
Tin consumption	World	221.4	210.5	199.7	202.2	212.7	214.3	224.7	226.9	236.5	241.7	232.7
	Asia	42.2	43.5	39.0	42.4	48.2	45.6	53.5	55.6	58.9	64.8	65.7
	Japan	30.9	30.5	28.7	30.4	33.3	31.6	31.5	32.6	32.2	33.5	33.8
	S. Korea	1.8	2.2	2.0	2.6	3.6	2.6	5.2	5.8	7.2	12.2	7.8
	India	2.3	2.8	2.1	2.3	2.1	2.3	2.9	2.7	3.0	3.0	3.0
	China	10.0	10.0	10.0	11.0	11.0	11.5	11.0	12.5	14.0	18.0	18.0
	Soviet Union	25.0	26.0	27.0	27.0	27.0	31.5	31.5	29.0	28.0	24.0	20.0
	Oceania	3.4	3.4	2.8	2.7	2.9	2.8	2.8	2.8	2.8	2.9	1.4
Cadmium production	World	18.5	17.2	16.9	17.7	20.1	19.6	19.5	19.7	21.8	20.8	19.9
	Asia	2.6	2.4	2.5	2.8	3.0	3.2	3.1	3.1	3.4	3.5	3.3
	Japan	2.2	2.0	2.0	2.2	2.4	2.5	2.5	2.5	2.6	2.7	2.5
	S. Korea	0.4	0.3	0.3	0.5	0.4	0.5	0.5	0.5	0.5	0.5	0.6
	India	0.1	0.1	0.1	0.1	0.1	0.2	0.2	0.2	0.2	0.3	0.3
	China	0.3	0.3	0.3	0.4	0.5	0.5	0.7	0.8	0.8	0.9	1.0
	Soviet Union	2.9	2.8	2.8	2.8	2.8	2.8	2.7	2.6	2.7	2.6	2.4
	Oceania	1.0	1.0	1.0	1.1	1.1	0.9	0.9	1.0	0.9	0.7	0.6
Cadmium consumption	World	16.7	16.3	16.0	18.2	17.4	17.8	18.9	19.3	19.9	20.4	19.4
	Asia	1.6	1.6	1.7	2.4	2.3	2.5	2.6	3.2	4.5	6.0	5.6
	Japan	1.1	1.2	1.1	1.9	2.0	1.9	2.0	2.3	3.7	5.1	4.8
	S. Korea	0.2	0.2	0.2	0.1	0.0	0.3	0.4	0.4	0.4	0.4	0.4
	India	0.1	0.1	0.2	0.2	0.2	0.3	0.2	0.3	0.2	0.3	0.3
	China	0.3	0.3	0.3	0.3	0.3	0.4	0.4	0.4	0.4	0.4	0.4
	Soviet Union	2.4	2.2	2.3	2.5	2.8	2.9	2.7	2.5	2.6	2.3	2.0
	Oceania	0.3	0.1	0.1	0.1	0.1	0.1	0.0	0.1	0.1	0.1	0.1

(Continued)

Annex 1. Mineral Consumption and Production by Selected Countries/Regions

('000 tonnes)

		1980	1981	1982	1983	1984	1985	1986	1987	1988	1989	1990
Magnesium production	World	319.2	397.6	251.2	261.1	328.7	321.2	324.1	326.4	346.1	354.2	357.6
	Asia	9.4	5.8	5.7	6.1	7.2	8.6	8.2	8.5	10.3	12.4	13.8
	Japan	9.3	5.7	5.6	6.0	7.1	8.5	8.1	8.2	10.0	12.1	12.8
	S. Korea	0	0	0	0	0	0	0	0	0	0	0
	India	0.1	0.1	0.1	0.1	0.1	0.1	0.1	0.3	0.3	0.3	1.0
	China	7.0	7.0	7.5	8.5	8.5	9.0	9.0	9.0	14.0	15.0	16.0
	Soviet Union	75.0	76.0	77.0	80.0	85.0	85.0	85.0	86.0	90.0	85.0	80.0
	Oceania											
Magnesium consumption	World	284.1	260.3	248.5	266.3	287.7	300.6	30[illegible].9	323.0	342.6	340.8	335.9
	Asia	20.6	20.0	21.9	23.7	24.8	28.5	2[illegible].2	26.8	29.7	28.9	34.1
	Japan	18.3	18.0	19.0	19.6	20.8	21.6	19.6	21.4	22.6	22.8	25.0
	S. Korea	0.5	0.4	0.3	0.6	0.6	1.8	0.8	1.1	1.5	1.1	1.5
	India	0.5	0.6	0.6	0.6	0.3	0.3	0.6	0.7	1.0	0.8	1.5
	China	10.0	10.0	10.0	12.0	13.5	14.0	15.0	16.0	18.0	16.2	16.0
	Soviet Union	60.0	62.0	62.0	65.0	70.0	76.0	78.0	80.0	80.0	80.0	77.0
	Oceania	3.1	2.5	2.4	4.9	4.3	3.3	4.8	4.1	4.3	3.8	3.8
Nickel production	World	733.4	705.4	631.6	680.8	758.1	780.2	762.6	793.1	837.0	861.8	860.7
	Asia	136.4	117.7	103.5	93.3	98.8	116.9	108.6	104.1	116.0	125.8	123.9
	Japan	109.3	93.6	87.3	82.3	89.4	95.2	93.0	94.2	100.6	105.8	100.3
	S. Korea	0	0	0	0	0	0	0	0	0	5.0	8.2
	India	0	0	0	0	0	0	0	0	0	0	0
	China	11.0	12.0	12.5	13.0	17.5	24.6	23.3	25.5	25.5	26.3	27.5
	Soviet Union	165.0	170.0	190.0	192.0	193.0	198.0	195.0	210.0	215.0	225.0	230.0
	Oceania	67.9	70.5	73.9	63.5	67.9	76.9	75.[illegible]	74.0	79.4	79.2	79.0

(Continued)

Annex 1. Mineral Consumption and Production by Selected Countries/Regions

('000 tonnes)

		1980	1981	1982	1983	1984	1985	1986	1987	1988	1989	1990
Nickel consumption	World	706.3	659.5	648.9	696.8	786.4	771.1	772.0	832.7	853.5	847.4	849.9
	Asia	138.5	126.1	127.0	142.9	174.2	162.9	172.7	200.9	207.4	219.9	224.7
	Japan	123.8	106.4	108.4	118.1	144.6	136.1	127.8	153.9	161.0	163.0	164.2
	S. Korea	3.0	3.9	2.8	2.6	3.1	3.5	6.6	5.4	6.0	18.3	24.0
	India	7.0	9.5	11.0	13.0	15.7	14.0	16.0	19.3	15.0	14.8	12.2
	China	18.0	18.0	18.0	18.0	19.0	21.0	23.0	24.0	27.6	26.7	27.5
	Soviet Union	130.0	130.0	138.0	145.0	140.0	138.0	137.0	135.0	130.0	120.0	115.0
	Oceania	4.8	3.6	2.8	3.0	3.6	3.7	3.3	3.6	3.5	3.5	1.7

Source: *Metallstatistik 1980-1990*, 78th Edition, 1991.

Annex 2. Energy Consumption and Production by Selected Countries/Regions

('000 tonnes of coal equivalent)

Consumption		1980	1981	1982	1983	1984	1985	1986	1987	1988	1989
World	Total	8,544,341	8,455,687	8,429,30[illegible]	8,590,329	8,906,128	9,142,328	9,314,882	9,614,855	10,012,893	10,176,643
	Solid	2,631,884	2,649,929	2,684,80[illegible]	2,761,880	2,876,560	3,028,594	3,072,958	3,166,348	3,287,863	3,311,959
	Liquid	3,777,799	3,645,368	3,566,60[illegible]	3,606,950	3,661,249	3,630,751	3,701,344	3,775,868	3,921,193	3,975,897
	Gas	1,833,733	1,840,082	1,844,421	1,861,279	1,973,115	2,055,569	2,097,200	2,206,409	2,312,465	2,393,325
Japan	Total	434,827	425,881	409,46[illegible]	421,515	461,221	460,245	444,922	445,252	480,086	495,896
	Solid	80,496	88,170	87,784	89,053	105,011	109,834	104,224	102,617	113,237	113,260
	Liquid	299,464	282,276	264,197	270,434	280,402	266,516	253,030	252,264	273,254	283,610
	Gas	33,246	33,352	34,491	37,003	49,730	53,311	56,254	57,209	59,705	64,378
India	Total	139,383	148,817	157,557	170,557	176,681	192,004	206,489	218,064	236,132	256,876
	Solid	95,095	100,985	107,064	116,970	118,550	129,437	138,598	148,056	160,295	172,049
	Liquid	36,470	39,364	41,284	43,387	46,736	50,700	53,493	55,450	59,003	65,168
	Gas	1,726	2,001	3,024	3,636	4,279	4,994	7,152	8,020	8,878	10,776
S. Korea	Total	52,343	56,370	55,611	57,371	62,206	65,439	65,732	76,367	87,407	93,114
	Solid	18,804	23,037	22,137	22,865	27,840	31,178	27,692	33,682	36,625	35,502
	Liquid	32,868	32,644	32,763	33,070	32,623	31,755	32,970	34,196	41,530	47,477
	Gas	0	0	0	0	0	0	99	3,003	3,888	3,757
Asia	Total	1,579,289	1,589,006	1,634,899	1,719,366	1,839,066	1,925,505	2,048,331	2,140,258	2,263,221	2,414,360
	Solid	697,278	717,204	759,064	813,964	893,257	962,089	1,011,626	1,054,607	1,089,558	1,171,919
	Liquid	721,267	708,189	706,668	731,287	747,704	744,781	757,488	783,778	841,162	887,616
	Gas	115,334	115,085	117,744	116,925	136,917	150,953	208,438	226,243	251,670	272,315
China	Total	562,776	556,617	584,860	613,265	666,448	708,228	766,788	799,789	817,530	892,478
	Solid	432,193	433,385	464,952	498,528	550,103	587,732	627,579	652,702	663,031	731,718
	Liquid	104,438	98,223	94,883	87,871	89,087	91,827	109,133	115,988	121,941	127,136
	Gas	18,948	16,917	15,841	16,213	16,511	17,196	18,320	18,660	18,970	19,988
Soviet Union	Total	1,473,133	1,508,504	1,566,902	1,618,259	1,667,462	1,716,076	1,757,535	1,823,749	1,946,776	1,875,232
	Solid	490,825	480,795	490,969	486,722	483,184	489,178	504,207	507,475	570,248	495,094
	Liquid	503,955	514,748	524,468	537,331	526,835	509,232	521,789	538,100	554,579	547,602
	Gas	450,692	484,050	522,563	561,437	618,092	674,277	688,811	732,465	772,015	783,780

(Continued)

Annex 2. Energy Consumption and Production by Selected Countries/Regions

('000 tonnes of coal equivalent)

Consumption		1980	1981	1982	1983	1984	1985	1986	1987	1988	1989
Oceania	Total	105,269	107,125	110,622	111,092	115,774	120,905	123,223	129,358	132,904	141,325
	Solid	40,060	41,227	42,285	41,661	43,088	46,268	45,730	49,702	50,680	54,760
	Liquid	47,761	45,849	45,784	46,498	47,696	47,969	48,082	49,658	51,123	54,092
	Gas	13,520	15,622	18,308	18,715	20,560	22,026	24,941	25,419	26,360	27,578
Production		**1980**	**1981**	**1982**	**1983**	**1984**	**1985**	**1986**	**1987**	**1988**	**1989**
World	Total	9,264,622	9,061,566	8,986,278	8,917,454	9,267,546	9,455,165	9,820,657	10,043,327	10,467,206	10,611,083
	Solid	2,625,747	2,632,157	2,707,382	2,719,402	2,839,091	2,964,788	3,075,498	3,153,610	3,259,662	3,266,289
	Liquid	4,497,149	4,247,313	4,079,733	3,981,898	4,043,603	4,010,770	4,178,208	4,198,589	4,389,515	4,445,910
	Gas	1,840,416	1,861,630	1,865,226	1,855,598	1,989,536	2,052,062	2,124,075	2,226,236	2,328,561	2,405,395
Japan	Total	42,261	42,217	43,410	43,774	44,701	49,078	49,257	48,456	47,462	47,115
	Solid	17,082	16,759	16,681	15,163	14,887	14,463	14,075	11,472	9,868	8,958
	Liquid	653	590	607	626	606	786	924	894	870	794
	Gas	2,905	2,785	3,034	2,960	3,130	3,245	2,844	2,929	2,833	2,714
India	Total	113,528	133,490	146,662	160,319	173,995	181,249	196,871	207,304	221,784	236,101
	Solid	92,040	103,289	108,677	114,101	122,559	126,611	137,699	149,407	159,437	167,870
	Liquid	13,666	21,726	28,769	36,008	40,027	42,767	44,788	43,425	45,643	48,720
	Gas	1,726	2,001	3,024	3,636	4,279	4,994	7,152	8,020	8,878	10,776
S. Korea	Total	12,910	13,744	13,930	14,487	15,786	17,320	19,909	21,437	20,986	19,743
	Solid	12,239	13,054	13,219	13,052	14,043	14,814	15,938	15,951	15,622	13,365
	Liquid	0	0	0	0	0	0	0	0	0	0
	Gas	0	0	0	0	0	0	0	0	0	0
Asia	Total	2,464,232	2,289,628	2,117,248	2,080,154	2,139,046	2,171,163	2,463,380	2,521,678	2,728,763	2,908,606
	Solid	620,477	635,415	673,475	715,800	777,471	828,220	892,236	925,851	942,742	1,017,132
	Liquid	1,681,059	1,488,518	1,273,517	1,184,094	1,161,178	1,123,925	1,288,806	1,295,321	1,458,142	1,547,211
	Gas	117,474	117,392	119,065	123,385	139,537	151,572	211,605	224,870	247,129	261,807

(Continued)

Annex 2. Energy Consumption and Production by Selected Countries/Regions

('000 tonnes of coal equivalent)

Production		1980	1981	1982	1983	1984	1985	1986	1987	1988	1989
China	Total	615,112	608,481	641,207	679,842	744,047	798,407	854,590	884,742	903,037	971,910
	Solid	434,959	436,327	467,728	501,521	553,170	591,468	637,961	662,169	675,367	742,114
	Liquid	154,046	147,176	148,487	151,500	163,705	[illegible]	186,697	191,629	195,293	196,357
	Gas	18,948	16,917	15,841	16,213	16,511	[illegible]	18,320	18,660	18,970	19,988
Soviet Union	Total	1,935,909	1,971,842	2,021,132	2,052,943	2,114,288	[illegible]	2,233,666	2,301,417	2,425,211	2,356,765
	Solid	510,430	502,200	503,262	499,426	496,738	505,526	524,155	527,487	599,443	514,752
	Liquid	878,586	886,760	892,215	881,939	876,788	851,523	875,696	885,696	883,696	868,993
	Gas	516,895	551,548	594,329	635,937	698,409	[illegible]	790,563	838,264	887,398	919,488
Oceania	Total	118,823	131,175	138,377	143,450	155,328	175,231	192,390	205,430	196,925	207,122
	Solid	70,817	81,321	86,086	91,911	97,087	108,948	120,901	134,292	124,414	136,727
	Liquid	30,490	29,806	29,737	28,602	33,250	39,614	42,084	41,111	41,404	37,915
	Gas	13,587	15,622	18,308	18,719	20,561	22,026	24,936	25,447	26,366	27,585

Sources: United Nations, *Energy Statistics Yearbook*, various issues; and World Bank, *World Development Report*, various issues.

REFERENCES

Bradshaw, M.J., 1988, "Soviet Asia-Pacific Trade and the Regional Development of the Far Soviet East," *Soviet Geography: Review and Translation,* Vol. 29, No. 4, pp. 367–393.

Bradshaw, M.J., 1990, "Soviet Far Eastern Trade," in Chapter 10 of *The Soviet Far East: Geographical Perspectives on Development,* Allen Rogers, ed., Routledge, London and New York.

Clark, Allen L., and Gyoo J. Jeon, 1990, "Metal Consumption Trends in the Asia-Pacific Region: 1969–2015," *Proceedings of the Minerals and Energy Forum,* Pacific Economic Cooperation Conference, Manila, Philippines.

Jeon, Gyoo J., 1989, "Innovative Methods for Long-term Mineral Forecasting," unpublished Ph.D. dissertation, University of Arizona.

Metallgesellschaft Aktiengesellschaft, 1991, *Metallstatistik 1980–1990,* 78th edition, Metallgesellschaft Aktiengesellschaft, Frankfurt am Main.

The Economist Intelligence Unit, 1990, *Country Report: USSR,* No. 2, United Kingdom.

United Nations, *Energy Statistics Yearbook,* various issues, Department of International Economic and Social Affairs, Statistical Office, United Nations, New York.

U.S. Bureau of Mines, 1988, *World Economy Minerals Yearbook,* U.S. Government Printing Office, Washington, D.C.

U.S. Bureau of Mines, 1988, *USSR Minerals Yearbook,* U.S. Government Printing Office, Washington, D.C.

World Bank, *World Report,* various issues, Washington, D.C.

Chapter 16

MULTILATERAL ENERGY COOPERATION IN NORTHEAST ASIA: A FOCUS ON OIL AND NATURAL GAS DEVELOPMENT

Keun-Wook Paik

INTRODUCTION

During the last few years, new geoeconomic patterns have been rapidly replacing past geopolitical alignments due to Sino-Soviet rapprochement, normalization between South Korea and the former Soviet Union, economic cooperation between South Korea and the People's Republic of China, the improving political relationships between the United States and North Korea and Japan and North Korea, and a somewhat improved inter-Korean relationship. More encouraging are the signs that multilateral economic relations are emerging, possibly leading to a "Northeast Asian Economic Circle." Once the Tumen River Basin Development Scheme under the auspices of the United Nations Development Plan (UNDP) gets under way, the full-fledged development of multilateral economic relations in the region will be a more likely prospect.

The trend towards multilateral economic cooperation may also usher in multilateral energy cooperation in Northeast Asia. If this occurs, it may be possible to think about a "Northeast Asian Energy Regime" or a "Northeast Asian Energy Charter," similar in structure to the European Energy Charter. In fact, energy relations in Northeast Asia have been undergoing their own peculiar brand of perestroika as communism retreats. The establishment of robust energy relationships between South Korea, the former Soviet Union, and China are a typical example of this new development.

This chapter will first examine bilateral energy relations, especially during recent years, and then explore the trend towards multilateral energy cooperation in Northeast Asia in the 1990s. Finally, the chapter will examine the need to establish a "Northeast Asian Energy Charter" as a parallel to the European Energy Charter.

J.P. Dorian et al. (eds.), CIS Energy and Minerals Development, 293–313.

BILATERAL ENERGY RELATIONS IN NORTHEAST ASIA

The former Soviet Union and the People's Republic of China.

The energy relationship between the former Soviet Union and China—disrupted for three decades because of political and territorial disagreements—has been resumed. Spurred on by the improved relationship after normalization between the Soviet Union and China in May 1989, Li Peng, the Chinese Premier, and Nikolai Ryzhkov, then Soviet Premier, signed a second protocol covering the period 1990-2000 when Li Peng visited Moscow in April 1990. The protocol encourages cooperation in nonferrous metals, petrochemicals, natural gas, agriculture, transportation, and public health fields, without specifying individual projects. In November 1990, the Soviet Union and China agreed that the former would help to develop two or three of China's oil fields on a turnkey basis. In return, Chinese construction workers would help develop two oil fields in the Soviet Western Siberian Tyumen Province. In May 1992, Xinhua News Agency reported that a team of Chinese geophysicists was about to take part in oil exploration in Russia's Tyumen region. In addition, in December 1990, a Chinese newspaper hailed the first achievement of Sino-Soviet energy cooperation, the discovery of Nanyang oil field in Henan Province, as being the result of the application of *Dianchang Chafen Fa* (Electric Field Proportional Distribution Method), invented by a Soviet scientist.

Taiwan and China.

Even though the Republic of China/Taiwan government has rejected any moves toward energy cooperation with the mainland, there have been a number of breakthroughs. In May 1990, China approved an ethylene project by Taiwan Plastic Corp. (TPC). Both sides signed an agreement for the construction of a 1.27 million tonnes/year ethylene plant in Fujian Province during 1991-93. At the end of 1990, the Institute of Oceanology at the National Taiwan University and China's Academia Sinica's South China Sea Institute of Oceanology in Guangzhou agreed to formal and regular exchanges of information on oil reserves and geophysical structures, ocean flows, and geological features in the Taiwan Strait. The move is understood to be the first step toward a joint oil exploration effort in the future, although there have been setbacks, such as the Taiwan Ministry of Economic Affairs' refusal of Beijing's offer of oil at a special price.

However, the Taiwan government found an alternative. During the fourth quarter of 1990, the Chinese Petroleum Corp. (CPC) expressed its interest in buying crude oil from the Soviet Union, even though there have been no relations between the Republic of China and the Soviet Union for decades. In addition, CPC sent a mission to inspect the Salyn oil field in Tyumen and officials of CPC and a Soviet association discussed a possible CPC role in development of the field.

South Korea and the former Soviet Union.

The establishment of diplomatic relations between South Korea and the Soviet Union in late September 1990 has paved the way for energy cooperation. Even though South Korean companies Honam Oil Refinery Co. and Yukong Ltd. had imported a total of 0.4 million barrels of oil products from the Soviet Union in March 1987, March 1988, and February 1990, the first imports of Soviet crude (volume 106,000 barrels, value US$3 million) were made by Honam Oil Refinery Co. through a spot purchase contract in late September 1990. In July 1991, Jindo Do entered a contract with a Soviet organization to directly import five million barrels of Soviet crude oil.

As far as oil and gas exploration and production-related investment is concerned, three cases have borne fruit. Hyundai Co. and the Kalmyk Republic signed an agreement on joint ventures in oil drilling and refining. Hyundai also decided to participate in the development of the North Buzachi oil field in Kazakhstan. In addition, Samsung Co. agreed to invest US$3 million in a project by MDSeis to expand oil production at four fields in Western Siberia. It is not certain whether Tongwon Consolidated Coal Mine Development Co. will set up a joint venture to explore the Okruzhnoye area onshore Sakhalin. Without compromises on management, profits remittances, and exchange rate application issues, the plan will not be realized.

Great potential for investment lies in the development of the Sakhalin offshore, Sea of Okhotsk, and Yakutsk areas. The recent warming of relations between Seoul and Moscow could give rise to opportunities for Korean firms to participate in oil and gas development projects in the Russian Far East. Since early 1989, Hyundai Co. has expressed an interest in laying a pipeline grid from Yakutsk to South Korea through North Korea, and its application to bid for the Sakhalin area could be interpreted as an initial step towards this goal. Unlike Hyundai's approach, Palmco Corp. has targeted the development of the gas-rich Lunskoye field since 1989. Encouragingly, Samsung Co. decided to cooperate with Palmco Corp., and Palmco/Samsung presented a feasibility study for the field to the Experts Commission set up under the leadership of Mr. Olkhovikov by the Russian Ministry of Economy and Finance on October 31, 1991. It is not certain, however, whether either side would secure a bridgehead in the Sakhalin contract.

South Korea and the People's Republic of China.

The China National Offshore Oil Corp. (CNOOC) officially offered a proposal to jointly develop offshore Yellow Sea areas to South Korea in May 1991. Considering that a series of Korean-American seismic surveys in the Yellow Sea in May and June 1991 and Gulf Oil's drilling operations in Korea's sea-bed mining Block II from February to June 1973 were suspended because of fierce protests and threats from China and North Korea, the proposal in and of itself should be regarded as great progress. Full-fledged offshore development cooperation between South Korea and China in the Yellow Sea and East China Sea, however, would not be possible without reaching an agreement on the delimitation of the continental shelf areas. Recent protests from China against Korean offshore exploration activities being carried out in Korean waters confirm this.

It is worth noting that Korean companies, including Yukong Ltd., Honam Oil Refinery Co., Kyung In Energy Co. and Kukdong Oil Co., began directly importing Chinese crude (mainly from Shengli and Chengbei oil fields) in mid-1988 when there were no diplomatic relations between South Korea and China. The volumes recorded were around 3.4 million barrels in 1988, 8.1 million barrels in 1989, and 7.2 million barrels in 1990. In addition, in August 1988 the China National Chemical Import and Export Corporation (SINOCHEM) requested that a Korean oil refinery, Kukdong Oil Co., process its heavy crude oil into a lighter product and deliver it back to China. And in January 1992, SINOCHEM opened a branch office in Seoul. The abovementioned cases imply that energy cooperation between South Korea and China will accelerate now that diplomatic relations have been established. As South Korea opened its trade office in Beijing in January 1991, and a Sino-South Korean trade agreement was signed in December 1991, it seemed only logical that establishment of diplomatic relations would follow in September 1992.

North Korea.

North Korea's role in the energy relations of the region has been insignificant; however, its special energy relations with the former Soviet Union and China are worth noting. During 1986–89, its crude oil imports from the Soviet Union and China reached approximately 1.9 and 1.2 million tonnes, respectively. (North Korean's current refining capacity is about 3.5 million tonnes.) A reduction of crude imports from its main suppliers has led to an unprecedented energy shortage, with the former Soviet Union requiring North Korea to settle its payments with hard currency at international prices beginning in 1991. In fact, in 1990 the Soviet crude oil supply to North Korea was 0.41 million tonnes, but during January–July 1991 the amount was only 0.041 million tonnes. To make matters worse, China has also asked North Korea to pay with hard currency; but it is not certain the request applies to the crude oil trade.

North Korea has already recognized the seriousness of its energy situation and has made efforts to find alternatives to alleviate its predicament. First, offshore oil field discoveries were made in 1988 and 1989 in the Yellow Sea; however, they proved to be uncommercial. Second, North Korea apparently has no objection to a plan for laying a pipeline across its territory to bring Russian natural gas to South Korea. Third, around 30,000 barrels of diesel oil were delivered to North Korea by a South Korean company, despite the United States' protest based on COCOM (Consultative Group Coordinating Committee) regulations. Fourth, in early 1989 North Korea's Seungli (Victory) Chemical Co. asked Lucky

Goldstar Co. to supply it with around 30 million barrels of crude oil annually. In return, Seungli suggested, payment could be arranged by supplying refined products and nonferrous metals, such as zinc ingots. North Korea has thus left no stone unturned in order to secure a stable oil supply and has been forced to take a positive attitude towards energy cooperation even with South Korea and Japan as energy shortages cripple its economy.

Japan and China.

Japan has become a major importer of Chinese crude since the mid-1970s. China has utilized its oil export card to make Japan turn away from Soviet East Siberian and Sakhalin oil and gas development and toward Tarim basin development. The Tarim basin in the Xinjiang Uygur Autonomous Region (northwestern China) is listed as a major area for oil exploration during the Eighth Five-Year Plan (1991–95). As of 1990, 27 oil wells had been sunk in the Tarim area, which has oil and gas reserves estimated at up to 12 billion tonnes. In April 1991, a senior Chinese delegation made a trip to Japan, Canada, and the United States in order to attract foreign investment to the area and Japan responded immediately. After the Japan National Oil Corporation (JNOC) agreed to provide eight billion yen (US$58 million) to fund a five-year seismic program for the oil-rich Tarim basin in July 1991, it was reported that Japan was to lend China about US$5.1 billion to help promote the development of oil, natural gas and coal resources. This was the third resource development loan from Japan, following loans of US$3.1 billion in 1979 and US$4.2 billion in 1984.

In addition to efforts at developing the Tarim basin, China embarked on an ambitious plan that is widely regarded as the first step towards the substantial development of the East China Sea. As of March 1990, 17 wells had been drilled in the East China Sea of which eight have resulted in commercial oil and gas flows, and six have had good oil and gas shows. Figure 16.1 shows the locations of those oil and gas fields in the East China Sea. In order to fully develop the East China Sea, however, China would need to settle the Senkaku Islands dispute with Japan. The Tarim basin and the East China Sea may prove to be the most important areas of multilateral energy cooperation in the region in the 1990s as their potential is confirmed.

Japan and the former Soviet Union.

Even though the oil trade between Japan and the Soviet Union had been very limited during the 1980s, the most striking development in bilateral energy cooperation in Northeast Asia during the 1990s will come from a compromise between Japan and Russia. Up to the present, the dispute over four of the Kurile Islands has proved the most formidable stumbling block to energy cooperation between Japan and the Soviet Union. For example, as shown in Figure 16.2, even though

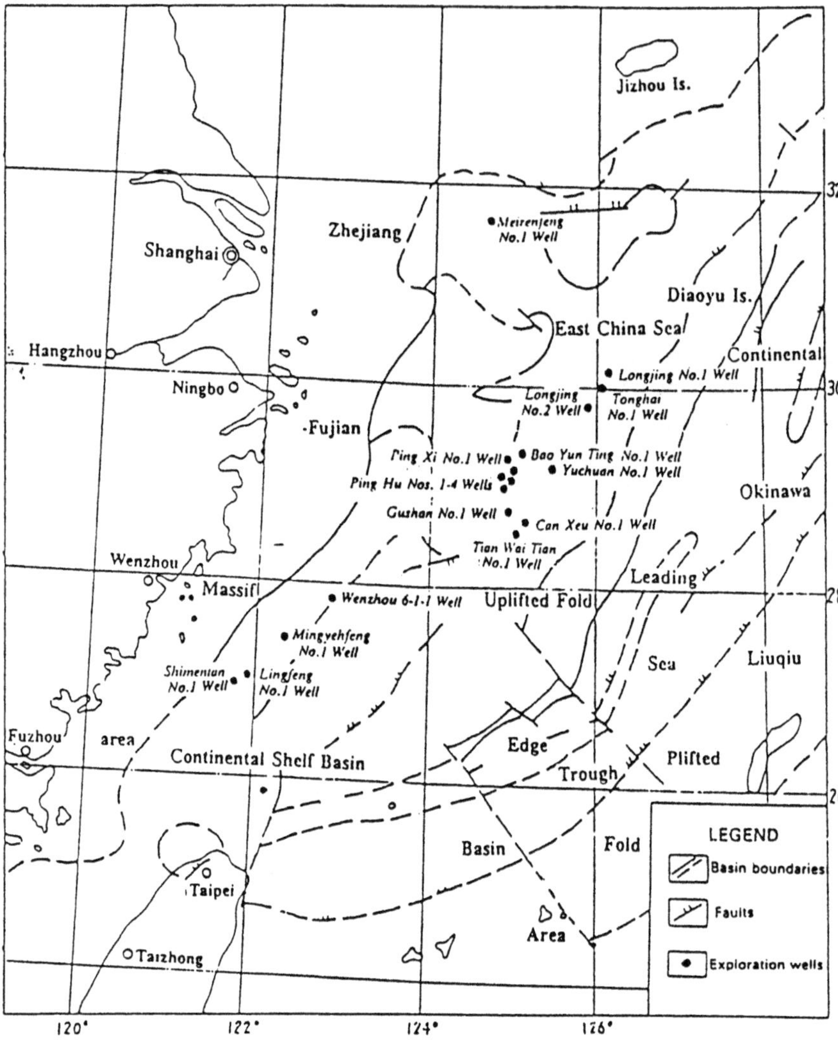

Source: *China Oil*, No. 22, 1989, p. 17.

Figure 16.1. Oil and Gas Fields in the East China Sea

SODECO (Sakhalin Oil Development Cooperation Co. Ltd.) had discovered the Odoptu-More and Chiavo-More oil fields in the Sakhalin offshore in 1977 and 1979, development of those fields did not occur during the 1980s.

Soon after the failed August 1991 coup, however, Japan began to soften its hardline policy towards the former Soviet Union. After the Soviets hinted at a deal on the Kurile issue (the Russian president, Boris Yeltsin, who suggested a five-phase solution to the issue in January 1990, recently made it clear that he would settle the issue within this century), Japan decided to allocate US$1 billion (out of US$1.8 billion worth of export credit insurance) to boost depressed

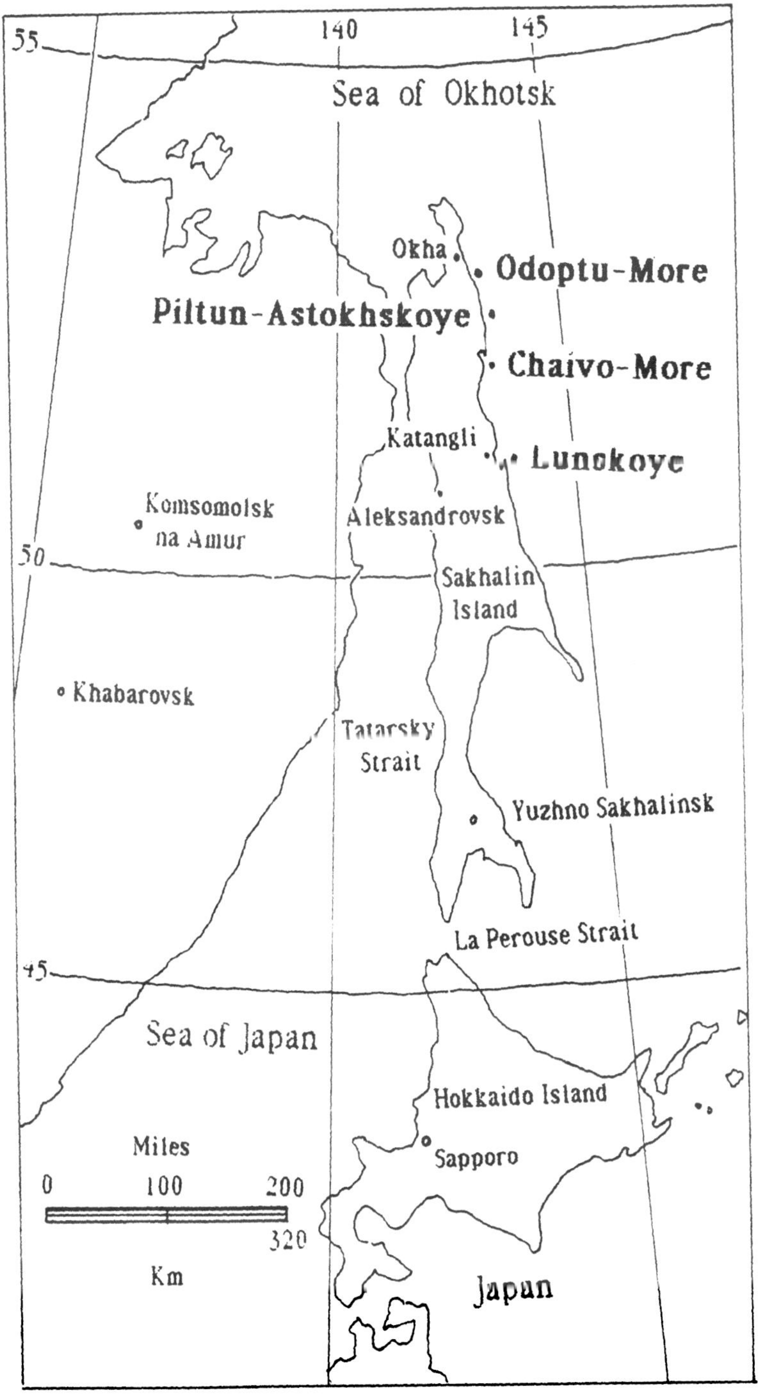

Source: *Oil and Gas Journal,* March 18, 1991, p. 35.

Figure 16.2. Oil and Gas Fields in Sakhalin Offshore

Russian production. The sheer number of Japanese companies, including SODECO, Mitsui, Idemitsu Kosan, Mitsubishi, C. Itoh & Co., Nisso Iwai Corp., and Shoseki Oil Development Co., applying to bid for offerings by Sakhalinmorneftegas confirms their interest in oil and gas development in the Russian Far East. After repeated reversals of the decision, the award of the Sakhalin tender went to a U.S.-Japanese consortium (dubbed 3M) made up of Marathon Oil Co. and McDermott International Inc. of the United States and Japan's Mitsui & Co. An official contract was signed on March 30, 1992. Japan's Ministry of International Trade and Industry (MITI) was reported to have begun exploring ways that SODECO could participate in the development. This indicates how Japan carries out its energy policy towards the former Soviet Union. Since the Japan National Oil Corp. (JNOC) is the main shareholder (49.9%) in SODECO, the Japanese government's influence is indirectly but firmly reflected in SODECO's decision making as well as its direction.

PROSPECTS FOR MULTILATERAL ENERGY COOPERATION IN NORTHEAST ASIA

An examination of bilateral energy relations in the region during the last few years reveals a totally different picture compared with the previous two decades, as shown in Figures 16.3 and 16.4. However, ambitious energy development plans, like the Russian Vostok Plan and Chinese Tarim and East China Sea development plans, could give rise to fundamental changes in regional energy trade patterns. These will reflect trends towards multilateral energy relations in the region. Figure 16.5 shows the anticipated multilateral and bilateral energy relations in Northeast Asia during the 1990s. The 20-year Russian Far East campaign, the Vostok Plan, serves as an example of the trends.

Despite all the uncertainties involved, it is worth mentioning the details of the Vostok Plan. As shown in Table 16.1, if the program goes ahead as planned this effort would by 2005 provide about 15.7 million tonnes/year of natural gas to Russia, 6.0 million tonnes/year to South Korea and Japan, and 1.3 million tonnes/year to North Korea. As shown in Figure 16.6, the core of the campaign will be to lay a 3,230 km gas pipeline from Sakhalin across Russian territory and North Korea to South Korea by 1995 and a 3,050 km line from Yakutsk to Khabarovsk by the year 2000. Total investments for the plan will amount to 28.2 billion rubles, including exploration costs of 10.4 billion rubles, plus US$9.7 billion.

In the long-term perspective, the Sakhalin contract is the first step toward the development of the Vostok Plan. The Sakhalin tender seems to be a turning point in the emergence of multilateral energy projects carried out with neighboring countries in the area. Even though 3M now has an official award, the current situation is somewhat confused because the award was made without apparent regard to the findings of the Special Committee set up by the Russian Parliament. On February 4, 1992, S.A. Filatov, the First Deputy Chairman of

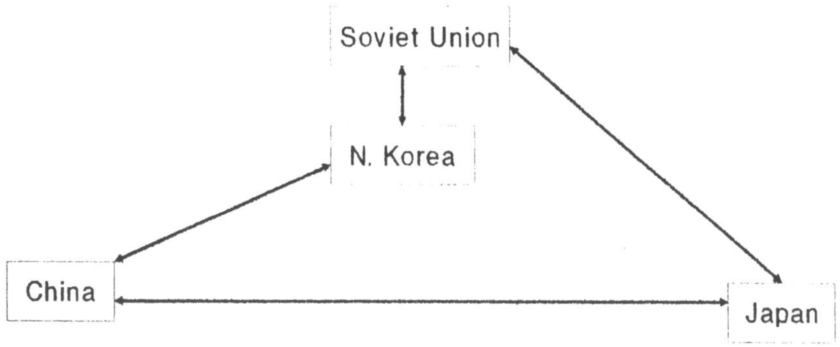

Figure 16.3. A Simplified Pattern of Bilateral Energy Relations in Northeast Asia until the Mid-1980s

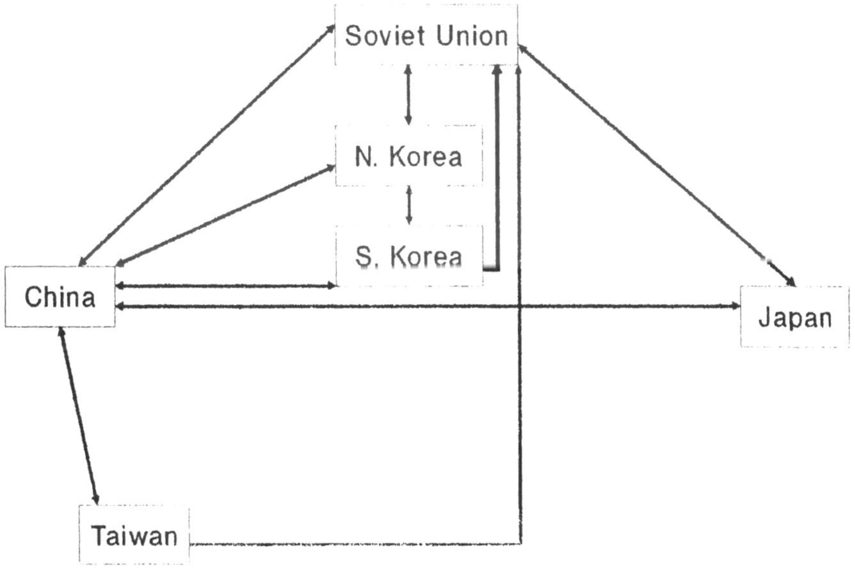

Figure 16.4. A Simplifed Pattern of Bilateral Energy Relations in Northeast Asia from the Late 1980s to the Present

the Russian Parliament, set up a new commission under Vladimir Shumeiko, also a Deputy Chairman of the Russian Parliament, to reconsider the Sakhalin tender decision. On March 26, 1992, the Shumeiko committee decided against an award to 3M. However, before its decision could be formally delivered to Mr. Gaidar, the Russian Deputy Premier issued an official contract to 3M on March 30, 1992 (see annex at the end of this chapter). Given the extent of the opposition to the 3M award, it cannot simply be assumed that it will be allowed to stand—at least not in its present form.

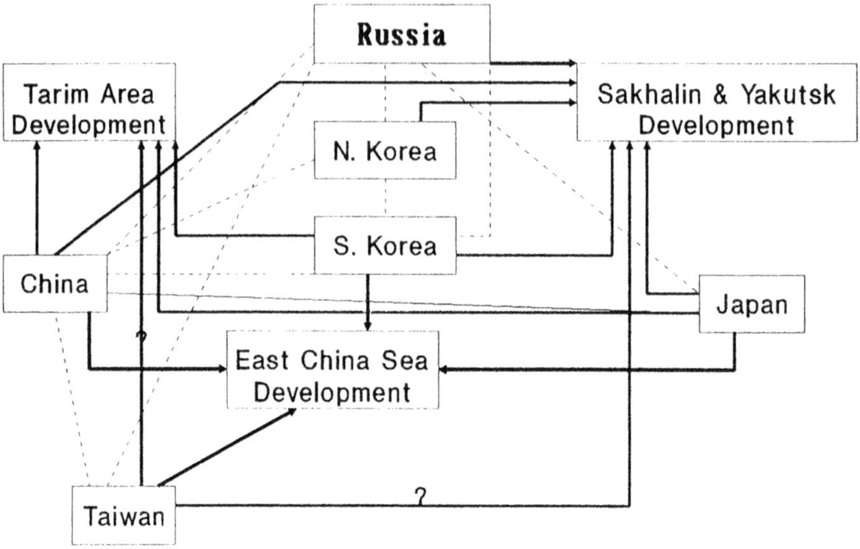

Figure 16.5. Anticipated Multilateral and Bilateral Energy Relations in North-east Asia during the 1990s

Table 16.1. Russian Far East Natural Gas Balance, 1995–2010 (million tonnes)

	1995	2000	2005	2010
Production	3.34	22.5	29.0	32.0
Sakhalin	3.34	14.9	15.7	14.7
Yakutsk	—	7.6	13.3	17.3
Domestic Consumption	2.34	10.9	15.7	18.7
Other requirement	0.07	2.5	3.3	3.5
Sakhalin	0.93	1.3	1.5	1.7
Yakutsk	—	2.0	2.3	3.3
Amur	—	0.6	1.3	1.7
Komsomol'sk	1.34	2.0	2.0	2.0
Khabarovsk	—	1.5	3.0	3.0
Primorsky	—	1.0	2.3	3.0
Export	1.00	11.6	13.3	13.3
North Korea	—	1.3	1.3	1.3
South Korea	1.00	6.0	6.0	6.0
Japan	—	4.3	6.0	6.0

Source: *Concept of Developing Yakutian and Sakhalin Gas and Mineral Resources of Eastern Siberia and the USSR Far East* (or the Vostok Plan).

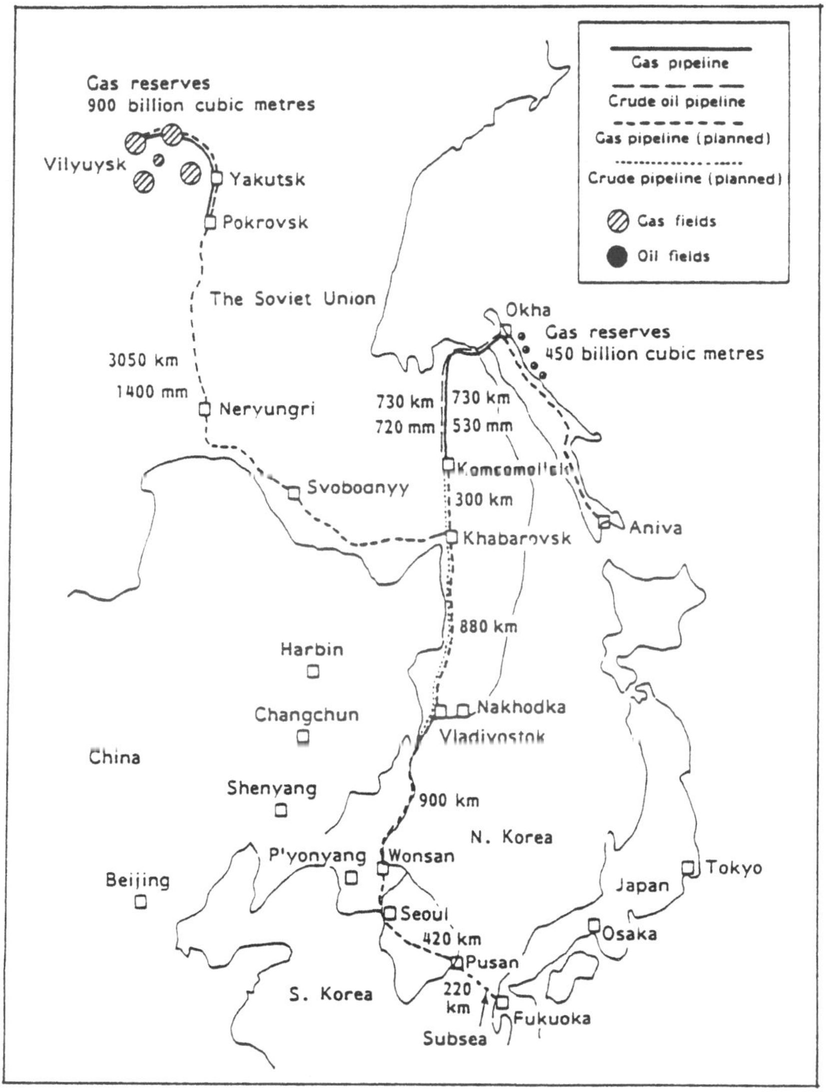

Source: Hyundai Resources Development Co.

Figure 16.6. Sakhalin and Yakutsk Natural Gas Project

An ideal arrangement for the Sakhalin tender would be to follow the example offered by the Azeri project deal. Amoco Eurasia, a subsidiary of Amoco Production Co. and the winner of the Azeri field bidding announced in June 1991, did not hesitate to accept other bidders as partners in consideration of the anticipated development costs of US$5–6 billion. Amoco decided to retain 45 percent of the Western stake in the Azeri field's development and the rest of the shares were divided as follows: Unocal (25%); BP, Statoil, and Ramco, an Aberdeen-based oil services company (20%); and McDermott (10%). However, it is doubtful that this sort of compromise can be reached for the Sakhalin tender.

The second best alternative would be a division of the original area embraced by the tender into blocks, with a wide range of foreign companies involved in the development. This approach is more realistic and desirable. If Korean participation is excluded from the Sakhalin project, the first phase of the Vostok Plan aimed at supplying one million tonnes of Sakhalin offshore gas to South Korea would collapse. A real multilateral energy project comprising all neighboring countries as participants as well as beneficiaries would arise when South and North Korea, China, and Japan all take part in the development of the Yakutiya gas fields and the establishment of a pipeline grid in the Russian Far East. If the Koreans can secure a major role in the Sakhalin project, it would not only be helpful in carrying out the first stage of the Vostok Plan but also in pushing the starting point of the Vostok Plan's second stage forward.

Strictly speaking, the decision of the Yakut-Sakha Republic in October 1990 to open 17 promising areas of southwestern Yakutiya to exploration and development by foreign companies was the first step taken toward realization of the Vostok Plan. In June 1991 the two most prolific structures (Chayandinskaya and Illeginskaya) were granted to ÖMV, the Austrian state oil company. A joint venture between ÖMV and Lenaneftegas Geologija, named "Takt," has the objective of raising proved reserves from the Yakutsk region's current natural gas reserves of 28 trillion cubic feet (tcf) to the 42–52 tcf required to justify a 2,500-mile pipeline to Korea and Japan.

A weakness in the original Vostok Plan was the omission of China from the list of Soviet natural gas beneficiaries. However, a Soviet article recently revealed that representatives of the Yakut Autonomous Republic have indeed conducted preliminary negotiations with the Harbin Gas Company (HGC) of China. The HGC had reportedly expressed its interest in purchasing five billion cubic meters of gas/year for Harbin District. In addition, it would buy, under an option arrangement, eight billion cubic meters for the Open Littoral Zones.

The Open Littoral Zones incorporates the Tumen River basin area at the junction of the Russian, Chinese, and North Korean borders. As shown in Figure 16.7, the area that stretches from the North Korean port of Shonjin to Yanji City in China and north to the Russian port of Vladivostok and the outlet of the Tumen River is at the heart of a region of 300 million people that is rich in natural resources, strategically located and close to major markets in Japan, South Korea, and China's Jilin and Heilongjiang Provinces. According to a United Nations Development Program (UNDP) mission report on Tumen River area development, the resources of this Northeast Asian region and their complementarity reinforces the concept of the Tumen delta area as a future Hong Kong, Singapore or Rotterdam with the potential for entrepôt trade and related industrial development. The development costs of the area could run as high as US$30 billion over a span of 20 years.

The UNDP report highlighted coal as a main energy source for the area, based on its abundance in the region, and suggested a possible way of eliminating its

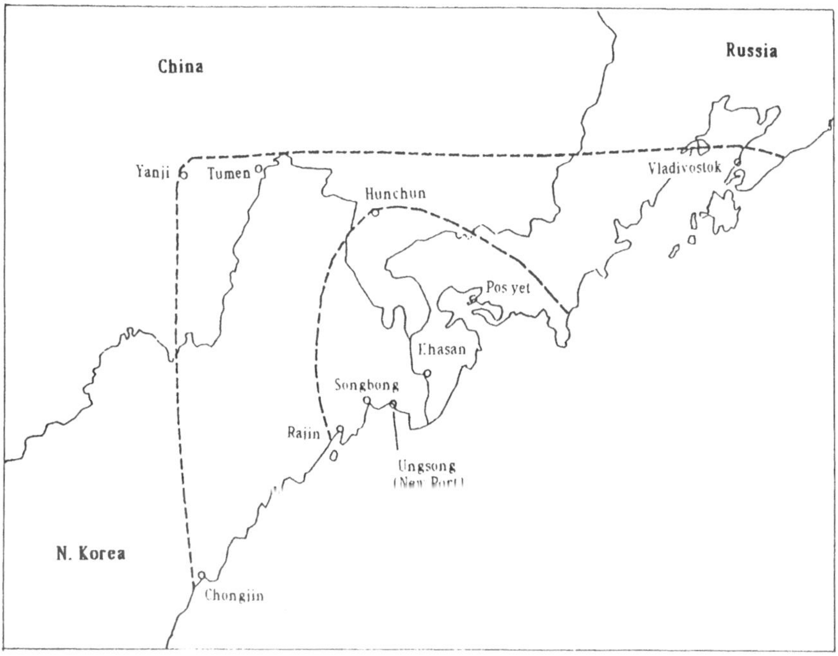

Source: M. Miller et al., "Tumen River Area Development - Mission Report," AFE 02. 04, UNDP.

Figure 16.7. Tumen River Delta Area: Development with Small/Large Delta Zone

main disadvantage—pollution—by manufacturing a synthetic fuel. However, the importance of natural gas as an energy source for the area's development should not be overlooked. Considering that the Tumen River area is located at the junction of the proposed Soviet pipeline grid to the Korean Peninsula, establishment of the pipeline grid introduces natural gas as an alternative energy source for the Tumen River area development. Once the UNDP program makes headway with the development of the Vostok Plan, the area would provide another market for the Russian gas exports.

The Asian Energy Community Plan proposed by a group of 32 Japanese companies indicates another possibility for integrating China as a recipient of Russian gas. As shown in Figure 16.8, this extensive plan proposes establishing a 26,900 kilometer natural gas pipeline grid that will link Yakutsk in the Russian Far East to Dampier in northwestern Australia, with lines through China, Korea, Japan and Taiwan, and the six countries of the Association of South East Asian Nations (ASEAN). Even though a feasibility study for a proposed US$10 billion natural gas pipeline project to connect all six ASEAN countries began at the end of 1991 and the project appears feasible in the long term, it is only a part of the Asian Energy Community Plan whose realization seems questionable in terms of economics, not to mention politics.

The crucial element in the plan is the establishment of a gas pipeline grid in Northeast Asia. Here, however, the Japanese plan leaves something to be desired.

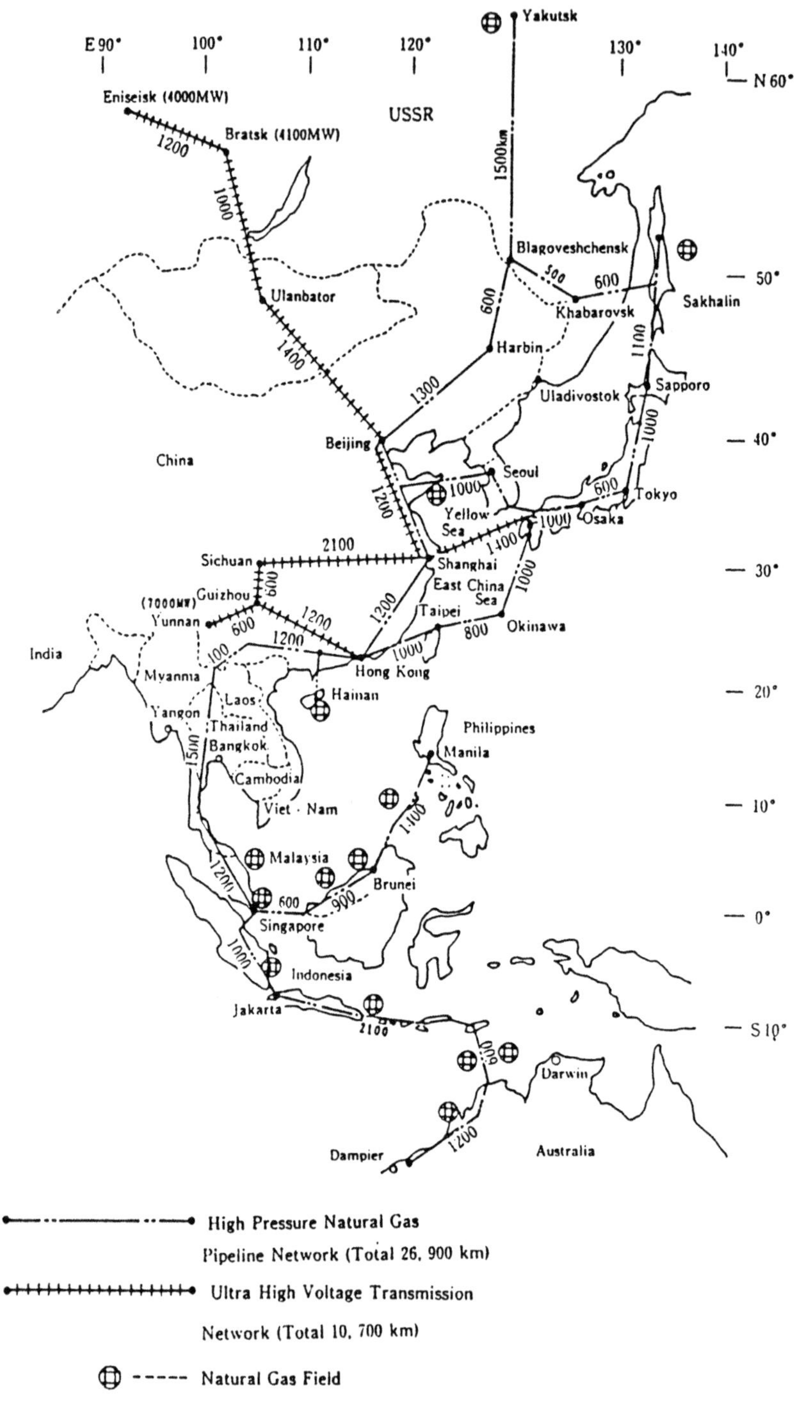

Source: Masaru Hirata, "Proposal of Asian Energy Community," University of Tokyo, July 1991.

Figure 16.8. Asian Energy Community Plan: Natural Gas Pipeline Grid

First, from the Russian point of view, it would not make sense to omit a gas pipeline from Khabarovsk to Vladivostok which runs through the Primorsky region and can be extended to the Tumen River area. In addition, a plan called the "Larger Vladivostok Free Economic Region" has been worked out by the United Nations Industrial Development Organization (UNIDO) for the development of the Russian Far East. This plan raises questions about the benefits of gas development. Second, from the Korean point of view, a gas pipeline from Vladivostok to Pusan through North Korea would provide a symbol of confidence and peaceful coexistence between the two Koreas. But a Yakutiyan gas pipeline to South Korea across the Yellow Sea after passing through Mainland China could constitute a conceivable alternative, though it would not, of course, contribute to the resolution of tensions prevailing on the Korean Peninsula. Third, the memory of Japanese imperial expansion in East Asia before the Second World War, epitomized by Japan's "Greater East Asia Co-Prosperity Sphere" has not been forgotten. This means that the Japanese plan is likely to cause misunderstandings—no matter how beneficial its purpose may be.

In short, as far as the proposed gas transmission routes in northeastern Asia are concerned, the Asian Energy Community Plan needs revision. If an ideal route, which took geographic, economic, and strategic factors in Northeast Asia into consideration, could be developed by consensus among the region's countries, it would make a considerable contribution to the area's economic development and political harmony, and make the prospects for the completion of an Asian gas grid more realistic.

If the Vostok and the Asian Energy Community Plans were to be combined, they would create an ideal gas pipeline system satisfying all the countries in Northeast Asia. The ideal system, however, cannot be easily designed as the countries in the area have had no experience in cooperation in such an important project. A special effort should be made to foster cooperation. The formulation of a 'Northeast Asia Energy Charter' with the establishment of a 'Northeast Asian Energy Regime (NAER)' would be the objective. This would generate a climate of confidence propitious for optimum utilization and allocation of resources and provide a framework to guarantee a safe energy supply, possibly contributing to the political stability and economic prosperity of the region.

CONCLUSIONS

The uneven distribution of major production factors among the Northeast Asian countries paradoxically indicates that mutual benefits can be derived from energy cooperation among the countries in the region, as shown in Table 16.2. In fact, Russia and China, with their huge reserves of oil and gas, certainly need outside capital, technology and equipment for resource exploration and development, while Japan, Taiwan, and South Korea, with their capital, technology and equipment, surely need to reduce their heavy dependence on Middle Eastern oil and diversify their energy supplies.

Table 16.2. Qualitative Comparison of Production Characteristics of Northeast Asian Countries

	Oil and Gas	Coal and Minerals	Labor	Capital	Tech-nology	Managerial Expertise
Russia	VR	VR	VP	VP	S	S
China	R	VR	VR	VP	S	S
Taiwan	S	S	S	R	R	R
N. Korea	A	R	R	VP	VP	VP
S. Korea	A	S	S	R	R	R
Japan	VP	S	S	VR	VR	VR

Note: VR = Very Rich; R = Rich; S = Short; VP = Very Poor; A = Absent

In other words, great potential for a vertical division of international labor in energy lies in the region, and if it were to be realized, the effects and benefits could be tremendous and reach into economic, social, and cultural spheres. Because of the strategic importance that energy and especially oil has had, benefits from energy cooperation would also be extended to regional politics. For example, if the intractable task of laying down a Russian gas pipeline to South Korea and Japan through North Korea can be settled through a pan-Northeast Asian agreement on energy, the agreement will make a great contribution to overcoming political difficulties arising from mutual mistrust among formerly hostile neighbors.

Even though the Minerals and Energy Forum (MEF) of the Pacific Economic Cooperation Conference (PECC) has provided a forum for discussion and consultation on key issues relating to minerals and energy sectors in the Asia-Pacific region since 1986, it has not been and will not be formalized into an international energy regime whose resolutions are officially binding. If any sort of coordinated official initiative on the energy policy in the region is to result, it will have to come from the APEC (Asia-Pacific Economic Cooperation) Conference as the first government-level conference in the Asia-Pacific area.

At present, however, it is uncertain whether APEC members will reach a consensus on giving priority to energy development in Northeast Asia in the foreseeable future. In this context, as an alternative, the establishment of a 'Northeast Asian Energy Regime', whose aim is not develop a self-sufficient energy system but to lay down guidelines for multilateral energy cooperation in Northeast Asia is conceivable.

In Northeast Asia, the significant improvement of bilateral energy relationships has raised real prospects for multilateral energy cooperation. Multilateral energy cooperation with neighboring countries, however, would not occur if each country in the area pursues a shortsighted policy that only leads to competition and confrontation. The establishment of a 'Northeast Asian Energy Charter' as a parallel to what has been created in Europe would mark the beginning of a new era of energy cooperation in Northeast Asia. Ultimately, it could lead to the formulation of an 'Asia-Pacific Energy Charter'.

ANNEX: SAKHALIN TENDER CHRONOLOGY

May 1991:

- Sakhalinmorneftegaz (Sakhalin Offshore Oil and Gas) Association revealed that the Soviet Ministry of Oil and Gas decided to set up a joint venture involving the association, and it, instead of the joint sponsors, the Soviet Oil and Gas Ministry and the Russian Republic's State Committee on Geology and Utilization of Fuel, Energy, and Mineral Resources, announced an international tender for development of the Lunskoye and Piltun-Astokhskoye fields.
- The tender covered a total offshore area of 17,000 sq. km or about 10 percent of the explorable Sakhalin shelf.

August 10, 1991: Deadline of proposal submission

- After the announcement of an international tender, six groups took part in the bidding. The six groups were as follows:
  1. 3M: McDermott, Marathon, and Mitsui;
  2. Exxon and SODECO;
  3. BHP, Amoco, and Hyundai;
  4. Mobil;
  5. Idemitsu Kosan; and
  6. Shell, Mitsubishi, Nisso Iwai, Showa Shell Sekiyu, and Shoseki Oil Development Co.
- Palmco and Ralph M. Parsons did not join in the tender.
- Examination of the proposals was based on the following criteria:
  1. Time schedule for commercialization.
  2. Time schedule for delivery of gas to consumers in Russia's Far Eastern Economic Region.
  3. Exploration and surveys in the tender zone.
  4. Evaluation of promising new areas in terms of time and the size of the deposit.
  5. Reasoning behind the engineering and technological solutions and their reliability and environmental safety.
  6. Preparedness to capitalize authorized funds.

September 20, 1991:

- Sakhalinmorneftegaz made a proposal to the Tender Organizing Committee (TOC) to recognize as the tender winner the 3M group and recommended a feasibility study to be jointly drafted.

October 5, 1991.

- Sakhalin Deputy Governors Viktor Sirenco and Valery Mozolevsky declared the tender void.

October 18, 1991:

- The Sakhalin government imposed new conditions of the 'Social Development Program'. These additional requirements involved US$10 billion in investment and loans for development of transportion systems, telecommunications, agriculture, and other industries. However, financing for the program would come from the proposed reduction in royalties from 12.5 percent under the tender terms to 1.5 percent and from granting the foreign partners a seven-year tax holiday for the repatriated profit.

- The government asked the foreign companies to submit their respective proposals by November 10, 1991, including the following provisions:
  1. Delivery of gas to the Sakhalin region was to be increased to three billion cubic meters (bcm) in 1995, five bcm in 1997, and eight bcm in 2005.
  2. The amount of gas for export was to ensure minimal profitability of the project, with the remaining gas to be delivered to Khabarovsk Territory.
  3. Cooperation options to be provided were to include—apart from joint venture concessions—production-sharing contracts and service contracts.
- The government suggested that the following construction should commence simultaneously with the beginning of the Sakhalin offshore project:
  1. Reconstruction of the two-lane road between Okha and Yuzhno-Sakhalinsk, a distance of 800 km, with branches to Poronaisk, Uglegorsk and Aleksandrovsk-Sakhalinskii;
  2. Railroad from Nogliki to Okha (provide an analysis of necessity and estimate of costs);
  3. Maritime ports in Ilinskii with ferry terminal or in Prigorodnyi for ships with a capacity of 100,000 tonnes or with a 150,000 tonnes a day turnover with a container terminal;
  4. Development and reconstruction of the maritime port of Korsakov for ships of 30,000 tonnes with a container terminal;
  5. Expansion of the Okha TETs with two power blocks and a capacity of 80 megawatts;
  6. LEP 220 kilovolts from Okha to Dagi, a distance of 180 km;
  7. Reconstruction of the airports in Okha, Nogliki, and Zonalnoe;
  8. Link oblast communications with international circuits; and
  9. Water supply system of Yuzhno-Sakhalinsk.

November 12, 1991:

- Russian President Boris Yeltsin approved the Sakhalin authority's decision to annul the original tender. However, Yegor Gaidar, Deputy Chairman of the Russian Council of Ministers, challenged Yeltsin's decision.
- The same day, by decision of the Sakhalin Governor Fedorov, the Sakhalin Tender Committee (STC) was set up, and seven groups of experts representing the subcommittees of the Executive Committee on energy, environmental protection, economics, construction, transport, communications, food, social issues, and public relations were constituted. STC was to submit to the Sakhalin governor its proposal on the winner of the revised tender by November 30, 1991.

November 18, 1991:

- Four days of presentations by the bidders were held in the oblast executive committee meeting hall.
- 3M (McDermott, Marathon, and Mitsui) offered to rebuild the Yuzhno-Sakhalinsk and Okha airports if it won and offered US$115 million as a development fund after contract signing.
- Mobil Oil flew in 30 tonnes of medical supplies and powdered milk from Anchorage (November 16, 1991).

November 28, 1991:

- The Sakhalin administration put forward additional requirements to foreign companies, asking them to submit draft contracts for a feasibility study by December 7, 1991 to determine the economic viability of the fields on the assumption that the foreign and domestic partners would select a mode of cooperation calling for a production-sharing contract.

December 3, 1991: Salmanov Commission

- A panel of experts presented some conclusions. The panel was constituted by decision of the Examining Council under the Chairman of the Russian Government and was headed by F. Salmanov, First Deputy Minister of Geology of the Soviet Union.
- The Salmanov Commission called for the tender area to be divided into several blocks.
- The panel of experts proposed dividing the zone as follows:
  1. **Block 1,** about 5,800 sq. km, including the Arkutun-Daginskoye oil and gas field and two conditioned structures, East Kaiginskaya and East Odoptinskaya. The panel included the Chaivo and Odoptu fields in the block. However, the tender organizers excluded two fields from the tender zones because they were included in the general agreement with Japan. The panel slated the block for transfer to the Exxon-SODECO group.
  2. **Block 2,** about 4,500 sq. km, including the Piltun-Astokhskoye gas field and the conditioned Lozinskaya and Bautinskaya structures. The block was slated for transfer to an ad hoc group of 3M-Idemitsu.
  3. **Block 3,** more than 7,500 sq. km, including the Lunskoye, Veninskoye, and Kirinskoye oil/gas/condensate fields and the conditioned Nabilskaya and south Lunskaya structures. The block was slated for transfer to the ad hoc group of BHP, Amoco, and Hyundai-Mobil.

December 23, 1991: Danilov-Danilyan Committee

- A committee convened by order of the Russian government headed by V.I. Danilov-Danilyan, the Russian Federation's Minister for Ecology and Natural Resources, reviewed the result of the Sakhalin tender.

January 27, 1992:

- Mr. Yegor Gaidar, First Deputy Chairman of the Government of the Russian Federation, in response to the recommendation of the Danilov-Danilyan

Commission, announced that 3M (McDermott, Marathon, and Mitsui) had won the Sakhalin tender for a feasibility study on the development of reserves estimated at 730 million barrels of oil and more than 400 billion cubic meters of gas. A formal agreement was due to be concluded on March 31, 1992.

- Before the announcement, the local officials had asked the Russian Federation to postpone tender results until plans from bidders for investing in the local infrastructure had been evaluated.
- After the announcement, Sakhalin Governor Valentin Fedorov, claiming that the opinions of local government officials had been ignored, resigned from the tender committee. The governor was concerned that if SODECO were left out of the development, the Japanese government that had financed earlier exploration in the area would demand repayment of loans that amounted to US$181.5 million at a six percent annual interest rate.
- As a result of such local opposition, 3M agreed to allow SODECO and Mobil Oil to participate in the feasibility study.

February 4, 1992: Shumeiko Commission

- S. A. Filatov, First Deputy Chairman of the Supreme Soviet of the Russian Federation, set up a Deputies' Commission, chaired by Vladimar K. Shumeiko, also a Deputy Chairman of the Russian Parliament, to reconsider the Sakhalin decision. As a result, the Russian Supreme Council suspended the award again, and the original bidders were reconvened. The new committee of experts (Russian Supreme Council) was composed of eight members, including officials from Sakhalin.
- Until February 20, 1992, the Deputies' Commission was asked to consider the decision of the Government Commission and Experts' Council of the Chairman of the Government of the Russian Federation and submit its conclusion to be considered by the Presidium of the Supreme Soviet of the Russian Federation.

March 26, 1992:

- The Shumeiko Commission concluded that the decision of the Government Commission, dated January 27, 1992, should be abolished.
- What the Commission has recommended is that the area of competition should be divided into at least three blocks, and it was envisaged that the winner-group in every block would have a 60 percent share of the rights for exploration and development, while the rights for the remaining 40 percent are to be distributed between winners in other blocks.

March 27, 1992:

- Based on conclusions by the Shumeiko Commission, Mr. S.A. Filatov advised Mr. Y. Gaidar to retract his January 27 decision in favor of 3M.

March 28, 1992:

- Deputy Prime Minister Yegor Gaidar authorized the negotiations and concluded an agreement with 3M to do a feasibility study for the exploration and development of the Piltun-Astokhskoye and Lunskoye fields.

March 30, 1992:

- 3M signed an agreement with the Russian Ministry of Fuel and Energy pursuant to Mr. Gaidar's decision on March 28.

Chapter 17

AUSTRALIA-COMMONWEALTH TIES IN THE MINERALS SECTOR

Alaster Edwards and Christopher Findlay

INTRODUCTION

The Russian Far East is sometimes cited as a potential mineral treasure house. It is well endowed with many minerals, including gold, platinum, diamonds, tin, tungsten, copper, lead, zinc, silver, coal and nonmetals. In terms of both their tonnages and grades, many of these deposits are world-class deposits.[1] Significant tonnages of tin and tungsten are already mined, but these do not attract much interest from Western companies because of the market structures for those minerals.

These resources are located in a region which has an area larger than that of Australia, but with a population of only about six million. The complementarities between the Russian Far East and the rest of Northeast Asia are therefore striking. Proximity to this large market adds to the potential of the region as a supplier of raw materials and products processed to some degree.

In principle, therefore, there is scope for the movement of capital and technology from the rest of the world into the Russian Far East. These opportunities exist not only for other Northeast Asian economies but also for resource-rich countries, like Australia, which have technologies or marketing systems that are relevant to the Russian Far East.

Despite these complementarities, there are few examples of foreign investment by Australian firms in mineral deposits in the Russian Far East.[2] The only publicly reported example of which the authors are aware is summarized in Figure 17.1. This agreement is significant as it is the first joint venture allowed in the precious minerals sector. The motivation for relaxing restrictions on foreign involvement in that sector appears to be the interest in increasing investment and gaining access to foreign technology.[3]

At this stage, relaxation of controls on foreign involvement does not extend to the international marketing of precious metals. If it did, these reforms could have an effect on world prices, since Russia is such a large potential supplier. Marketing systems are discussed in more detail below.

J.P. Dorian et al. (eds.), CIS Energy and Minerals Development, 315–323.

It is reported that an Australian consortium, Star Technology Systems (STS), has secured permission from the Russian government for the first joint venture in the mining of gold and silver. The parties in the joint venture are STS, Lenzoloto and the regional administration of Irkutsk. Originally the proposal was opposed by the dominant gold producer in Russia, Rossamazzoloto, but the deal was approved after the Australian government, at the request of the Russian government, endorsed STS as a credible partner for the venture. An option to buy the STS interest in this project has been sold to the Australian-listing company Central Mining Corporation.

The plan is to develop alluvial and open-pit mining of gold and silver. The investment by STS is reported to be US$150 million to pay for equipment, technology and technical assistance. In return, STS will receive the right to 31 percent of the extra gold output from the project. Press reports suggest that the target is to increase output to 42 tonnes a year, which is four times the previous level. In that case, STS would have claim to about 9.8 tonnes of gold a year, valued at approximately US$117 million at world prices.

However, uncertainties remain. Reports suggest that

- STS has promised to sell its gold to the Russian banking system but, according to press reports, the price to be paid has not been determined; and
- the rate of taxation to be paid by STS has not been fixed.

Sources: *Australian Financial Review*, 14 April 1992, 22 April 1992, 27 July 1992; *The Bulletin*, 20 October 1992.

Figure 17.1. Recent Press Report of Australian Foreign Investment in Russian Far East

In a previous paper, the authors reviewed some of the factors which are limiting the development of the capital and trade flows between the rest of the world and the Russian Far East. In this chapter, the authors aim to update and extend that review.

Some of these issues have also been considered in a general setting in papers presented at meetings of the Mineral and Energy Forum (MEF). For example, in the Dallas 1990 meeting of the MEF, Douglas Ritchie reviewed some of the issues that a mineral explorer must take into account when evaluating projects in various countries (Ritchie, 1990). Some of the rankings of variables stressed by Ritchie in the 1990 meeting are summarized in his chapter, Chapter 14, in Table 14.1 and Figure 14.4. Table 14.1 shows that the right to repatriate profits, the absence of price controls and the absence of limits on foreign equity were the three most important factors. Figure 14.4 shows another set of factors graded on a scale from critical to negotiable.

All of these factors are important to international mining companies:

- security of title for both mining and exploration;
- the right to mine if successful in exploration;
- management and equity control of the project;
- the right to remit profits freely;
- the right to access international markets;
- international arbitration of disputes; and
- a clearly defined fiscal environment.

Although current reforms are attempting to address some of these issues, the current situation in Russia still falls short of meeting the requirements of the major Western companies.

ENTRY STRATEGIES

The prospective investor in the minerals sector must decide in which commodity he is prepared to invest, in which locations and in what form, i.e., existing mine, explored deposit or grassroots exploration.

Infrastructure issues

Of all the commodities listed in the introduction, many Western companies would argue that only in the cases of gold, platinum and diamonds is the economic viability of the project less dependent on the quality of the infrastructure. This is because of the low weight/volume of the items compared to their value. In the absence of infrastructure, high unit-value items like gold or diamonds can be flown out of the mining region. Other commodities require the development of transport infrastructure before they can be developed. In the Russian Far East, commodities such as coal or industrial minerals are more likely to be considered for development if they are located near existing high quality infrastructure such as the Trans-Siberian—BAM railway corridor, or if they are near the coast (although in that case, the problem is coping with the winter ice).

However, even if lower unit-value items are located adjacent to transport infrastructure, developers would still face the problem of paying for access to those services. Obtaining information about the costs of the transport system (actual or potential) is difficult in Russia. This complicates the negotiation of rates for high volume users like mining projects. Furthermore, once the project began and became committed to a particular infrastructure, the ventures would be concerned about the problem of the "obsolescing bargain."

If, instead of using an existing transport infrastructure, a mining venture considered building its own, then the tax treatment of the investment in the infrastructure would be an issue. The infrastructure could be the responsibility of the Russian side or the foreign joint venture partners. In a previous paper, the authors reviewed some of the tax issues involved.

Access to deposits

Having considered these options and decided where the economical sites for projects were located, the next step for the potential investors is to obtain more information about potential investment targets in those areas. The usual options are:

- to invest in a proved deposit;,
- to select an area of proved geological potential which has not been fully explored but which has significant prospects (or 'blue sky'); or
- to explore a new area.

The accepted practice in the world mineral exploration industry is that if an organization is attempting to seek a joint venture partner, then it makes available to the potential joint venture partners all of the data that has been collected that is relevant to the project. This data is considered as part of the vendor's contribution to the venture and is recognized in the equity structure of any subsequent joint venture. Generally, potential joint venture partners examine the data under an agreement that they will not pass on the data to anyone else. If a decision is made to engage in a project, then the new partner will fund any additional exploration required. This will become part of the equity of the new partner in the project. The new partner may also fund feasibility studies and possibly even construction.

The new partner may also make a cash payment to the vendor. This amount represents both a payment for the right of access to the deposit (in which case, it would be passed on to the relevant level of government) and also some compensation for the capital costs incurred by the vendor in the exploration to date. The amount paid will depend on the expected profitability of the project.

Investment in known deposits would appear to be a minimum-risk strategy for a foreign venture partner to follow. In order to facilitate this, in 1989 the Vniizarubezhgeologia (VZG), or All-Union Research Institute of Geology of Foreign Countries, prepared a list of some 121 deposits that were described as open for joint ventures. The deposits were offered along with the results of some exploration undertaken to date. They were also offered at a fixed price.

The effects of a fixed fee as a source of collecting mineral rents have been examined by Ross Garnaut and Anthony Clunies Ross (1983). They argue that auctioning mineral rights would be 'ideal' when there is certainty about the productive power of the deposit, stability in government policy, and a high degree of competition among investors. However, they also point out that generally there is little knowledge about the value of a deposit at the time when exploration rights are being given out. So they conclude that "there is little value in auctioning exploration rights as a principal method of collecting rents."[4]

Unfortunately, in the Russian case, few of these deposits on offer had the potential to support an operation that would be considered viable in Western economic terms. A number of conditions significantly eroded the investors' confidence in the desire of the VZG to attract serious investors:

- data packages that were necessary to assess these potential projects were offered for sale at unrealistic prices and with unrealistic conditions;
- there were no guarantees offered for the right to mine or for security of tenure; and
- the deposits selected for joint venture involved commodities of low unit-value in remote areas or deposits which had technical problems.

The VZG apparently failed to recognize the fact that Western companies would only invest in the Soviet Union if they saw opportunities that would provide greater returns and thus compete with those in other countries, many of which represented lower levels of political and financial risk.

The lack of tenure was also a serious issue. In the minerals sector, exploration usually means the company has first claim on any deposits found. Removing that right to mine obviously reduces the incentive to explore.

In addition, there were some problems arising from the process of assessing reserves and the interpretation of that work. Under the old Soviet system, Soviet exploration organizations strove to maximize the estimated size of a deposit subject to the regulations of the State Reserves Committee. These regulations, which were often rigorously implemented, were framed with a view to the confidence limits imposed by existing technologies and computational techniques. The aim of the regulations was to ensure that the value of the deposit was estimated as precisely as possible, and the deposit was only handed over to a production organization for development after the requirements of the Committee had been satisfied. The explorers were not involved with the deposit once production commenced.

As a result of this system, the Soviet explorers did not have the constant pressure to justify the exploration expenditures against the potential profits to be derived from the exploitation of the deposit. The time frame in which they operated was established by the need to comply with the regulations rather than the need to determine the value of the deposit as cheaply as possible in the minimum amount of time. Furthermore, their primary aim was to establish the geological reserve according to the Committee's regulations, and this geological reserve did not necessarily correspond with the mining reserve.

Potential investors commonly hear the phrase "there is a guaranteed reserve," which implies that the reserve has been assessed according to the standards of the Committee. However, Western companies have found that the methods of reserve estimation commonly employed in the West have, in some cases, produced reserve estimates that differed significantly from the classical Soviet techniques. The reconciliation of the "guaranteed reserve" with the reserves estimated by other techniques can be a cause of mistrust between the political partners.

From the Western investor's point of view, there is also uncertainty as to whether it is the government's intention to fully recover the exploration expenditures from the project or whether these costs will be written off. If complete cost recovery is envisaged, then this will have a significant impact on the viability of many of these projects.

DECENTRALIZATION AND THE MINING LAW

The authors have previously argued that:

> . . . the next stages of the reform will be characterized by the development of much more complex relations between the center and local governments

> in terms of revenue raising, expenditure and responsibility for the profits and losses of local enterprises. The complication for foreign investors is that they will be dealing with enterprises and their local authorities whose powers and responsibilities will change over the life of a project. Of course, this is highly likely in long-lived resource projects. The main issue in developing these projects is not the absence of rules, regulations and guidelines but rather arriving at some expectation of path of the reforms, especially, as argued below, in relation to intergovernmental relations.

Critical to the future of the Russian mining industry has been the question of the division of control of resources between the federal government and the governments of the autonomous republics and regions. This matter has been addressed in the new constitution of the Russian Federation and will provide the regional authorities with a significant control over the resources situated within areas of their jurisdiction.

The increased role of the regional governments has also been recognized in the new Law of Mineral Resources. Under this law, which separates the administration of the mineral resources of the Russian Federation from those organizations that are participating in the mineral industry, licenses must be approved by both the regional government and the State Committee on Geology and Mineral Resources. In addition, the distribution of the financial returns to the district, republic or regional governments and the Russian Federation from the exploitation of mineral resources is defined.

A significant feature of this law is the provision for the allocation of licenses to develop known deposits by means of auctions or contests. While tendering of deposits minimizes the risk borne by the owners of the deposits and provides some safeguards as to propriety of the process, the process is greeted with some misgivings by many Western companies who observed the Sakhalin petroleum tender process. The rules of tendering would have to be clearly defined and strictly adhered to in order to maintain the interest of Western investors.

Unfortunately, the procedures for tendering have not been defined, and those authorities charged with administering the process are, in some cases, unsure as to how they should proceed. As a result, some multistaged schemes have been outlined for administering the tenders, commencing with the purely technical evaluation, and with a separate competition for the project. Such multiphase schemes fail to recognize the complex interdependence of marketing and financing arrangements and technical and engineering studies. They also fail to recognize the fact that major Western investors may not accept partners chosen for them by the authorities, and that they will demand to choose their own financial partners. If Western investors are to be attracted, they must see a clearly defined path for their investment.

MARKETING ARRANGEMENTS

Price controls

All precious metals and precious stones must be sold to the Committee for Precious Stones and Precious Metals of the Ministry of Finance. This committee controls all the precious metal refineries and sorting of stones. In contrast to the situation in the past, the price of gold to be offered by the Committee will now be uniform across Russia. The purchase price has been below world prices, but at the end of February 1992, the purchase price was fixed at 1058 rubles per gram or about US$10 per gram (at market exchange rates) which, depending on quality, is closer to the world price of Russian gold of US$11–12 per gram. The previous purchase price of 508 rubles per gram was about half of the world price. There were reports in February 1992 that miners in Yakutiya had been on strike to demand that the purchase price for gold be raised to world market levels. The Magadan region did not send any gold mined in 1991 to the Russian State Treasury because of the dispute over gold prices. The Russian government responded by revising its purchase price. The outlook for gold mining depends on the rules for adjusting the gold purchase price to allow for inflation and the falling value of the ruble.

Other minerals are subject to export licensing arrangements and export duties. The export licensing system applies to coal, mineral fertilizers, copper, zinc, tin, primary aluminum, and nonferrous scrap, for example. The export duties have recently been reduced.[5] The licensing system is probably the binding constraint on export volumes. It means that because supplies are diverted to the domestic market, prices are lower in Russia than on world markets. There is evidence that prices are lower, for example, for lead (28%), zinc (8%), copper (42%), tin (35%), nickel (59%) and aluminum (64%).[6]

These systems of price control lower the profitability of projects. Either the project sells to the domestic market at these lower prices or, if it is able to get a license to export, then it still pays an export duty.

Foreign exchange arrangements

The regulation of the foreign exchange system in Russia will also have an impact on the evaluation of projects by foreign investors. According to the regulations of January 1992, all enterprises doing business in Russia must sell part (typically 40%) of any foreign currency earnings to the Russian Central Bank. They may retain only part of their total earnings. The ruble price they receive for the foreign exchange they sell will be determined by the bank. This could be called the official exchange rate.

There is another rate which could be called the market exchange rate. This is the rate at which importers can buy foreign currency to import goods and at which foreign enterprises can buy foreign currency in order to remit profits.

This price is determined on the basis of "supply and demand"; some reports indicate that the official rate is "about half" the market rate.

The application of the official exchange rate to export earnings on top of the export duties is a further discouragement to exports.[7] This is offset to some extent by the right of exporting enterprises to retain about 60 percent of their foreign exchange earnings—these earnings would be valued at the market rate so the rate applying to exports is a weighted average of the official rate and the market rate. From the point of view of foreign investors, the higher rate applying to repatriated profits also lowers the returns on a project.[8]

It appears that the 40 percent rule does not apply to joint ventures with a greater than 30 percent foreign investment or to enterprises wholly owned by foreign investors (although there may be a requirement, which also covers other enterprises, to sell 10% of their earnings of foreign exchange to the Russian Central Bank at the market exchange rate). Therefore, foreign investors who have sufficient equity, who operate an export-oriented project and who can deal directly with the world market escape to some extent the disincentives provided by the foreign exchange system. However, those enterprises also pay higher export duties than others which do make contributions to the Russian currency reserve. Gold producers, who receive about 75 percent of the value of their output in rubles (the balance in foreign currency), could not escape the problem of having to pay the higher market exchange rate to obtain foreign exchange for repatriated profits.

CONCLUSIONS

Western mining companies are becoming increasingly global in their outlook. They are profit maximizers, but they are also averse to risk. They are able to consider the profit and risk combinations of a variety of projects in different economies. If in one jurisdiction there are uncertainties about the mineral exploration and exploitation regime and constraints on access to international markets, two key issues in the Russian Federation which are stressed in this chapter, then the evaluation of projects in that jurisdiction will be downgraded. This will occur despite expectations about the quality of the mineral deposits. These issues have been relevant to the evaluation of projects in the Russian Far East by Australian mining companies. They have contributed to the low level of joint venture development in the region.

ENDNOTES

1. For further information, see V. I. Ivanov and N. M. Baikov (eds.), 1989, *The Mineral and Energy Resources of the Russian Far East: Material for the Forum on Minerals and Energy of the Pacific Economic Cooperation Conference,* Moscow, Vladivostok and Khabarovsk: The Institute of World Economics and International Relations, USSR, Academy of Sciences.

2. There are reports of an investment by Hancock Resources in a steel plant at Komsomolsk-Na-Amurye, but this plant will use Australian ore. In this chapter the authors concentrated on projects involving investment in mineral deposits in the Russian Far East.

3. The Russian government is said to have explained to an IMF mission that the shortage of capital was holding back investment in heavy machinery in the mining sector and thereby reducing production levels. There are some reports that in the gold sector yields have been falling for the same reason.

4. The Russian strategy of selling rights to further explore may have reflected oil industry experience. Garnaut and Clunies Ross (1983) point out (p. 92) that the auction system has worked in some oil projects where there were a large number of potential bidders, where the risk of policy changes was small and where the risk borne by the bidders for any one project was small, because the project was small relative to their whole portfolio.

5. Enterprises "belonging to foreign investors or with a (foreign) share of more than 30 percent in the authorized capital of a joint venture" appear to have to pay higher export duties. Export duties are calculated as absolute amounts (measured in ECUs) per ton of export.

6. These comparisons are based on an exchange rate of 110 rubles to the U.S. dollar (the so-called market rate at the Russian Federation Central Bank); Russian prices are "Moscow exchange prices." All data are in February 1992 prices. Since then, the rise in the general price level in Russia has led a devaluation of the exchange rate to over 240 rubles to the U.S. dollar. Note that energy prices have also been below world prices so the recently announced price rises will also affect the profitability of mineral projects (see *Australian Financial Review,* 13 July 1992).

7. It may still be the case that in the minerals sector, the licensing system is the binding constraint on export volumes.

8. This system is similar to that operating in China, except that in China the extent of the market in foreign exchange is much larger. The low official rate discourages exports, lowers the supply of foreign exchange and drives up the price in the swap markets. Reform of the system would raise the official rate (i.e., depreciate the official rate), encourage exports and, at the same time, lower the local currency price of foreign exchange in the swap markets as more foreign exchange came into the market.

REFERENCES

Garnaut, Ross, and Anthony Clunies Ross, 1983, *Taxation of Mineral Rents,* Claredon Press, Oxford, pp. 91–92.

Ritchie, Douglas, 1990, "Investment criteria for international exploration and mining companies in the Asia-Pacific region," paper presented to the Fourth Minerals and Energy Forum, November, Dallas, Texas.

Chapter 18

PROSPECTS FOR SINO-CIS COOPERATION IN NONFERROUS METALS

Youhua Chen

BACKGROUND

China and the former Soviet Union are the most mineral-rich countries in the world; however, the per capita value of minerals in the nations is quite low. The two countries share several thousand miles of borders. Their cooperation in the area of mining will not only benefit each other but will also be important to global markets.

Official economic relations between the Soviet Union and the People's Republic of China (PRC) began in 1952, when the government of China decided to undertake long-term economic planning and adopted the Soviet model of development. At that time, only the Soviet Union was willing and able to provide technical and financial assistance to the new Chinese state. China's First Five-Year Plan (1953–57) was successful, in part, because of Soviet aid to nearly 150 investment projects, for which the Soviets provided designs, equipment, engineers and loans. All this came to an abrupt halt in 1960, when severe political differences resulted in the Soviet withdrawal of assistance from China.

Economic relations and cooperation between the two countries resumed in the late 1970s on a very limited scale. In the 1980s, three obstacles hindered the development of closer ties: the Soviet invasion of Afghanistan, Soviet support for Vietnam's occupation of Cambodia, and the Soviet military presence along the border with China. With the eventual Soviet withdrawal from Afghanistan and the reduction of forces along the border, as well as the Vietnamese withdrawal from Cambodia, relations between the two Asian superpowers have improved.

The normalization of relations between the Soviet Union and China in 1989 marked the emergence of a new pattern of superpower relations in Asia. Both

J.P. Dorian et al. (eds.), CIS Energy and Minerals Development, 325–332.

nations were struggling through periods of transition from one economic system to another. From a Soviet perspective, normalized relations with China put it in a better position to improve ties with other nations in Asia.

China's trade with the former Soviet Union declined sharply in 1991, because instability left the new republics unable to produce or pay on time. Although it was one of China's smaller trading partners, the Soviet Union had enjoyed slow but steady growth in trade with China since the late 1980s. In 1991, however, two-way trade dropped 12.1 percent from the 1990 total to US$3.9 billion. China's exports to the former Soviet Union were US$1.8 billion, down 18.5 percent, and its imports from the republics had decreased to US$2.1 billion, a decline of 2.7 percent.

Border trade between the two countries played an increasingly important role after it was resumed in 1983, when it reached a value of six million rubles. Across the border areas, the Soviet Union supplied China with steel, cement, chemical fertilizers, machinery, cars and trucks and rolling stock. In turn, China exported to the Soviet Union such commodities as mineral ores, consumer goods, grains, fertilizers and textile products.

A recently completed railroad linking Xinjiang (Alashankou) to Kazakhstan is likely to expand trade across the border areas, particularly shipments of mineral raw materials. Completed in August 1990, the new railroad links China with Central Asia and several Middle Eastern and European nations and cuts freight costs between China and Kazakhstan by a reported 30–50 percent. The former Soviet Central Asian republics may be the greatest beneficiaries of the growing cross-border trade. In addition to the railway line through Xinjiang to the Kazakhstan border, there are also weekly flights between Alma-Ata and Urumqi, China.

More than 100 protocols and agreements for joint venture activities, labor exchanges, and scientific and technological cooperation were signed in 1989 involving, in part, iron and steel, building materials, nonferrous metals and geology, and the pace of cooperation between the two countries is accelerating. In October 1989, the first Sino-Soviet joint venture since the suspension of economic ties nearly 30 years earlier was announced in the form of a project to construct a 13-km pipeline to take coal from the Jiangquhe slurry manufacturing site in Binxian County, Shaanxi Province, to the Weihe power station near Xiangyang City. The pipeline reportedly has cut transportation costs by two-thirds.

NONFERROUS METAL RESOURCES OF THE CIS AND CHINA

In comparing the nonferrous metals of China and the Commonwealth of Independent States (CIS), four categories can be distinguished according to reserves and the supply-demand balance:

- **Category 1 includes lead, zinc and mercury. Both China and the CIS have adequate resources of these metals to be self-sufficient.**

Both the CIS and China have huge lead reserves. During the past decade, the Soviet Union produced 5,395,000 tonnes of lead concentrates and 7,850,000 tonnes of refined lead (including secondary lead), according to *World Metal Statistics*. It consumed about 7,700,000 tonnes of lead during the same period. From 1981 to 1990, the Chinese production of lead concentrates was 2.42 million tonnes, balanced by a consumption of 2.41 million tonnes of refined lead.

It is unlikely that these two countries will be short of lead in the near future. This takes into consideration lead recycling and the limits of lead use in some areas.

For zinc, the reserves of China and the CIS are among the largest in the world. While the per capita holdings are very low, the two countries are able to meet their needs with domestic production. At one time in the early 1980s China was short of slab zinc. It imported about 800,000 tonnes of zinc during the decade. However, as a result of the policy "to vigorously develop the lead and zinc industry," the production of slab zinc surpassed consumption in the late 1980s. In contrast, Chinese mine production of zinc exceeded consumption during the entire decade. The zinc production of the Soviet Union was slightly lower than its consumption, but there were still only small amounts of imports during in the 1980s.

The former Soviet Union was the world leader in mine mercury production. Its consumption level was also among the highest in the world. China's mercury production had always exceeded its consumption, and most of its production was exported for hard currency.

- **Category 2 includes those metals in which both countries are deficient; an example is aluminum.**

Owing to geological conditions, the quality of bauxite resources in both China and the CIS is not very high. Most of the bauxites are low-grade diaspores. Only a small fraction of the mines produce gibbsite bauxites. It is estimated that the Soviet Union imported at least 1.5 million tonnes of alumina every year in the 1980s. There is also a large gap between the production and consumption of aluminum in China. From 1981 to 1990, net aluminum metal imports amounted to 1.6 million tonnes, in addition to large amounts of alumina imports.

- **Category 3 includes metal resources abundant in China but in short supply in the CIS. These metals include tungsten, tin, and antimony.**

In terms of metal content in the resources, both China and the CIS have abundant reserves of tungsten, tin and antimony. However, given the qualities of the deposits and the supply-demand relationships in the countries, there are some shortages in the final products. Tungsten mine production in the former Soviet Union was the largest in the world. Yet, the country still depended on imports because of high consumption levels. Most of the minerals in the CIS deposits are sheelites, with only a small percentage of wolframite, and most of the mines are of below-average grade. Imports came mainly from China, which possesses abundant high-grade tungsten minerals. From 1981 to 1990, the Soviet Union bought about 62,000 tonnes of tungsten concentrates from China.

China is rich in tin. The mines are located primarily in Yunnan, Guangxi, Guangdong, and Hunan in the central-southern part of the country. The Gejiu mine in Yunnan Province and the Dachang mine in the Guangxi Autonomous Region are collectively known as "The Capital of Tin." In the 1980s, China produced about 232,800 tonnes of tin concentrates and 205,000 tonnes of refined tin. Total domestic consumption during this period was only 134,000 tonnes. Net exports for the decade totaled 72,300 tonnes.

During the 1980s, mine and refined tin output of the Soviet Union amounted to 156,000 tonnes and 170,000 tonnes, respectively. The country consumed about 280,000 tonnes of refined tin, yielding an apparent shortage approaching 110,000 tonnes.

As with the above two metals, China possesses huge antimony reserves, while the country's consumption of the metal is low. Most antimony production in China was exported, which constituted about 60 percent of the world supply. Prior to 1973, the Soviet Union was able to supply itself with antimony; after that, it had to depend partly on imports.

- **Category 4 consists of metals in which China is deficient, while the CIS possesses abundant resources. These metals include copper, nickel and gold.**

As the third most abundant metal after iron and aluminum in the country, copper resources in China are relatively abundant. However, the sizes and grades of the mines do not take advantage of the abundance of the resource. Most mines are below average in scale and grade. As of yet, none of the copper mines can be described as very large on a global scale. In addition, some of the mines are located in remote areas with difficult geographical conditions, such as Tibet and Qinghai, and thus, are not favorable for mining. Each year China has spent large amounts of hard currency on imports of copper.

Although the grades of copper mines are not high in the CIS, the country does boast favorable mining conditions, unlike China. Kazakhstan and Russia account for most of the copper output. According to *World Metal Statistics,* during the 1980s maximum copper production of the Soviet Union exceeded consumption by 1.2 million tonnes. It exported 569,000 tonnes of copper to Western countries during this period, with peak exports of 171,000 tonnes in 1990. It is important to note that the significant export levels were achieved before the huge Udokan mine began operations.

As with copper, China is well-endowed in nickel resources, yet the economic reserves are limited. The development of the nickel industry is constrained by unfavorable mining conditions and the lack of workable reserves. China has been forced to import large amounts of nickel every year, even though consumption remains relatively low (China's Ni/steel consumption ratio was only 0.051%, or half of the world average during the past ten years). The consumption level will surely rise in the course of economic development and with an expansion of steel-making capacity. A second stage of expansion in Jinchuan can be expected during the Eighth Five-Year Plan (1991–1995), raising China's total nickel production capacity to 40,000 tonnes. It is expected that total output will reach 60,000 tonnes by 2000, yet it will still be difficult to meet domestic consumption.

CIS nickel reserves are not much larger than China's; however, the mining conditions are far better and the recoverable reserves are much higher. The former Soviet Union produced large amounts of nickel ores and metal, and consumed relatively little of the metal. In the 1980s, the country imported small amounts of concentrates to feed its huge steel-making capacity, yet the imports were far exceeded by exports of refined nickel.

The CIS has an abundance of high quality gold resources. Around two-thirds of total Soviet gold production was located in Russia, with the remaining output coming primarily from Uzbekistan and Kazakhstan. While only 10 percent of China's gold reserves occur in surficial placer deposits, it is estimated that two-thirds of CIS gold resources are in placers.

SINO-CIS COOPERATION IN THE AREA OF NONFERROUS METALS

There are ample opportunities for cooperation between China and the CIS in the nonferrous metals industry. The discussion below highlights the various forms of cooperation now being pursued.

Sino-CIS trade cooperation

Both China and the CIS have abundant supplies as well as shortages of various nonferrous metal resources, depending on the specific commodity. It is possible for the two countries to alleviate their shortages through trade. China can continue to import copper and nickel products from the CIS, while in turn, the CIS can buy tungsten, tin and antimony from China.

In the 1980s, nonferrous metals trade between China and the Soviet Union was very limited. Except for tungsten and concentrates of some less-common metals such as tantalum and niobium, statistics on Sino-Soviet trade of nonferrous metals before 1989 are not readily available. Indeed, it was not until 1989 that China began to import aluminum, copper and nickel products from the Soviet Union. The amount of imports has increased every year since then. Yet, no data are available on Chinese exports to the Soviet Union, other than those before 1988 which refer to shipments of several thousand tonnes of tungsten concentrates annually. It is surprising, however, that China, as one of the main mineral suppliers in the world, did not establish a foothold in the Soviet tin and antimony market.

As with trade of other commodities, trade of nonferrous metals tends to be diversified, ranging from intergovernmental trade to activities involving a combination of intergovernmental and private enterprise trade. The recently booming border trade between the two countries is an example of this.

It is reported that only 45 percent of the nonferrous metals imported by China are bought with the country's hard currency reserves. Barter, which constitutes the majority of border trade, accounts for most of the remaining payments for goods, while cash in consumers' hands makes up the remainder. In 1991, most of the aluminum products, almost all of the magnesium, and a large amount of nickel imported by China were bought from the Soviet Union through barter. About 40 percent of the imported copper, valued at US$270 million, was obtained through barter in 1991.

Sino-CIS border trade is attracting more and more attention. Along the border with Russia in northeast China there are several key trade areas, including Heihe, Suifenhe and Manzhouli. In northwest China, there are also such trade areas along the border with Kazakhstan, Kyrgyzstan and Tajikistan. As mentioned earlier, border trade is expected to increase in the future, following the opening of the Eurasian continental rail bridge at Alashankov in Xinjiang. This type of trade is flexible and quite profitable, and thus, very attractive. It will continue to expand as a primary trade form. In fact, China has designated the above areas as open trade ports, and more than 100 cities have sent trade administrative bodies to Heilongjiang Province in China to coordinate future trade activities.

Nonferrous metals are principal among the commodities traded at the Sino-CIS border. A significant share of the Chinese imports of copper, nickel and aluminum were bought from the former Soviet Union through border trade, with the payment being either in cash or consumer goods, which are in need in Russia. In addition, with barter trade, China can pay for its imports with supplies of tungsten, tin and antimony, which are also in short supply in parts of the CIS.

Sino-CIS cooperation in prospecting and mining

Along the Pacific metallogenic belt, China is well-endowed in mineral reserves in Heilongjiang Province, while Russia has abundant reserves in its Far East region. There are also potential resources in Hami, the Junggar basin and the Altai mountain range in Xingjiang Autonomous Region, while there are abundant nonferrous metal resources on the other side of the border in Kazakhstan and Uzbekistan. In all of these areas, it is quite possible that China and the CIS could work together extensively in prospecting activities.

China and the new independent CIS states can also cooperate in geological surveying and economic feasibility evaluation. During a recent visit to Uzbekistan, senior officials from the Geological and Mineral Economic Institute of China signed a memorandum of understanding with the Central Asia Institute of Geology and Mineral Resources in Uzbekistan in which both sides agreed to cooperate in seven areas. The third and the fourth areas of cooperation include research on expanding the mineral resources base and an exchange of computerized geological data and information about establishing additional computerized data networks.

China and the CIS will also hopefully be able to cooperate in mining. While the Sino-Soviet Union Intergovernmental Group for Nonferrous Metals Cooperation was holding discussions in Moscow in 1990, the possibility of establishing a joint venture in the Duobaoshan copper mine in Heilongjiang Province in China was examined. A technical-economic evaluation of such a project was completed during the conference. Located in Nengjiang County near the Xiaoxinganning mountain range, the mine has proved reserves of about 2.5 million tonnes of copper in a porphyry deposit and unknown potential reserves. If this mine were put into operation, the copper-deficient Chinese market could certainly be much better supplied. In the northeastern part of China, there is another area that has the potential to support a copper mine: Unugetou Mountain, which has reserves of about 640,000 tonnes of copper.

In the memorandum mentioned above, the seventh area of cooperation concerned the mining of medium- or small-scale gold-bearing placers in Uzbekistan planned to commence in 1992. The Chinese will be responsible for the mining and dressing technology. It is estimated that gold reserves in Uzbekistan account for about 40 percent of those of the CIS. After gaining control over the mineral resources of its territory, the republic seems quite willing to cooperate with foreign countries in the exploitation of its gold mines located in both placer and vein deposits. In fact, Uzbekistan has already received investment from a U.S. firm in its largest gold mine, Muruntau. Among the 16 known gold mines in the republic, only seven are in operation. China and Uzbekistan could cooperate in mining any of the other nine mines.

As the CIS develops its mining industry, China can offer effective cooperation in the form of joint ventures, technology sales and labor. For example, the huge Chinese geological and mining labor force could play a role in the development of the Russian Far East, which contains abundant reserves of critical non ferrous metals.

Sino-CIS cooperation in mining technology

A variety of opportunities for cooperation in this area could be realized; indeed, some examples of successful cooperation in mining technology have already occurred. In the field of exploration, for example, China utilized reversed-flush sampling technology from the Soviet Union. This technology brought favorable results when applied to the exploration of beach gold deposits in Yantai, Shandong Province, on the east coast of China. China is also likely to acquire noncore drilling technology for rock powder sampling. The technology could be used for nonstandard core sampling in complex mining conditions, making it possible to use drilling instead of underground excavation.

The mining experiences associated with the Udokan copper mine in Russia can also be of great value to China in applying such mining techniques in areas where natural conditions are not favorable. The former Soviet Union has successfully processed pseudonepheline and alunite to make up for a lack of bauxites. This process would certainly be useful to China while it is trying to expand its aluminum base. Another technology that might be attractive to China is bacteria leaching used in extracting gold, copper and other resources from low-grade concentrates.

With such a diversity of geological conditions, China, too, can provide the CIS with necessary excavation technology, for example, for mining small- to medium-scale gold mines. There is also the possibility of cooperation in areas where both countries have limited experience, including the processing of diaspores and low-grade copper ores.

COOPERATION IN INFORMATION EXCHANGE

International markets are today playing important roles in the Chinese and CIS economies. There are therefore good prospects for cooperation in the exchange of information on technology and markets which will benefit both sides. The Techno-Economic Research Center of the China National Nonferrous Metals Industry Corporation (CNNC), for example, seeks to learn more about the nonferrous metals industry and market in the CIS, and can provide information on China in return. Cooperation in this field can assist in expanding the many areas of cooperation described in the preceding pages. It might be useful to set up a market information network to be shared by both countries.

SUMMARY

As has been demonstrated, areas of potential cooperation between China and the CIS in nonferrous metals are quite numerous, and it is not possible to describe all of them in this chapter. Clearly, there is a promising future for Sino-CIS mining cooperation in nonferrous metals, and both countries will be making efforts to facilitate increased cooperation in the years ahead.

Chapter 19

EVOLVING COMMONWEALTH-JAPAN ECONOMIC RELATIONS

Hiroshi Takahashi

INTRODUCTION

In July 1986, during a visit to Vladivostok in the Soviet Far East, Soviet President Gorbachev proclaimed his country's desire to become more actively involved in the economic growth and dynamism of Asia. He affirmed that the Soviet Union wanted to re-establish itself as an Asian power, willing and capable of participating in the phenomenal growth of the region. Soviet-Asian relations began to improve noticeably that year, though Japanese-Soviet ties remained strained due to the long-lived territorial dispute over four islands north of Japan.

Soviet policy towards Asia in 1990 represented a continuation of the ambitious directives put forward by Gorbachev in 1986 and again during a speech in 1988 at Krasnoyarsk. Despite growing domestic turmoil and economic deterioration, 1990 was marked by high-level government exchanges between Moscow and several Asian nations, and the expansion of industrial relations and trade. Having firmed ties with China in 1989 and later with South Korea, Gorbachev solidified his Asia-Pacific standing with a historic trip to Japan in April 1991.

One of the issues discussed at the April summit was trade relations and how to improve them. In 1991 Soviet-Japanese trade amounted to about US$5.4 billion. In 1990, Japan's trade with the Soviet Union had declined sharply as a result of Soviet delays in paying their bills and Japanese concerns over the fragility of Gorbachev's hold on power (Table 19.1). By June of that year, Soviet businesses were behind in payments to Japanese companies by at least US$250 million, prompting Japan's trade ministry to urge some companies to halt shipments to the Soviet Union at that time. Though the delays in Soviet payments were deemed temporary and largely the result of domestic economic problems, they tarnished the Japanese government's image of economic reform in the Soviet Union and fueled doubts that Soviet-Japanese trade would expand in the near term.

J.P. Dorian et al. (eds.), CIS Energy and Minerals Development, 333–340.

Table 19.1. Japan's Trade with the Soviet Union, 1970–1991 (million US$)

	Export	Import	Total	Balance
1970	341	481	822	-140
1975	1,626	1,170	2,796	457
1980	2,778	1,860	4,638	918
1982	3,899	1,682	5,581	2,217
1985	2,751	1,429	4,180	1,321
1989	3,082	3,005	6,086	77
1990	2,562	3,350	5,914	-788
1991	2,119	3,313	5,432	-1,195

Source: SOTOBO's *Monthly Bulletin*

Even faced with payment problems, Japanese business leaders continue to evaluate opportunities to participate in Commonwealth industrial activities, including those related to mining, metallurgy, and energy production. Although the former Soviet Union contains the world's largest mining industry and is essentially self-sufficient in most mineral commodities, it is slowly depleting its minerals base. As production costs continue to escalate in the Commonwealth, Asia will likely become an increasingly important source of raw materials and products by way of trade and joint venture activities. Economically robust nations like Japan and South Korea may also become providers of financial and technical assistance to the former Soviet Union for improvement and expansion of its mining industry.

Since 1988, Japanese delegations have visited many parts of the former Soviet Union to assess areas for investment in minerals and energy development. Many Japanese industry officials are aware of Soviet scientific potential and are now taking steps to initiate cooperation with a number of former Soviet republics in several new areas. Moscow, too, has encouraged Japan to assist in renovating Russian steel and nonferrous metal facilities in a bid to boost overall productivity. In 1991, Japanese energy experts visited the Soviet Union to investigate possibilities for Japanese investment or technology transfer to the country to help increase its waning energy supply. Today, the Russian Federation is seeking overseas investment to help explore for new energy sources and build facilities in its Siberian and Far East regions, but the Japanese government has been generally reluctant to consider such ventures until the Northern Territories issue is resolved.

Although Japanese companies are willing to do business with the new Commonwealth, the private sector can do little without a resolution of the Northern Territories issue. Meanwhile, the Japanese business community is increasingly concerned about being left behind other nations such as Germany, Italy, Switzerland and the United States that have already begun engaging in joint venture activities and providing large loans to the former Soviet Union.

Clearly, resolution of the territorial dispute would mark a turning point in bilateral relations between Japan and the Russian Federation that would yield political and economic benefits to both countries. While it is uncertain as to whether in the near future President Yeltsin will offer territorial concessions in exchange for economic assistance from Japan, the resolution of the Kurile Islands dispute will undoubtedly set a tone for future relations between Japan and Russia which, in turn, will have broader implications for the rest of Asia.

PRESENT AND HISTORICAL TRADE BETWEEN JAPAN AND THE FORMER SOVIET UNION

The volume of trade between Japan and the Soviet Union in 1991 decreased by 8.1 percent and totaled US$5.4 billion. Trade between the two countries in the second part of the 1980s was characterized by a stagnation of Japanese exports to the Soviet Union. Consequently, the trade surplus with the Soviet Union that had been ongoing since 1975 was transformed into a trade deficit by 1990, reaching US$1.2 billion dollars in 1991. Although the trade volume between the two countries does not represent a significant share of Japan's world trade, some industrial sectors, like steel and engineering and construction equipment, generated significant profits from exports to the Soviet Union, while imports of timber, fish and nonferrous metals from the Soviet Union are also of some importance to Japanese industry (Table 19.2).

The trade structure between the two countries has been evolving. Traditional trade had been in goods connected with Siberian development, including items related to Japanese-Soviet joint venture projects. Japanese export goods associated with Japanese-Soviet projects included construction equipment, cranes, trucks and other related items (Table 19.3). In addition, Japan also exported various types of steel pipes for gas and oil development in Siberia.

As for imports, Japan imported wood and coal from the Soviet Union during the development history of a number of Siberian projects. Table 19.4 exhibits the structure of Japan's imports from the Soviet Union during the 1975–1991 period. Throughout the trade history of the two countries, the activities primarily responsible for promoting Japanese-Soviet trade have been joint venture projects and Soviet oil and gas development.

Japanese-Soviet trade reached its first peak in 1982, an increase caused by the two activities mentioned above. In addition, until the early 1980s large plant construction was an important basis for trade. These contracts were financed by official credits. For example, Japanese companies were contracted to build seven plants (mainly chemical plants) financed by official credits in 1975–1977. The value of the contracts reached more than US$1 billion. Subsequent to these

Table 19.2. Japan-Soviet Share of Japan's World Trade, 1979-1991 (%)

	Export	Import
1979	2.4	1.7
1980	2.1	1.3
1981	2.1	1.4
1982	2.8	1.3
1983	1.9	1.2
1985	1.6	1.1
1986	1.5	1.6
1987	1.1	1.6
1990	0.9	1.4
1991	0.7	1.4

Source: SOTOBO's *Monthly Bulletin*

Table 19.3. Structure of Japan's Exports to the Soviet Union, 1975-1991 (%)

	1975	1980	1985	1990	1991
Food	0.0	0.0	0.0	0.4	1.0
Steel	33.8	34.8	31.4	14.7	9.7
General machines	9.9	18.3	22.8	26.1	28.6
Electric machines	5.2	2.8	6.9	18.1	16.0
Transport machines	9.2	7.5	7.2	5.1	11.0

Source: SOTOBO's *Monthly Bulletin*

plant contracts there was only one other contract signed in 1981 and financed with official credits.

Problems with Japanese-Soviet trade in the early 1980s

Japanese-Soviet trade stagnated in the first part of the 1980s after peaking in 1982. Stagnation was caused by the rise in East-West tensions after the Soviet invasion of Afghanistan as well as turmoil in Poland. At this time the Japanese government followed the lead of the West and limited official credits to the Soviet Union. The attitude toward the Soviet Union in business circles was overwhelmingly negative.

In the early 1980s, the Japanese economy's demand for raw materials diminished in comparison with demand in the 1970s largely because of structural changes in the economy. The "crisis consciousness" of the Japanese economy with respect to supplies of raw materials that had fueled the passion for Siberian projects in the 1970s had disappeared by the 1980s.

Table 19.4. Structure of Japan's Imports from the Soviet Union, 1975–1991 (%)

	1975	1980	1985	1990	1991
Fish	1.2	1.9	7.1	9.6	11.6
Cotton	14.6	5.8	2.0	0.6	1.3
Lumber	35.7	38.0	24.5	15.6	13.7
Coal	14.0	6.5	13.8	13.8	9.8
Pig iron	Negl.	N/A	0.3	6.2	7.2
Nonferrous metal	13.8	17.7	16.1	29.9	35.1
Gold	N/A	2.6	10.8	7.1	6.6

Negl. = Negligible.
N/A = Not available.
Source: SOTOBO's *Monthly Bulletin*

Japanese-Soviet trade in the Gorbachev era

The attitude in Japan toward the Soviet Union changed after Mikhail Gorbachev became the leader of the country in 1985. Several positive developments subsequently occurred in the foreign relations between Japan and the Soviet Union. In January 1986, for example, Soviet Minister for Foreign Affairs Eduard Shevardnadze visited Japan and the trade agreement of 1986–1990 was signed. Subsequently, a general exhibition of Japanese industry was held in Moscow for the first time in nearly a decade.

Trade between Japan and the Soviet Union increased in value by US$1 billion in 1986. However, the increase was not as significant as it appeared. If the value were recorded in Japanese yen, it actually decreased by 12.7 percent. The strong yen negatively affected exports.

The volume of Japanese-Soviet trade recovered in the latter half of the 1980s and reached a second peak in 1989. In contrast to the 1970s, Japanese imports mainly drove the volume upward. Japanese imports from the Soviet Union doubled from 1985 to 1989, while exports increased only 12 percent. Japan's exports increased slowly for several reasons:

- almost all Japanese-Soviet Siberian projects were completed;
- the backlash from the "Toshiba machine tool case" (in violation of COCOM rules) had a negative effect on the willingness of Japanese exporters to sell certain items to the Soviets;
- the Soviets cut investments, especially for oil and gas development; and
- a strong yen made Japanese goods more expensive.

Japanese exports to the Soviet Union depended heavily on Soviet investment policies. Japan traditionally exported steel pipes for oil and gas development as well as machinery and components for large plants. However, the sudden decrease in revenues to the Soviet Union from oil exports as a result of lower oil prices in the international market from 1985–86 limited the country's ability to invest in oil development projects and purchase imports from Japan.

Japanese exporters had anticipated that Mr. Gorbachev would accelerate investment after he came to power. Yet, the Soviet leader was forced to revise this policy due to shortages of foreign currency revenues.

In contrast to exports, Japanese imports from the Soviet Union increased in the latter 1980s, including, primarily, nonferrous metals, gold, and fish. The significance of timber imports declined. A strong yen provided companies with an incentive to import, and general domestic demand also rose. Some Japanese companies considered an increase in imports in the short run to provide a means to generate an export increase to the Soviet Union in the long run because the superpower needed foreign currency.

In the late 1980s, Japan was a relatively small but significant market for Soviet mineral commodities. Japan's imports from the Soviet Union were dominated by timber, coal, and nonferrous and rare metals. More recently, however, nonferrous metals became the most significant Japanese import from the Soviet Union, accounting for 35.1 percent of Japan's imports from the country in 1991. Platinum and aluminum are the primary import items, as well as steel.

Recent problems in Japanese-Soviet trade

While tremendous near-term opportunities in Japanese-Soviet trade were envisioned as a result of Soviet President Gorbachev's visit to Japan in 1991, problems that led to decreases in trade between the two countries were also developing. Delays by the Soviets in paying their bills occurred more and more frequently by the end of the 1980s. By March 1992, delays in payments reached a value of about a half billion dollars. This problem created an export disincentive for most Japanese companies. While the sharp decrease in foreign currency revenues is a primary reason for the payment delays, it is impossible to ignore the effects of organizational disorder in the newly formed CIS. As the Soviet Union collapsed, this disorganization intensified. For example, Japanese companies can no longer expect to make any normal transactions through the former central bank, Vneshekonombank.

Japanese anxieties about the dissolution of the Soviet Union and Boris Yeltsin's rise to power led to a sharp decrease in trade in the fourth quarter of 1991. Trade between Japan and the new Commonwealth in January–February 1992 fell by 40 percent in comparison to the same period the previous year. There were three primary reasons for this significant drop in trade: organizational turmoil, a strict Russian tax regime, and a prohibitive foreign currency policy toward export enterprises. Japanese companies have been trying to import more goods from the former Soviet Union in order to improve that country's foreign currency position; however, the strict tax and currency policy have impeded such efforts. Moreover, the Commonwealth economic and financial systems are changing rapidly. Before they take further action, Japanese companies want Russia to establish and follow a stable course.

A new phenomenon in Japanese-Soviet trade

It is worth noting a new phenomenon in Japanese-Soviet trade that occurred in the second half of the 1980s: consumer cross-border trade. It is not possible to attach a value to this trade as the goods are transported across borders by individual consumers. The cross-border trade was negligible in the past, yet visitors to Japan from the Soviet Union increased by 50 percent in 1988–1990. Today, the main items being purchased by Russians on such consumer visits are used automobiles. As a consequence, there are many used Japanese cars in Russia's Far East region. It is said that more than 50 percent of the automobiles in cities like Vladivostock, Nakhodka, and Yuzhno-Sakhalinsk are Japanese-made. Japanese car makers have even set up a joint venture service station in Vladivostok. In addition, Russians have been buying electrical appliances at Japanese ports. A shop specifically aimed at Russian customers was opened recently at the port of Otaru in Hokkaido in the northern part of Japan. Japanese businessmen believe that a number of Russians hold large amounts of foreign currency, and some companies want to establish businesses to cater to these Russians.

THE JAPANESE-CIS JOINT VENTURE SITUATION

By mid-1992, there were around 100 Japanese-CIS joint ventures, accounting for only a small percentage of all joint ventures registered in the CIS. The Japanese are being criticized by the Russians for such a small number of joint ventures in comparison with the trade volume between the two countries. However, these critics do not consider that such joint ventures accounted for a quarter of all joint venture exports in 1990 and that the level of activity of Japanese-CIS joint ventures is indeed high. Such enterprises generally begin operating as soon as the required documents are signed. In addition, the share of Japanese-Russian joint ventures is highest among all joint ventures registered in the Far Eastern region.

JAPAN'S INTEREST IN THE RUSSIAN FAR EAST

Japanese government and business officials seek to develop closer economic relations with Russia for geopolitical, territorial, security, and bilateral reasons. Although Japan and the Russian Far East are geographic neighbors, trade has been limited and travel restricted. Today both nations are, however, working towards normalizing relations.

The Russian Far East is considered by many to be Moscow's last frontier. It is attractive to foreign businesses because of its unique combination of natural resources and advantageous geographic position. This vast area covers 6.2 million km^2 and stretches along the northwest Pacific rim, bordering on China, North and South Korea, Japan and the United States. Its rich endowment of natural resources accounts for much of Russia's minerals output, a third of its timber resources, and a majority of Russia's fish harvest. The Russian Far East is multinational, with many more ethnic groups in the region than anywhere else in the Soviet Union. Like Siberia, the Far East has a history of autonomy.

In contrast to the earlier Soviet policy, the Russian government is now anxious to open the area to foreign investors and increase economic ties with nations in the Asia-Pacific region. As trade among Asia's rapidly growing nations increases, the Russian Far East's economy will probably become more and more active. Today, interest in trade deals, joint venture projects and large-scale economic linkages is growing among both existing and potential trading partners in the region. The nearest and largest potential investor in the Russian Far East is Japan. For reasons having do with conflicting territorial claims as well as economic realities, the Japanese have adopted a slow and deliberate strategy.

Eventual resolution of the Northern Islands territorial dispute would mark a turning point in bilateral relations that would yield political and economic benefits to both countries. The time has come for relations to improve, as there are pressures on both governments to seek resolution of the issue. The Japanese government is ready to assist Russia with the conversion of military enterprises to civilian activities and provide much-needed economic aid and investment to help support a struggling economy.

Chapter 20

THE NEW ZEALAND-SOVIET UNION/CIS ECONOMIC RELATIONSHIP: PROBLEMS AND PROSPECTS IN A TIME OF CHANGE*

John H. Beaglehole

INTRODUCTION

The economic relations of New Zealand and the former Soviet Union since the mid-1980s have reflected the dramatic changes in the political and economic environment of the Soviet Union, culminating in the disintegration of the Union at the end of 1991. The ending of the Cold War and wide-reaching reforms have completely altered the political and economic environment within which New Zealand-Soviet Union relations had been conducted. Despite earlier steps towards the liberalization of foreign economic relations, there were no tangible changes in the administration of external economic activity and the state monopoly on Soviet foreign trade even by the end of 1988. The principal landmark in the decentralization of foreign economic relations was the decree of 1988 which made possible direct access for enterprises and organizations to foreign partners as of April 1989. Complementing this and subsequent decrees and legislation in this field was the formulation of policies with regard to foreign economic relations of economic programs both at central and republican levels in 1990 and 1991. Equally significant was the parallel development of political and economic decentralization, following the growing assertion of economic and legal sovereignty by republican and subrepublican authorities.[1]

The late 1980s and early 1990s also brought changes in the trade environment. New Zealand-Soviet Union/Commonwealth of Independent States (CIS) trade

*An earlier version of this chapter was published in the journal *Soviet Union (Union Sovietique)*, Moscow, Russia, copyrighted by Charles Schlachs, Jr., Publisher, California.

J.P. Dorian et al. (eds.), CIS Energy and Minerals Development, 341–358.

has inevitably reflected the trends influencing Soviet foreign trade in general. The last decade has seen substantial fluctuations in Soviet trade relations. At the beginning of the 1980s, there was a dramatic rise in Soviet imports, especially of agricultural products from nonsocialist countries. Soviet imports peaked in 1981 and 1982 and declined through the oil price collapse of 1986–87. From 1987 on, Soviet imports from the West began to climb again, but still only reached 1981–83 levels. By the late 1980s, the Soviet Union faced increasing hard currency problems combined with a deteriorating economic situation domestically. The attempts it had made to cut back on nonsocialist imports in the mid-1980s could not be sustained, as domestic economic difficulties pushed the Soviet Union in the opposite direction. As a result, in 1989 food imports increased by almost 25 percent, with a corresponding increase in foods' share of total imports.

While desperate to save hard currency, the Soviet Union was also under pressure to maintain imports of food and consumer products as a result of the continuing decline of agricultural output in the late 1980s and 1990. These imports were sustained, in part, by revenues from massive gold sales and higher earnings resulting from a moderate recovery of oil prices.[2] However, by mid-1990 there were signs of trouble over repayments. New Zealand then began to face a new experience in its relationship with the Soviet Union—the inability to pay. This coincided with a situation where Soviet-New Zealand trade reached the boom levels of the early part of the decade.

On the whole, the Gorbachev era was a good one for New Zealand's economic relations with the former Soviet Union. The period saw a substantial increase in trade between the two countries, and indeed a position was reached where fears were expressed in New Zealand that the country had become unduly dependent on the Soviet market for its major commodity exports, a point brought home by the debt crisis of 1990–91. That debt crisis together with the final collapse of the Union, however, undermined the established framework and structure of New Zealand-Soviet Union economic relations, thereby posing serious problems and creating new uncertainties.

FRAMEWORK OF ECONOMIC RELATIONS

The changes that occurred in the former Soviet Union since the late 1980s radically changed the framework within which New Zealand's economic relations with the Soviet Union were conducted.

Prior to the recent developments in the former Soviet Union, New Zealand's trade with the nation was conducted within the traditional framework of trade with centrally planned economies. In 1963, a trade agreement had been signed, and a decade later a joint trade commission was established. A fisheries agreement followed in 1978 which provided for representation in New Zealand of the Soviet Ministry of Fisheries and Sovrybflot. The Soviet merchant marine also had a representative in New Zealand.[3] Trade with the Soviet Union was conducted by trading companies, although 65–70 percent of wool was traded by

auction. As a result of the launching of economic reforms in the Soviet Union, it was recognized in New Zealand that new conditions were being created that would permit private purchasing, and thus, a new organization, 'Sovenz', with responsibility for conducting most of the agricultural trade with the Soviet Union, was established by the Dairy Board. In 1987, Sovenz took over both the Meat and Dairy Board licenses and acquired representation in Moscow.[4] New Zealand's economic relationship was conducted with foreign trade organizations, in particular Prodintorg, which acted as an agent for the Soviet Union's food imports, and Exportljion, and later Novoexport, for wool. With regard to fisheries, New Zealand's agreements for licensed fishing were with the Ministry of Fisheries and its agents Sovrybflot and Dalryba.

The trade and economic links then reflected the established pattern of agreements with central purchasing agencies. During the second half of the 1980s, however, the regionalization of the market was becoming evident. To a limited extent, there had been some degree of devolution of trade in certain border areas for many years—for example, in Europe through Fintorg and in the Far East through Dalintorg. In 1987, New Zealand was included for the first time in the list of border traders in the Far East, a status already enjoyed by Australia, and initially some trade was conducted through Dalintorg. Also, agreements in fishing were reached with Dalryba, the Far Eastern agent of the ministry. Gorbachev's emphasis on the importance of the Soviet Far East in the context of Soviet relations with the Asia-Pacific region undoubtedly influenced New Zealand's decisions on the directions of economic links with the Soviet Union.

Changes in Soviet legislation authorizing direct economic access and preferential provisions for the Soviet Far East encouraged in New Zealand's case the expansion of relations with the region and the adoption of new approaches to the conduct of economic relations with the Soviet Union.[5] With the erosion of power at the center and the assertion of claims of sovereignty, particularly economic sovereignty in the republics, New Zealand business, like the foreign business community in general, was forced to identify the new centers of decision making and establish new links. In practice, some decisions were being made and claims to power and control were being asserted at republican and also subrepublican levels. New Zealand firms responded to these new conditions and changes by establishing direct relations with the government of Sakhalin. It was already evident by 1989 that decisions were effectively being taken by the ispolkom (executive councils) at krai and oblast levels which were trying to establish direct links and create favorable conditions for foreign enterprise. It had been a constant complain of New Zealand traders under the monopoly system of foreign trade that they had been unable to gain access to end users. Under the new conditions things were rapidly changing. By 1990, Sovenz was reaching agreements with local authorities in Sakhalin, Altai and Kemerovo, and through them with individual enterprise groups. The involvement of Sovenz in the Soviet Far East has resulted in the establishment of two offices there, in Sakhalin and in Khabarovsk, with a third planned for Kamchatka. There is a possibility in the longer term of New Zealand setting up honorary consular representation in the region, and at least one other New Zealand company, Amalgamated Marketing, has decided in principle to establish an office there.

By mid-1990, it was evident that the Soviet central trading organizations were under serious threat. Prodintorg was touting itself to become an agent for food imports for the republics, and was also involved in serious arguments with the central government over the shortage of hard currency—a further reflection of the greater autonomy of the republics and regions, which were now able to retain some of their hard currency.[6] The same uncertainties were felt in the Novoexport wool trade organization which reflected what had by then become a crisis in New Zealand-Soviet Union trade relations. Officials generally were becoming uncertain about their future as each republic developed its own trade organization, and began to carry 'two cards' as they sought business opportunities outside their official capacity. The regional branches of central organizations, Dalryba and Dalintorg, were also experiencing the effects of greater autonomy and decentralization. Those enterprises, such as Kuzbass, that formerly supplied Dalintorg with export products had begun to acquire their own licenses, and Dalintorg was finding that its relations with krai and oblast authorities were not as good as they were a few years earlier. Soviet state traders, however, strove to protect their trading monopoly, and newcomers, such as Sovenz, at one stage faced some serious problems when they began to establish direct links with regional organizations and authorities. As of early 1992, Prodintorg still seems to have survived, possibly because the need for central purchasing has persisted as a result of the food aid program, coupled with the lack of knowledge of markets on the part of more remote areas of the Commonwealth; however, it is probably doomed as the process of decentralization and marketization continues. This outcome was made further inevitable, it seems, by President Yeltsin's announcement that Prodintorg was "a sufficiently corrupt" organization, and that it would be banned henceforth from carrying out monopoly operations of purchasing food for St. Petersburg (British Broadcasting Company, January 17, 1992). The Russian Ministry of Agriculture has assumed responsibility for agreements and contracts with foreign governments, and New Zealand has already sent draft plans to the ministry for cooperation in all aspects of agriculture in Russia. Roles are now being played also by newly established bodies, such as Mosgortorg, and a number of specialist institutes in the agriculture field.

The authority over fishing in the Russian Far East has been taken over by Russia's Ministry of Fisheries. Before the breakup of the Soviet Union, fishing companies which had established direct links with New Zealand companies began to bypass Dalryba. This accorded with changes in New Zealand's own fisheries policy, as charter fishing became the principal form of foreign fishing in New Zealand waters. Nevertheless, Dalryba remains and it sees itself as a broker retaining the power to make fish allocations, thereby limiting the autonomy of the local fishing companies. A new factor in this struggle for power and resources in the Russian Far East is the role of the local authorities, which claim to exercise supervisory control. With the recent visit of a delegation from the Baltic states, there has also been some indication that non-Russian fishing fleets may be interested in the New Zealand region.

The collapse of the former Soviet Union and the establishment of the CIS have created a new environment for the whole framework of agreements between the CIS and New Zealand. Russia has declared itself as legal heir and successor of the Soviet Union, and has assumed responsibility for all existing international agreements, including trade and fishery agreements. This still leaves uncertain the extent to which these treaty obligations are in fact being fulfilled, as recent concern about Soviet overfishing in the Antarctic suggests.[7] New Zealand has recognized Russia, and its ambassador is formally accredited by Uzbekistan, Kazakhstan, the Ukraine and Belarus.[8]

THE STRUCTURE OF THE ECONOMIC RELATIONSHIP

The effects of the changes on the New Zealand-Soviet Union relationship since the introduction of perestroika have varied. While overall the relationship has grown closer, the changes have also brought some negative results reflecting certain characteristics of the New Zealand-Soviet Union trade structure.

As a market, the Soviet Union began to acquire significance for New Zealand during World War II with large purchases of wool. However, major growth in trade came only in the 1970s, reaching the first peak in the early 1980s and again in the late 1980s and 1990. The rapid growth was the result, in part, of major changes in the Soviet Union's trade, foreign and domestic policies. The 1970s saw a massive increase in Soviet trade with the world, including a dramatic increase in agricultural imports combined with a windfall in energy prices. This occurred at the same time that New Zealand also sought to diversify its markets, particularly as a result of Great Britain's entry to the European Economic Community (EEC). The Soviet Union's involvement in the South Pacific and the waters surrounding New Zealand was also increased by the rapid development of its sea transport capabilities and its increased interest in fisheries in the region.

As a result, there developed to a limited extent a convergence of interests between the Soviet Union and New Zealand. New Zealand, as a specialized producer of agricultural products, was already familiar to the Soviet Union as a major supplier of wool. The New Zealand-Soviet Union/CIS relationship is first of all characterized by the marginality of New Zealand's trade with the former Soviet Union. Even in the best years, New Zealand accounted for only 0.2 percent of the Soviet Union's total trade. The Soviet Union's share of New Zealand trade overall has in recent years exceeded one percent, but represents well over two percent of its exports. New Zealand also faces in the Soviet Union a market which is itself a very large producer of the commodities which New Zealand sells to that market. A further feature of the trade relationship is that trade has been confined to a narrow range of products. In the post-World War II years, exports to the Soviet Union consisted almost entirely of wool; however, during the 1970s trade became slightly more diversified as a result of regular purchases of meat from New Zealand, and in the 1980s, dairy products assumed increased importance. During those two decades, imports from the Soviet Union also began to be more diversified.

During the second half of the 1980s, imports from the Soviet Union increased dramatically in proportional terms, but from a low base. The value of the total rose from 5.2 million rubles in 1986 to 15.9 million rubles in 1989. The very earliest imports from the Soviet Union consisted of some wood products and luxury items. During the 1960s and 1970s, chemical fertilizer came to account for a large part of Soviet exports to New Zealand; by the 1980s, these were balanced by car and tractor imports. In the second half of the 1980s, fish imports made up the largest share, representing 38 percent of imports from the Soviet Union in 1989, a reflection of the expansion of charter fishing.[9]

A persistent feature of the trade relationship has been the large trade imbalance because of the very limited volume of Soviet exports to New Zealand. The ratio of imbalance has fluctuated widely, but in 1990 stood at 15 : 1 in New Zealand's favor. This is a situation typical of the former Soviet Union's relations with commodity suppliers—the consequence of the Soviet Union's inability to sell manufactured products in significant quantities to countries that do not require its energy or primary products. The Soviet Union continued to complain about the gross trade imbalance over the years, but without any significant results. Some opportunity for modifying the imbalance was provided, however, by the introduction of the Exclusive Economic Zone (EEZ) in New Zealand in 1978, together with the development of Soviet interest in fishing in New Zealand waters, diversification of the economic relationship in the late 1980s, and the expansion of the Soviet merchant fleet.[10]

While New Zealand exports to the Soviet Union have been confined to a narrow range of products, the significance of that trade for both countries has been much greater than the overall level of trade would suggest. This can best be illustrated by a more detailed analysis of the trade of the individual commodities involved.

COMMODITY TRADE

The success of the New Zealand wool trade with the former Soviet Union is based primarily on the substantial wool requirements of the nation coupled with certain characteristics of the wool industries in both countries. The size of its population and the nature of its climate account for the fact that the former Soviet Union is the world's largest consumer of wool. While it possesses the second largest population of sheep in the world, the former Soviet Union is at the same time the world's second largest importer of wool. The dependence on imports is explained by the deficiencies of the Soviet wool industry which are related to the climatic conditions facing the industry, the high priority given to other sectors of agriculture, and the relative inefficiency of Soviet livestock agriculture. In view of the concentration in the Southern Hemisphere of much of the world's wool export production, the growth of the New Zealand-Soviet Union relationship was therefore a natural development.

Australia and New Zealand could provide a variety of wool requirements. The merino-based Australian wool industry is the major exporter of fine wools. New Zealand, with an industry based predominantly on cross-breeds, serves much of the world's import requirements for semicoarse grades. As a result, a symbiotic relationship has developed between the Soviet Union, on the one hand, and Australia and New Zealand, on the other, with the latter supplying the Soviet market on a complementary rather than competitive basis. New Zealand's own position in that market is further affected by the character of the Soviet sheep industry, which is directed towards the development of fine wool capacity. As in the case of New Zealand's other commodity trade with the Soviets, the wool trade with the Soviet Union has also reflected the Soviet concern with the limited availability of hard currency and, therefore, the avoidance of payment for value added. As a result, the Soviet Union has in almost all cases insisted on the import of the greasy as opposed to the scoured wool product.

As already indicated, there is a long history of New Zealand wool exports to the Soviet Union, but dramatic increases only began to appear in the late 1960s, after which growth reached a new plateau. By the end of the 1970s, again a new level was reached and was maintained throughout most of the 1980s until the collapse of 1990–91. Throughout the Gorbachev years, exports to the Soviet Union represented an average of over 10 percent of New Zealand's wool exports, and in 1989, the Soviet Union became New Zealand's largest single wool market. In terms of wool, the Soviet Union is more dependent on New Zealand than vice versa. Whereas overall the Soviet Union has been taking about 10 percent of New Zealand wool exports, these accounted for over 20 percent of Soviet wool imports during the 1980s. If the semicoarse category alone is considered, however, imports from New Zealand have on average represented over 60 percent of Soviet imports in this category by value in the 1980s; by quantity, however, the percentage was far higher, averaging 87 percent from 1986 to 1989 (USSR Ministry of Foreign Economic Relations, 1989).

On the Soviet side, the wool trade was conducted by a central trading organization. As central power in the Soviet Union collapsed and the effects of decentralization in foreign economic relations began to be felt, a crisis arose in the wool trade with the Soviet Union which is still unresolved. The trade has depended on allocations of hard currency and the consistent priority given to wool imports. In New Zealand's case, the Soviet trading organization had always paid cash, no credit lines had been required, and the Soviet Union was regarded as a totally reliable trading partner. By early 1990, however, payments, were falling behind, and by mid-1990, it was admitted by Novoexport that payments could not be made, as it had not received its hard currency allocations. Much of New Zealand-Soviet Union diplomacy from then on was to be concerned with achieving a solution to debt payments. The problem was also inevitably to affect the rest of the commodity trade between the two countries. The crisis forced New Zealand traders to consider approaching the government and the banks for credit lines, a step that had never been taken in the past for this market. Given the uncertainties existing in the Soviet Union/CIS market and the tight economic circumstances at home, the New Zealand government turned down appeals for credit.[11] Some wool exporters succeeded in obtaining small lines of bank credit, but the bulk of the debt remained and still remains to be paid off. Faced with its credit problems, the former Soviet Union pulled out of the wool market, and this decision seriously affected the world wool situation and wool prices.

As in the case of wool, the development of the New Zealand-Soviet Union relationship in meat and dairy products also reflects a convergence of interests. This arose from a changed economic environment faced by both countries in the 1970s. The commitment made by the Soviet leadership in the Brezhnev period to improve nutritional standards and to make available a greater and more stable supply of agricultural products, combined with the continuing failure of Soviet agriculture to meet plan targets, resulted in major trade policy changes. The result was the rapid rise in agricultural imports which affected all agricultural trade.

The opportunity for New Zealand to supply the Soviet Union meat market, as in the case of wool, arose from the character of the New Zealand sheep industry. The main sheep products marketed by this New Zealand industry are wool and lamb, but mutton, a by-product, has also found a good but specialized market overseas. New Zealand's role in the Soviet Union meat market has been the result of an ability to supply on a reliable basis what for the Soviet consumer is a high-quality product at a low price. New Zealand became a regular supplier in the 1970s. The Soviet market came to represent a large proportion of the New Zealand mutton trade, averaging 33 percent of mutton exports in 1982–86, and since then, averaging over 20 percent.

The most recent commodity to establish itself in New Zealand-Soviet Union trade is milk products, especially butter. The Soviet Union/CIS market is of crucial importance to New Zealand butter exporters. It accounts for over half of the accessible international market for New Zealand butter, with Russia taking almost all of the Soviet Union/CIS butter imports and about 90 percent of New Zealand's butter exports to that market. Initially, butter became a significant factor during the first peak period of New Zealand exports to the Soviet Union in the early 1980s, when the Soviet share of New Zealand butter exports reached 31 percent. The trade was affected by the general decline of Soviet trade in the mid-1980s and again by the disposal of EEC surpluses in 1987–88. However, from then on the relationship acquired greater importance for both partners. In 1989, New Zealand became the Soviet Union's largest butter supplier, and in the following year, the Soviet Union emerged as New Zealand's second largest butter market. At the end of that year, New Zealand and the Soviet Union agreed to their largest deal of the decade.

Despite the priority accorded to food imports, however, the debt problem also involved New Zealand dairy sales. By November 1990, the Dairy Board was owed

Table 20.1. Principal New Zealand Exports to the Soviet Union (NZ$000) (% in parentheses)

	1986	1987	1988	1989	1990
Mutton	13,473	29,726	8,156	30,645	15,052
	(7.4)	(16.0)	(5.6)	(9.4)	(4.5)
Butter	40,253	17,027	0	132,651	159,340
	(22.0)	(19.3)		(40.7)	(46.5)
Wool	128,833	138,579	136,989	162,727	167,873
	(70.6)	(74.7)	(94.4)	(50.0)	(49.0)
Total	182,559	185,377	145,145	326,023	342,265

Source: New Zealand, Overseas Trade.

NZ$114 million. With the agreement on a NZ$260 million butter deal at the end of the year, the Board has since faced serious payments problems.

The diversification of exports to the Soviet Union has altered the relative importance of these products in the total trade picture. However, with the collapse of the Soviet wool market in 1990–91, the commodity distribution of the trade will have changed further (Table 20.1).

The crisis that erupted in 1990 over Soviet payments highlighted what had always been a problem in the trade relationship—that of hard currency. It had affected the character of New Zealand commodity exports and had produced the ongoing criticisms by the Soviet Union of the continuing imbalance in the trade relationship. One consequence of this had been the search for various ways in which the relationship could be diversified. The scope for such an approach was considerably widened by the liberalization and decentralization of foreign economic relations by the Soviet Union, as it was at the same time increasingly acknowledged that the future of a relationship based on cash for goods was limited.

Over the last few years, the economic relationship has developed beyond the traditional pattern of exchange in three directions: countertrade, turnkey projects and joint ventures. Countertrade has long been an established feature of trade with command economies, and New Zealand itself had conducted a limited bilateral barter trade within its meat and diary contracts. However, complex countertrade arrangements are more difficult to arrange and more risky. Nevertheless, examples of this are occurring, the most publicized case being that of the payment of part of a butter deal with coal, which was then sold by Sovenz to third parties.

Turnkey projects built by New Zealand labor and resources constitute a form of exports, but also leave behind a permanent resource for Soviet domestic requirements, particularly in food. Sovenz has built or is proposing to build ten plants, including meat works and cheese processing plants, in various areas of the CIS at a total value of about US$60 million. Amalgamated Marketing, which the Soviet Union; and finally, the need for Soviet commercial law, management discipline and pricing practices to be developed and modified, so as to provide a business culture more in accordance with Western experience. Emphasizing as it did the problems of doing business in the Soviet Union, it took a very cautious view of possibilities, and concluded that it was premature to hold onto the prospect of major attractions in large-scale, capital-intensive projects. Instead, it called for more modest, medium-sized ventures dependent on management or trading skills rather than capital.

Since 1989, changes in the former Soviet Union have accelerated, and the new situation evident by 1991 led to a second review of the situation by the government of New Zealand. The result, ***Doing Business in the Post Revolution Soviet** Union: A Strategy for New Zealand,*[12] prepared by Stuart Prior, is a report with recommendations for adaptation by New Zealand to the changing conditions, and urges the need for New Zealand "to lift our game significantly. . . if we are to be a serious player in the rapidly changing commercial field." It also sees this to be a matter of urgency, with "the present window of opportunity" remaining open only for a few more years.

The Prior report emphasized that New Zealand had to adopt a new strategy in its relations with the Soviet Union. This strategy should comprise four elements: risk reduction, targeting of effort, profile raising and niche location. The report argued that the end of the traditional relationship based on bulk purchasing between central agencies was in sight. This meant that New Zealand had to identify opportunities that were arising in the separate republics.

In the targeting of republics and regions, the report considers New Zealand's future in the market to lie in Russia, east of the Urals, and Kazakhstan. More specifically, with regard to Russia, it identifies Moscow, because of the size of the consumer market, southern Siberia (the Altai, Kemerovo, Novosibirsk), the Russian Far East, and, in Kazakhstan, Alma Ata and the northern agricultural provinces. Besides targeting the direction of trade, there is the need to identify the goods and services for which local authorities have already shown interest in New Zealand involvement. These areas are identified as agro-technology, farming practices, food industry, some aspects of engineering, marketing and training.

Equally important, the report states, is recognizing that the republics and authorities want the kind of cooperation that will increase their capacity to earn hard currency and to achieve greater autonomy for themselves. What is called for, the report argues, is a change in the nature of the economic relationship, and the need to look beyond turnkey projects and consider partnership with local companies. This, as it points, out, involves a significantly greater commitment to that market than hitherto.

At the time of this review, the payments crisis was already over a year old, and was coloring the whole discussion of the relationship with the then Soviet Union. The report's response was that the new conditions of trade and the new strategy the report was recommending required cooperation between the New Zealand government and the private sector. The issue of export guarantees had been widely discussed in 1990 and early 1991 as a result of the large debts owed

Table 20.2. Soviet Licensed and Charter Fishing in the New Zealand Region

	Years				
	1986–87	1987–88	1988–89	1989–90	1990–91
Licensed allocations (tons)	20,107	7381	0	0	0
No. of Soviet vessels in New Zealand zone					
a) Charter	10	22	22	40	64
b) Foreign licensed	15	14	0	0	0

Source: Ministry of Agriculture and Fisheries.

Until the recent changes in the former Soviet Union, such proposals had faced a number of problems, not the least of which was the issue of national security. The proposal has been kept alive over the years, but is now in a position to be judged solely by economic criteria. Agreement has been reached on most of the issues regarding Aeroflot access. However, the problem is that there is no longer the same level of interest on the part of the Russians, Aeroflot being registered as a Russian company. It seems that New Zealand's belated positive response on this issue was influenced by the fact that Australia has recently reached an agreement with Aeroflot for commercial flights to Australia.

Proposals for the provision of ship repair facilities, including a dry dock, in New Zealand now finally seem to be dead. Concern with saving expenditures in hard currency is also reflected in recent discussions regarding fuel requirements for the fishing fleet. Unable to gain access to New Zealand-based fuel without payment in hard currency, the fleets requested the right to bring their own tankers into New Zealand waters, subject to the provision that they do not use onshore fuel facilities. As a result of changes in the CIS, however, Russian companies now have to pay for fuel in hard currency, and there is therefore no longer the same advantage in bringing their own fuel down to New Zealand.

PROBLEMS AND PROSPECTS

Despite the obvious difficulties of the CIS market, it is too important for New Zealand to ignore. The need to examine what is involved and what the prospects are has resulted in two reviews of the market in the last three years.

In 1989, a New Zealand official and business mission undertook a visit to the Soviet Far East led by the chairman of the country's largest enterprise, Fletcher Challenge. The mission's report concluded that very real obstacles existed to the early achievement of development in this region of the Soviet Union (New Zealand Committee for Pacific Economic Cooperation, 1989). The mission saw these to be the scale of investment needed to provide the infrastructure necessary; the need for the new reforms to be fully understood at all levels in is heavily involved with Soviet fishing interests in New Zealand, has already built one cold storage plant and plans another. The construction of these plants has involved the employment of New Zealand labor and resources, and some small New Zealand firms have benefited substantially from this development as a result.

The legislation on joint ventures opened up another avenue for economic relations. New Zealand had limited experience with joint ventures in fishing, authorized under the New Zealand-Soviet Union Fishing Agreement. With regards to joint ventures in the Soviet Union, New Zealand as a small country would not be able to be a significant investor. More generally, the experience of joint ventures in the former Soviet Union, despite large registrations, has not satisfied expectations. Nevertheless, New Zealand established a joint venture initially in Sakhalin and now has five such ventures. The joint ventures are very small in scale, with a total value of only about NZ$2 million, and, following a common pattern of such investments in the Soviet Union, are in service industries and other fields which can earn hard currency.

In the course of the arguments over the last decade or so on trade imbalances, one new factor that became very important was Soviet fishing. Soviet active interest in the South Pacific fisheries emerged in the mid-1970s. With the introduction of EEZ provisions in 1978, the Soviet Union signed a fisheries agreement that provided for Soviet-licensed fishing. It was clear at that time that the purpose of the agreement from New Zealand's point of view was as part of a package to enable New Zealand to gain wider access to the Soviet market. Access to New Zealand fishing zones for the Soviet Union was subsequently severely restricted, as New Zealand joined the widespread international reaction to the Soviet invasion of Afghanistan. When relations were normalized in 1983, the New Zealand fishing situation had changed as a result of the adoption of policies favoring the development of a New Zealand fishing industry. However, in practice, New Zealand continued to discriminate against the Soviet Union until 1988. By then all foreign license quotas had been cut back, and the pattern of Soviet fishing in the area had radically changed. The encouragement of the local fishing industry and the adoption of marine conservation policies meant that the future of foreign fishing was becoming more uncertain. All fishing quotas were allocated to New Zealand fishing companies. Therefore, access to fishing, especially for valuable species, increasingly became available only through charter contracts with a New Zealand company. For Soviet fishing companies, an important additional factor was that licensed fishing involved the payment of hard currency, the charter being essentially a barter agreement. The change is illustrated in Table 20.2.

The result was that by 1991 the number of Soviet fishing vessels in New Zealand waters had substantially increased, despite the growing difficulties involved in distant fishing. The economics of distant fishing had led to consideration over many years of the possibility of broadening the fishing relationship into other areas. In particular, the Soviets and some New Zealand interests had lobbied for Aeroflot access to New Zealand to facilitate crew changes, and had sought more substantial ship repair facilities for the Soviet South Pacific fishing fleet.

to New Zealand for commodity payments. At the time, the government rejected appeals for provision of export credits. The Prior report argues that the new strategy proposed does not require major New Zealand investments, that foreign exchange does exist if it can be tapped into, and that through involvement in manufacturing joint ventures in the former Soviet Union, it should be possible to develop exports that can be sold for foreign exchange. It takes the view that while the debt question tends to be the focus of present discussion on the Soviet market, it would be a mistake to make future involvement dependent on repayments which seems to be the current mood. In this context, the report supports government-backed export credit guarantees, subject however to a number of clear conditions and requirements. The government is asked in the report to realize that the present state of change in the Soviet Union is highly unusual, and that it is important for New Zealand companies to secure a foothold now. This is, in turn, related to a major theme of the report that with the fragmentation of the economy it is essential for New Zealand to raise its profile.

What emerges from the report is that the erosion and what subsequently was to be the total collapse of central authority has created both new problems and new possibilities.

For New Zealand, the Soviet market has been important for many years and remains so. New Zealand traders had begun to adapt to the new circumstances in response to the growing regionalization of trade and the need to diversify the nature of the bilateral relationship. Using what was perceived as a competitive advantage, New Zealand especially targeted the Soviet Far East, and as economic decision making began to be decentralized, direct access was sought from local authorities. By the time of the collapse of the Soviet Union, Sovenz was operating at three levels: dealings with the central foreign trade organization (Prodintorg), direct ties with the Russian Federation and direct access to individual regional governments (Figure 20.1).

Russia accounts for the import of almost all food products to the CIS, and that relationship is therefore crucial to New Zealand. This has not, however, precluded the establishment of other links, especially to Central Asia. New Zealand has sold meat to Central Asia, an area in which sheep meat is readily acceptable, and large sales of butter are made in the Russian Far East market. While Sovenz had begun to adapt to decentralization, the wool industry found itself in total dependence on the traditional pattern of central purchasing.

The monetary and reserves crisis in the Soviet Union/CIS has, however, affected all commodity traders. The failure to make payments is a new experience for New Zealand in the Soviet trade and has brought real fears of overexposure. The wool trade had come almost to a standstill during 1991, except for a small quantity covered by bank credit, and in the dairy trade very little is being traded until there are signs of movement on debt repayments. New Zealand is owed about NZ$400 million, half of that for butter and milk shipped in 1991. The contract called for a series of payments throughout 1992, starting in February, but the first payment was not made. The Russians know, however, that failure to pay will mean that they will be denied further credit for their highest priority—

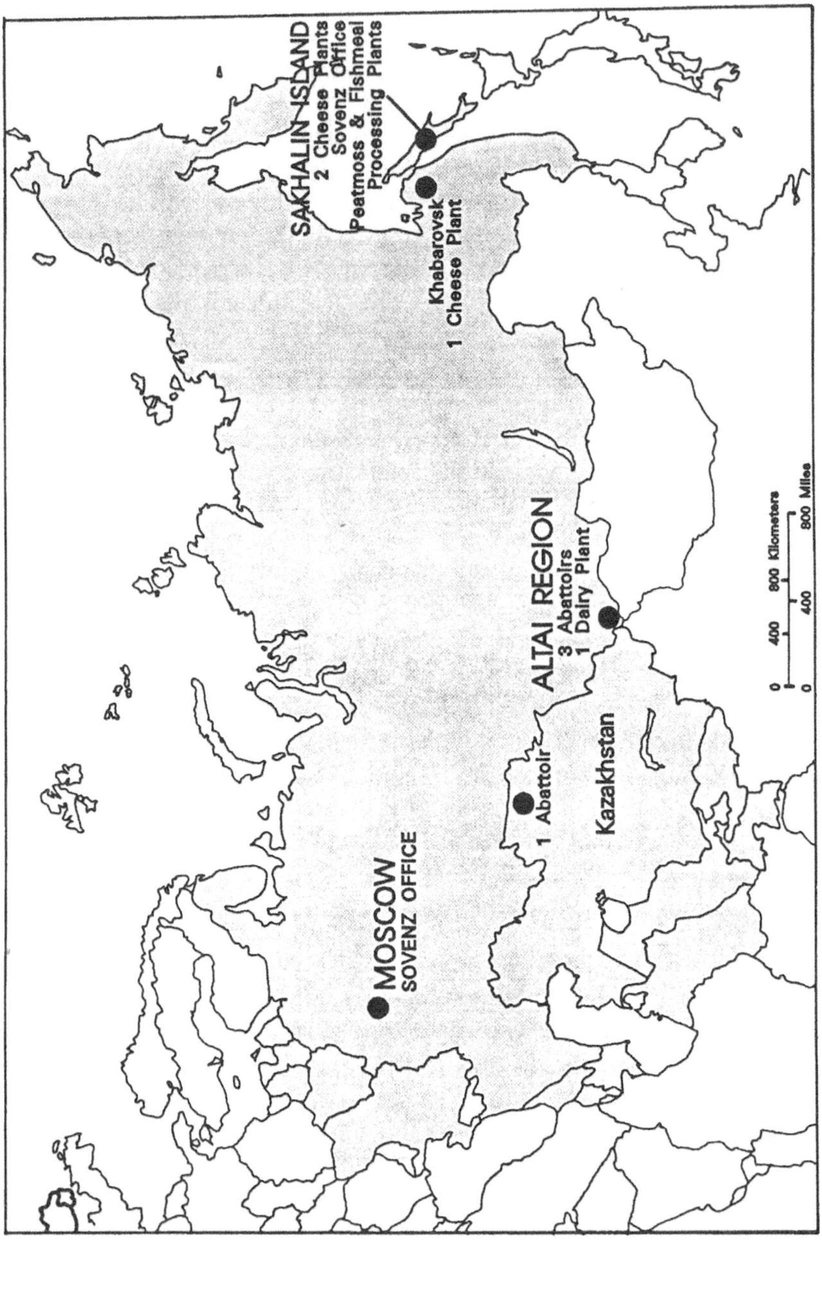

Figure 20.1. Sovenz in the Soviet Union

food. With regard to wool, New Zealand cannot afford to see the closure of mills in Russia, and serious efforts have been made to restore the flow of New Zealand wool to that market. In May 1992, these efforts resulted in an agreement in principle by the Wool Board to supply 16,000 tonnes of wool at a cost of NZ$100 million, with funds being made available through a countertrade package put together by Sovenz.[13]

The debt problem has been compounded by the breakup of the former Soviet Union. Agreement has been reached by the former Soviet republics on the sharing of existing debts, but there is no guarantee of payments by these states, which are facing a deteriorating situation, and the agreement does not cover future debts.[14] The Vnesheconombank of the Soviet Union is now officially defunct and has in effect been taken over by the Russian Vnesheconombank. While one has collapsed, the other has not yet established international confidence, and Germany's Deutsche Bank has been asked to form a coordinating committee to negotiate on the rescheduling of Soviet commercial debt.

Decentralization in foreign economic relations and the establishment of sovereign states have resulted in the application of different rules by the various states and the need for the establishment of new contacts. While the movement to marketization accompanied political decentralization, the relationship is by no means necessarily a complementary one. One of the features of the independence of the new states has been the adoption of autarkic economic policies, including the prohibition of exports. However, while such policies are being adopted, the need for food is so great that it seems such policies are not allowed in practice to prevent payment for food imports by way of export of resources.

The present situation has brought changes in the mechanics of trade with the regions and states. Much of the trade is now being conducted through European trading houses. These firms are planning to pay for Russian raw materials with food, and those, such as the New Zealand traders, that are experienced in supplying food packages may have the opportunity to benefit.[15]

Nevertheless, New Zealand is clearly concerned about the effect of massive credits to the CIS. At a meeting of New Zealand officials and business representatives in December 1991 to consider the coordination of effort and approaches to the CIS, discussion was dominated by the debt issue and the question of export guarantees. At the Washington Conference, New Zealand's Deputy Prime Minister supported the general view that aid should not distort moves towards free-market economies in the new states. This reflected New Zealand's concern with the possible effects on trading opportunities of gross oversupplies of dairy products, and New Zealand continues to urge EEC members to consider the effect of food aid on traditional trade.[16]

The changes in the former Soviet Union, as already indicated, also affected the Soviet/Russian Far East fishing operations. The future of these companies is uncertain for a number of reasons. First, the conflict of authority with regard to ownership, control and access to fishing resources continues to be unresolved. The industry furthermore faces the task of major restructuring which will involve unemployment and massive capital formation at a time when the industry is losing the subsidies it had enjoyed in the past. One response by Russian companies may be to look for partnership with New Zealand companies in New Zealand and encourage foreign participation in Russia. These Russian companies operate out of areas which have become important markets for New Zealand products, and the hard currency earnings of Russian fishing companies in New Zealand waters have facilitated the development of a relationship based on mutual interest. The role of Russian fishing in New Zealand waters is important for New Zealand. Russian charter vessels harvest about 30 percent of the New Zealand catch, and as they tend to dispose of most of it in New Zealand in return for hard currency, they make a very important contribution to New Zealand's fish exports.[17] However, the future of Russian fishing in New Zealand waters is uncertain. Because there are problems with the supply and cost of fuel and allocations, most of the Russian Far East fleets are tied up (British Broadcasting Company, January 3, 1992). Some are still coming to New Zealand, but recently in reduced numbers. Furthermore, the situation in New Zealand is also changing, as European fleets are beginning to show an interest in the New Zealand region, and Russian fleets will face increasing competition.

CONCLUSIONS

As the Prior report argued, under the new conditions New Zealand has to work harder to find new niches and raise its profile in the CIS. With the publicity attached to the food aid program, some felt that New Zealand's failure to contribute was an opportunity missed to help establish a fund of good will, a view taken also by Russian officials in New Zealand.

The overall result of developments in the Soviet Union/CIS has been to create uncertainties and a more complex and sensitive environment. The greatest uncertainty, however, still lies in the political outcome in the CIS. Already two parallel groups are emerging in the CIS. One group represents a movement towards the restoration of some of the economic integration with has been lost over the last two years. This group, led by Russia, may bring in turn a revival of some degree of central purchasing, and may make it easier to identify where the power to make decisions lies. One indication of this is that recent agreements between the CIS member-states include a protocol for establishing an Inter-State Commission on Food Imports (British Broadcasting Company, March 18, 1992). However, even if central organizations are retained, their function is likely to be one of coordination, with regions and enterprises generating the method of payment and determining their requirements.

In the longer term, as a major exporter of food New Zealand should be able to remain a valuable supplier to the CIS. Of the need for such supplies there can be no doubt. The current situation in the Russian livestock industry of declining production and falling livestock population should, in principle, ensure favorable export opportunities for New Zealand. Equally beyond doubt, however, is that the CIS market is now one in which New Zealand has never faced so many uncertainties as trade with that market in general faces serious decline (British Broadcasting Company, January 24, 1992).

ENDNOTES

1. For a review of these developments prior to August 1991, see the author's "Decentralization and Soviet Foreign Economic Relations," a paper presented at the Conference on "The Economies of Eastern Europe," at the Center of Soviet and East European Studies, University of Melbourne, Australia, August 1991.
2. For details of gold sales see the interview with G. Yavlinskii in *Russkaya Mysl',* November 22, 1991, p. 7.
3. FESCO is now the only former Soviet liner service on the New Zealand run.
4. Much of the meat trade with the Soviet Union had until then been conducted by one New Zealand company, Amalgamated Marketing Limited, whose involvement in Soviet trade can be traced back to 1957.
5. A factor in this expansion was the containerization of wool shipping. This resulted in New Zealand wool being carried by FESCO to the Soviet Far East instead of by the Baltic Line to Europe.
6. Discussion in this paragraph is based on interviews with officials in various organizations in Moscow and the Soviet Far East in September–October 1990.
7. See Greg Ansley, "Sea Rustlers," *New Zealand Herald,* January 18, 1992, Section 2, p. 1.
8. The absence of formal recognition of other states does not mean New Zealand does not recognize them, nor does it preclude doing business with them. New Zealand has also recognized the Baltic Republics. The Soviet Embassy is now known as the Embassy of the Russian Federation. It is however authorized to act on behalf of any of the other members of the CIS if requested by them.
9. The Soviet figures are calculated from *Vneshnie Ekonomicheskie Svyazi SSSR V. 1989 g. Statisticheskii Sbornik,* 1989, and for previous years. This volume changed its name in 1989.
10. Most of the NZ-Soviet Union trade is carried in Soviet vessels.
11. See, in particular, the reporting in the New Zealand *National Business Review,* in May 1990. Hopes were raised by Australia's announcement to set up a A$500 million credit line. The following month the New Zealand government said it would not follow in Australia's steps. On the case for credits, see *ibid,* November 8, 1991, p. 7.
12. The report was prepared for the Ministry of External Relations and Trade and the Trade Development Board by Mr. Stuart Prior of the External Assessments Bureau.
13. It is hoped that the NZ$40 million existing debt can be paid off to New Zealand creditors with the funds made available to Russia by the G7.
14. For shares of the debt, Plan Econ Report, *Toward the Resolution of the External Debt of the Former Soviet Union,* January 17, 1992, p. 10.

15. Opportunities may also arise in cooperation with countries bordering on the CIS. See, for example, the proposal of the Iranian Ambassador to New Zealand for joint ventures with Iran in agricultural industries in the Muslim states of the CIS, *New Zealand Herald,* February 28, 1992, p. 9.

16. The EEC is shipping large quantities of food, including butter and beef (reportedly being sold at 28 cents a kg), to Moscow, negating Russian interest in buying elsewhere. See *National Business Review,* March 27, 1992, p. 5.

17. Russian vessels represent about 37 percent of the total charter fleet for the 1991–92 season.

REFERENCES

British Broadcasting Company, *Summary of World Broadcasts,* SU/SO211, January 3, 1992, p. A/8.

British Broadcasting Company, *Summary of World Broadcasts* SU/280, January 17, 1992, pp. B4/5.

British Broadcasting Company, *Summary of World Broadcasts* SU/1286/C2/2, January 24, 1992.

British Broadcasting Company, *Summary of World Broadcasts,* SU/1332, C2/1, March 18, 1992.

New Zealand Committee for Pacific Economic Cooperation, 1989, *Visit to the Soviet Union,* Auckland, New Zealand.

USSR Ministry of Foreign Economic Relations, 1989, *Vneshnie Ekonomicheskie Svyazi SSSR V. 1989 g. Statisticheskii Sbornik,* Moscow.

Appendix A

MINERAL PRODUCTION STATISTICS OF THE FORMER SOVIET UNION

J.P. Dorian et al. (eds.), CIS Energy and Minerals Development, 359–366.

Table A2. Regional Distribution of Soviet Gas Production, 1980–1991 (billion cubic meters)

	1980	1985	1988	1989	1990	1991	% of 1991 Total
USSR (total)	435.2	642.9	770.0	796.1	814.8	810.5	100
Russian Federation	254.0	462.0	589.8	615.8	640.6	642.9	79.3
European Russia	41.3	30.3	29.3	22.4	19.7	17.5	2.2
Komi Republic	19.4	17.9	13.6	10.7	8.3	8.0	0.98
North Caucasus	14.1	7.5	6.2	5.8	5.6	4.5	0.55
Volga	7.8	4.9	9.5	5.9	5.8	5.0	0.62
Urals	51.0	48.9	46.6	45.4	43.3	43.0	5.3
Siberia	161.7	382.8	518.8	548.0	577.8	582.4	71.9
West Siberia	156.4	375.8	510.8	539.6	569.3	574.0	70.8
Tyumen' Oblast	156.2	375.4	505.3	539.2	568.9	573.7	70.8
Gas industry enterprises	145.0	350.6	474.9	507.6	536.6	550.6	67.9
Urengoy	50.0	258.2	300.0	288.3	287.9	295.1	36.4
Yamburg	—	0.1	85.0	128.9	158.8	167.9	20.7
Medvezh'ye	71.0	73.8	73.1	72.6	72.1	71.1	8.8
Vyngapur	16.0	18.1	17.6	17.5	17.6	16.6	2.0
(unidentified)	8.0	1.9	0.0	1.7	0.2	0.0	0.0
Oilfield gas	11.2	23.3	29.7	30.2	32.3	27.2	3.35
Noril'sk area	4.2	6.6	5.1	6.5	5.1	5.2	0.64
Tomsk Oblast	0.2	0.4	0.2	0.4	0.4	0.3	Negl.
Other Siberian	1.6	1.8	2.9	3.4	3.2	3.2	0.39
Yakutiya-Sakha	0.8	1.0	1.3	1.4	1.4	1.4	0.17
Sakhalin	0.8	0.8	1.6	2.0	1.8	1.8	0.22

Turkmenistan	8.0	6.0	5.7	5.8	5.6	5.4	1.05
Ukraine	7.5	5.8	5.4	5.4	5.3	4.9	0.95
Uzbekistan	1.3	2.0	2.4	2.6	2.8	2.8	0.5

Note: Figures are rounded.

— No production reported.

a. Soviet oil statistics include both crude and condensate (natural gas liquids).

b. The Bashkir ASSR is included in the Urals.

c. Does not include production in Tyumen' by Tatar and Bashkir associations.

Source: Matthew J. Sagers, "Review of Soviet Energy Industries in 1991," *Soviet Geography*, Vol. XXXIII, No. 4, April 1992, pp. 237–268.

Table A1. Regional Distribution of Soviet Oil Production, 1980–1991[a] (million tonnes per year)

Former republic/region	1980	1985	1988	1989	1990	1991	% of 1991 total
USSR (total)	603.2	595.3	624.3	607.3	570.8	515.0	100.0
Azerbaijan	14.7	13.1	13.7	13.2	12.5	11.7	2.3
Belarus	2.6	2.0	2.1	2.1	2.1	2.1	0.4
Georgia	3.2	0.5	0.2	0.2	0.2	0.2	Negl.
Kazakhstan	18.7	22.8	25.5	25.4	25.8	26.6	5.2
Kyrgyzstan	0.2	0.2	0.2	0.2	0.2	0.1	Negl.
Russian Federation	546.7	542.3	568.8	554.9	516.2	461.1	89.5
European Russia	153.9	105.5	90.0	83.6	79.6	72.7	14.1
Kaliningrad Oblast	1.3	1.5	1.3	1.2	1.21.6	0.31	
Komi ASSR	20.4	19.4	17.1	16.1	15.7	12.1	2.3
North Caucasus	18.8	10.8	9.6	9.0	8.4	8.0	1.6
Volga	113.4	73.8	62.0	57.3	54.3	51.0	9.9
Urals[b]	77.6	66.0	61.4	64.0	59.5	57.5	11.2
Siberia	315.2	370.7	417.3	407.3	377.4	330.9	64.3
West Siberia	312.7	368.1	415.1	405.1	375.2	328.8	63.8
Tyumen' Oblast[c]	307.9	361.1	405.7	394.7	365.0	319.0	61.9
Oil Industry Enterprise	302.8	352.7	394.0	382.9	353.1	308.4	59.9
Gazprom	—	2.3	6.0	6.9	7.3	7.5	1.45
Tomsk Oblast	4.8	7.0	9.4	10.4	10.2	9.8	1.9
Tomskneft'	9.8	13.1	14.6	14.9	14.7	1.8	0.3
Sakhalin	2.5	2.6	2.2	2.2	2.2	2.1	0.4
Outside Russia	56.6	52.8	55.5	55.1	54.6	53.9	10.5
Tajikistan	0.4	0.4	0.3	0.2	0.1	0.1	Negl.

Outside Russian Federation	181.2	180.9	180.2	180.3	174.2	167.6	20.6
Ukraine	56.7	42.9	32.4	30.8	28.1	24.4	3.0
Azerbaijan	14.0	14.1	11.8	11.1	9.9	8.6	1.06
Kazakhstan	4.3	5.5	7.0	6.7	7.1	7.9	0.97
Uzbekistan	34.8	34.6	39.9	41.1	40.8	41.9	5.2
Turkmenistan	70.5	83.2	88.3	89.9	87.8	84.3	10.4
Other republics	0.9	0.7	0.5	0.6	0.5	0.5	Negl.
Georgia	0.3	0.1	Negl.	Negl.	Negl.	Negl.	Negl.
Kyrgyzstan	0.1	0.2	0.2	0.2	0.2	0.1	Negl.
Tajikistan	0.2	0.4	0.3	0.2	0.1	0.1	Negl.
Belarus	0.3	0.2	0.3	0.3	0.3	0.3	Negl.

Table A3. Regional Distribution of Soviet Coal Production, 1980–1991 (million tonnes of gross mine output)

Republic/region	1980	1985	1988	1989	1990	1991	% of 1991 total
USSR	716.4	726.4	771.9	740.3	703.3	629.5	100
Hard coal	553.0	569.3	599.9	576.8	543.2	484.5	77.0
Anthracite	73.1	71.3	72.3	68.0	62.7	55.7	8.8
Coking coal	190.2	196.9	211.0	198.8	191.0	153.1	24.3
Steam coal	289.7	301.1	316.6	310.0	289.5	275.7	43.8
Lignite	163.4	157.1	172.0	163.5	160.1	145.0	23.0
Deep-mined	445.5	421.8	430.1	404.3	377.8	314.2	49.9
Strip-mined	270.9	304.6	341.8	336.0	325.5	315.3	50.09
Georgia	1.9	1.7	1.4	1.2	1.0	0.8	.01
Kazakhstan	115.4	130.8	143.1	138.4	131.6	130.0	20.7
Karaganda	48.6	49.8	52.3	50.5	48.7	47.2	7.5
Ekibastuz	66.8	80.5	89.7	86.6	81.9	82.0	13.0
Other	0.0	0.5	1.1	1.1	1.0	0.8	0.1
Kyrgyzstan	4.0	4.0	4.0	4.0	3.7	3.5	0.6
Russian Federation	391.4	395.2	425.5	410.0	395.3	353.3	56.1
European Russia	86.3	80.2	81.2	75.3	71.5	64.5	10.2
Moscow basin	25.4	19.3	8.0	15.6	13.3	11.3	1.8
Pechora basin	28.6	29.8	31.5	29.2	29.3	27.3	4.3
Donbas (East)	32.3	31.1	31.7	30.5	28.9	25.9	4.1
Urals	38.8	28.0	26.4	25.0	23.9	21.9	3.5
Siberia	266.3	287.0	317.9	308.8	299.9	266.9	42.4
Kuznetsk basin[a]	145.0	146.2	159.2	157.5	150.4	125.4	19.9

Kansk-Achinsk	34.8	41.4	50.6	48.2	52.3	50.0	7.9
South Yakutiya	2.1	11.9	15.5	17.3	16.9	15.5	2.5
Neryungri	2.5	8.5	14.0	14.9	14.5	14.0	2.2
Ukraine	197.1	189.0	191.7	180.2	164.8	136.0	21.6
Donets basin (Ukraine)[b]	173.3	167.5	168.5	157.5	145.0	125.0	19.9
Donets basin (Total)[a]	204.0	198.7	200.3	188.0	173.9	150.9	24.0
Uzbekistan	5.7	5.0	5.5	6.2	6.5	5.9	0.9
Tajikistan	1.0	1.0	0.7	0.5	0.5	0.3	Negl.

Negl. Negligible

a. Includes ministry and nonministry production. For Donets, includes both Ukrainian portion and that produced in Rostov Oblast in the RSFSR.

b. Ukrainian portion for the Donets basin only.

Source: Matthew J. Sagers, "Review of the Soviet Energy Industries in 1991" *Soviet Geography,* Vol. XXXIII, No. 4, April 1992, p. 260.

Table A4. Regional Distribution of Soviet Steel Production, 1980–1991 (million tonnes)

Republic/region	1980	1985	1988	1989	1990	1991	% of 1991 total
USSR	147.9	154.7	163.0	160.1	154.4	133.1	100
Azerbaijan	0.8	0.9	0.8	0.8	0.7	0.6	0.45
Belarus	0.3	0.8	1.1	1.1	1.1	1.1	0.8
Georgia	1.3	1.4	1.5	1.4	1.3	1.0	0.75
Kazakhstan	6.0	6.2	6.8	6.8	6.8	6.4	4.8
Latvia	0.5	0.6	0.6	0.6	0.6	0.5[a]	0.3
Lithuania	Negl.	Negl.	Negl.	Negl.	Negl.	Negl.	Negl.
Moldova	Negl.	0.2	0.7	0.7	0.7	0.6	0.45
Russian Federation	84.4	88.7	94.1	92.8	89.6	77.0	57.9
European Russia	26.0	30.0[a]	33.1	34.0[a]	34.0[a]	30.0[a]	22.5
Urals	42.0	43.0[a]	44.8	44.0[a]	42.0[a]	35.0[a]	26.3
Siberia	16.0	16.0[a]	16.0	15.0[a]	14.0[a]	12.0	9.0
Ukraine	53.7	55.0	56.5	54.8	52.6	45.0	33.8
Uzbekistan	0.8	0.9	1.0	1.1	1.0	0.9	0.7
Other republics	Negl.	Negl.	Negl.	Negl.	Negl.	Negl.	Negl.

Note: Figures are rounded.

Negl. Negligible

a. Figures are estimates only.

Sources: Narkhoz SSSR, various years; year-end statistical reports in the republic newspapers for 1991; Narkhoz RSFSR 1988, pp. 332–333; Promyshlennost' SSSR, 1991; and *Soviet Geography,* April 1992, p. 339.

The GeoJournal Library

1. B. Currey and G. Hugo (eds.): *Famine as Geographical Phenomenon.* 1984 ISBN 90-277-1762-1
2. S. H. U. Bowie, F.R.S. and I. Thornton (eds.): *Environmental Geochemistry and Health.* Report of the Royal Society's British National Committee for Problems of the Environment. 1985 ISBN 90-277-1879-2
3. L. A. Kosiński and K. M. Elahi (eds.): *Population Redistribution and Development in South Asia.* 1985 ISBN 90-277-1938-1
4. Y. Gradus (ed.): *Desert Development.* Man and Technology in Sparselands. 1985 ISBN 90-277-2043-6
5. F. J. Calzonetti and B. D. Solomon (eds.): *Geographical Dimensions of Energy.* 1985 ISBN 90-277-2061-4
6. J. Lundqvist, U. Lohm and M. Falkenmark (eds.): *Strategies for River Basin Management.* Environmental Integration of Land and Water in River Basin. 1985 ISBN 90-277-2111-4
7. A. Rogers and F. J. Willekens (eds.): *Migration and Settlement.* A Multiregional Comparative Study. 1986 ISBN 90-277-2119-X
8. R. Laulajainen: *Spatial Strategies in Retailing.* 1987 ISBN 90-277-2595-0
9. T. H. Lee, H. R. Linden, D. A. Dreyfus and T. Vasko (eds.): *The Methane Age.* 1988 ISBN 90-277-2745-7
10. H. J. Walker (ed.): *Artificial Structures and Shorelines.* 1988 ISBN 90-277-2746-5
11. A. Kellerman: *Time, Space, and Society.* Geographical Societal Perspectives. 1989 ISBN 0-7923-0123-4
12. P. Fabbri (ed.): *Recreational Uses of Coastal Areas.* A Research Project of the Commission on the Coastal Environment, International Geographical Union. 1990 ISBN 0-7923-0279-6
13. L. M. Brush, M. G. Wolman and Huang Bing-Wei (eds.): *Taming the Yellow River: Silt and Floods.* Proceedings of a Bilateral Seminar on Problems in the Lower Reaches of the Yellow River, China. 1989 ISBN 0-7923-0416-0
14. J. Stillwell and H. J. Scholten (eds.): *Contemporary Research in Population Geography.* A Comparison of the United Kingdom and the Netherlands. 1990 ISBN 0-7923-0431-4
15. M. S. Kenzer (ed.): *Applied Geography.* Issues, Questions, and Concerns. 1989 ISBN 0-7923-0438-1
16. D. Nir: *Region as a Socio-environmental System.* An Introduction to a Systemic Regional Geography. 1990 ISBN 0-7923-0516-7
17. H. J. Scholten and J. C. H. Stillwell (eds.): *Geographical Information Systems for Urban and Regional Planning.* 1990 ISBN 0-7923-0793-3
18. F. M. Brouwer, A. J. Thomas and M. J. Chadwick (eds.): *Land Use Changes in Europe.* Processes of Change, Environmental Transformations and Future Patterns. 1991 ISBN 0-7923-1099-3

The GeoJournal Library

19. C. J. Campbell: *The Golden Century of Oil 1950–2050.* The Depletion of a Resource. 1991 ISBN 0-7923-1442-5
20. F. M. Dieleman and S. Musterd (eds.): *The Randstad: A Research and Policy Laboratory.* 1992 ISBN 0-7923-1649-5
21. V. I. Ilyichev and V. V. Anikiev (eds.): *Oceanic and Anthropogenic Controls of Life in the Pacific Ocean.* 1992 ISBN 0-7923-1854-4
22. A. K. Dutt and F. J. Costa (eds.): *Perspectives on Planning and Urban Development in Belgium.* 1992 ISBN 0-7923-1885-4
23. J. Portugali: *Implicate Relations.* Society and Space in the Israeli-Palestinian Conflict. 1993 ISBN 0-7923-1886-2
24. M. J. C. de Lepper and H. J. Scholten: *The Added Value of Geographical Information Systems in Public and Environmental Health.* 1993 ISBN 0-7923-1887-0
25. J. P. Dorian, P. A. Minakir and V. T. Borisovich (eds.): *CIS Energy and Minerals Development.* Prospects, Problems and Opportunities for International Cooperation. 1993 ISBN 0-7923-2323-8

KLUWER ACADEMIC PUBLISHERS – DORDRECHT / BOSTON / LONDON